Aging Interventions and Therapies

Aging Interventions and Therapies

Suresh I S Rattan
University of Aarhus, Denmark

World Scientific

NEW JERSEY • LONDON • SINGAPORE • BEIJING • SHANGHAI • HONG KONG • TAIPEI • CHENNAI

Published by

World Scientific Publishing Co. Pte. Ltd.

5 Toh Tuck Link, Singapore 596224

USA office: 27 Warren Street, Suite 401-402, Hackensack, NJ 07601

UK office: 57 Shelton Street, Covent Garden, London WC2H 9HE

British Library Cataloguing-in-Publication Data
A catalogue record for this book is available from the British Library.

AGING INTERVENTIONS AND THERAPIES

ISBN 981-256-084-X

Typeset by Stallion Press
Email: sales@stallionpress.com

Printed in Singapore by World Scientific Printers (S) Pte Ltd

Contents

Introduction

During the last few years, tremendous progress has been made in our understanding of the biological basis of aging. Biogerontology is now a respectable and well-established field, and several universities, medical institutes and research centers throughout the world offer full-fledged courses on this topic. Most importantly, biogerontologists are now in a position to construct general principles of aging and explore various possibilities of intervention using rational approaches. While not giving serious consideration to the claims made by charlatans, it cannot be ignored that several researchers are making genuine attempts to test and develop various means of intervention for the prevention and treatment of age-related diseases and for achieving healthy old age.

Aging Interventions and Therapies takes status of the molecular, cellular, hormonal, nutritional, medical and lifestyle strategies being tested and applied for the prevention, intervention and treatment of age-related diseases. The articles included in this book are specially written for this compendium by internationally leading researchers and practitioners at the universities, research institutes, hospitals, clinics and cosmetics and pharmaceutical industry. Furthermore, a philosopher and bioethicist addresses the ethical questions related to anti-aging and longevity promoting research, biodemographers discuss the practicality and impracticality of human lifespan extension, and a theoretical biologist-cum-biofuturologist reviews the futuristic technologies in development which will have serious bearings on aging intervention.

The target readership of *Aging Interventions and Therapies* is the undergraduate and graduate students in the universities, medical and nursing colleges, and post-graduates taking up studies and research projects on different aspects of aging and anti-aging. Practicing clinicians who would like to know about the latest developments in the field of aging intervention and therapy will find this book very useful. The topics covered in this book are also highly relevant for pharmaceutical, cosmeceutical and nutrition and healthcare industry for an easy access to accurate and reliable information in the field of aging research and intervention.

Suresh I.S. Rattan, Ph.D., D.Sc.
Editor
Danish Centre for Molecular Gerontoloy,
Department of Molecular Biology,
University of Aarhus,
Denmark.

1

The Ethics of Aging Intervention and Life-Extension

Steven Horrobin
Faculty of Law
University of Edinburgh, UK
Emails: s.horrobin@sms.ed.ac.uk, s_horrobin@hotmail.com

INTRODUCTION

Why include an ethical chapter in a scientific text? What is the relevance of ethics to the pure research and practice of biomedical gerontology? If these questions are relevant, what are then the main issues, and how may they be addressed? My purposes in writing this chapter are two-fold: Firstly, I wish to answer the above questions, and thereby to convince the reader that certain philosophical and ethical questions and issues are prior to, coincident with, and consequent upon the research and practice of biogerontology, and should be seen as inseparable, necessary and beneficial components of the discipline. Secondly, I wish to provide the reader with a basic guide to approaching and dealing with these perhaps somewhat unfamiliar aspects of the field.

While the ethics of pure scientific research may be interesting in themselves, they are on the whole not germane to the kind of concern that is popularly expressed both by the general public, and by ethicists, when aging research is discussed. Therefore I shall, for the purposes of this paper, assume that research implies application, and that there exists an intention to intervene in the processes of aging, and so focus upon the ethical implications of this. There are two possible motivations for this intervention. The first is to mitigate the disabilities, infirmities, discomforts and impairments of the aging process. The second, which according to some thinkers in the field neither ought to nor can be disentangled from the first, is to obviate the aging process partially or altogether, and thereby achieve life-extension itself. Since life-extension is implied by both motivations, either as a goal or an effect, and since intervention in the pathologies of aging themselves is less controversial, I shall concentrate for the main part of what follows primarily upon the ethics of life-extension *per se*, though the ethics of aging intervention will of necessity be discussed *inter alia*.

The Problem of Value: Life-Extension and Death Postponing

In order to assess the ethical aspects of the notion of life-extension, one must first address the problem of value. If there is no value to be gained in the extension of life, or to put it another way, if life-extension has no value in itself, then a defense of its pursuit becomes difficult, or impossible, in the face of any risk or disvalue that may be posited. In order to properly assess the value of life-extension, we must first examine the nature of the action itself, taking into account not only effect, but motivation, in an attempt to find a value that underpins it. We shall see that this turns out to be the instrumental value of living.

O Death, Where is Thy Sting?

Can death in itself be argued to be something so negative that its very occurrence may be used as a justification for life-extension? If biogerontologists seek to extend life, are they mainly, as well, or at all seeking to stave off death itself? It has been argued that life-extension is death postponement.[1] While the postponement of death may explain a person's motives in seeking to extend a life, it is not the case that "life-extension" and "death postponement" are one and the same concept. I may be tired of life, and find no instrumental or intrinsic[a] value in its extension, but nevertheless wish to postpone my death. Such a desire is not motivated by a wish to extend life but rather from some notion of the disvalue of death. Perhaps death in such a case is feared as something negative in itself as a kind of anti-life, just as darkness is conceived in Milton as a kind of anti-light, or "darkness visible".[2] Or else perhaps it is feared, as in the case of Hamlet, simply because it is an unknown quantity:

> To die — to sleep —
> No more; and by a sleep to say we end
> The heartache, and the thousand natural shocks
> That flesh is heir to. 'Tis a consummation
> Devoutly to be wish'd. To die — to sleep.
> To sleep — perchance to dream: ay, there's the rub!
> For in that sleep of death what dreams may come
> When we have shuffled off this mortal coil,
> Must give us pause. There's the respect
> That makes calamity of so long life.[3]

Equally but very differently, I may conceive death as being nothing to be feared at all. This may be so if I accept, based on available empirical evidence, that death is not even comparable to the unconsciousness of dreamless sleep, but rather is a total oblivion and non-existence in which there is no longer an "I" upon whom suffering or disvalue may alight. The Greek philosopher Epicurus argued for just this conclusion:

> Make yourself familiar with the belief that death is nothing to us, since everything good and bad lies in sensation and death is to be deprived of sensation... For there is nothing to be feared in living for one who has truly comprehended that there is nothing to be feared in not living ... So [death] is nothing to the living and nothing to the dead, since with regard to the former, death is not, and as to the latter, they themselves no longer are.[4]

---

[a]For those unfamiliar with the use of these terms in philosophy, what is meant by instrumental value is value for some further end or purpose, as opposed to intrinsic value that is valuable in and of itself.

In the latter case, postponing my death would appear to have no particular import from the perspective that death is a bad thing in itself, since death is *no thing in itself*. So in the case of a person who accepts this view, action taken that happens to postpone death could only be fairly said to be motivated by the intention to extend life, since motivation to postpone death on its own would be unintelligible.

We do not have, and probably will never have, evidence as to the possibility or nature of death as an experienced state of being. Thus we should bracket any concerns along these lines, and go with Epicurus' view on the badness of death,[b] instead concentrating upon what is valuable in life, as the only reasonable justification for efforts to extend it.

Some of you are perhaps scratching your heads and wondering if it really can be that death is not bad in any way. How can that be? Are all our fears about and distaste for death unfounded? And what of prohibitions concerning death? If death is not bad in itself, could such a view not remove the badness from murder, provided it is conducted suddenly, and without expectation or pain? More relevantly here, if death is not bad, then why make special efforts to prolong life?

Well, of course, there is a way in which death may still be accounted as bad. On this view death is not bad *in itself*, but rather the badness of death becomes relative to what it negates, namely the continuance of life. Death is bad because of what we lose by it. So if we want to assess the ethics of life-extension, we must consider the value of life, rather than the disvalue of death.

The Value of Living: Life-Extension and the Relative Badness of Death

There are two possible modes in which life may be said to have value: the intrinsic and the instrumental. While some notion of the intrinsic value of life may be germane to the question of whether or not we should

[b]Though not, necessarily, with his other views on the matter, as Epicurus contends also that "the right recognition that death is nothing to us makes the mortality of life enjoyable, not by adding infinite time to it, but by removing the desire of immortality." (*ibid.*) However he here seems to feel that desire for immortality, or life-extension, may be located in fear of death. As I suggest, this is the wrong notion in any case.

actively take a life, it does not appear to tell us anything about why a continuing life, so life-*extension* should be valuable. If life has intrinsic value, it is neither diminished nor increased by added time. The value of continuing life derives from the value of *living*. The value of living should be understood as instrumental value. It is about what we can *do* with a life, not whether we are alive at all.

So then, we must establish a basic category of creature for whom life may be said to have instrumental value, such that extension of a life is valuable. It appears trivial that non-conscious life cannot have instrumental value. What about conscious life? For conscious life to have instrumental value to a creature, a creature must have the capacity to value. There are certain features which, added to consciousness, allow us a capacity to value. These other features of a conscious being are to be summarized as follows: a capacity for *self*-conscious, rational, autonomous will. These features put together constitute a basic category, the possession of whose characteristics is commonly known in bioethics as *personhood*.[c] So we are left with a question of what about living, then, is valuable to persons?

Being a person is not simply being an entity — it is an ongoing, time extended *process*. This process is composed of desires, wishes, hopes, preferences, thoughts, plans, actions, experiences, emotions, memories, etc. These and their kind are the goods in a person's life, and they constitute the value of living.

Of course, there are also evils in living, and things such as pain, sorrow, remorse, fear, etc. that contribute to disvalue in living. Can these things cancel each other out? Is there a basic value in living that is always there no matter what misfortunes a person suffers? The words of the eminent philosopher Thomas Nagel are useful at this point:

> The situation is this: there are elements which, if added to one's experience, make life better; there are other elements which, if added to one's experience, make life worse. But what remains when these are set aside is not merely *neutral*: it is emphatically positive. Therefore life is worth living even when the bad elements of experience are plentiful, and the good ones too

[c]For the purposes of this essay, I shall not enter into further discussion of the subtleties of the ascription of personhood in boundary or marginal cases.

> meager to outweigh the bad ones on their own. The additional positive weight is supplied by existence itself... like most goods, this can be multiplied by time: more is better than less.[5]

So the process of living may be regarded as essentially good. But what is it that makes or allows it to be good? The present and backward-directed elements of the process of being a person, such as experience and memory, have forward-looking counterparts: hopes, desires, plans, and so on. Hoping, desiring, and planning are intrinsically future-directed. Hoping for, desiring, or planning our past is meaningless or futile. These aspects of the process of personhood involve projection into a multiplicity of possible *futures*. The temporally extended process that is definitive of a person's life involves both the existence of these future-directed elements and their objectives, and the possibility of those objectives being realized, thereby becoming the objects of the counterpart elements of experience and memory. On this view, then, there is no point in time at which the continuation of a person's life may be said not to be valuable, since these forward-directed elements are intrinsic to the process of being a person. As such, the process of being a person is intrinsically open-ended. Nagel expresses a similar view:

> The situation is an ambiguous one. Observed from without, human beings obviously have a natural life-span and cannot live much longer than 100 years. A man's sense of his own experience, on the other hand, does not embody this idea of a natural limit. His existence defines for him an essentially open-ended possible future... Viewed in this way, death, no matter how inevitable, is an abrupt cancellation of indefinitely extensive possible goods. Normality seems to have nothing to do with it, for the fact that we will all inevitably die in a few score years cannot by itself imply that it would not be good to live longer.... If the normal life-span were a thousand years, death at 80 would be a tragedy. As things are it may just be a more widespread tragedy.[6]

In this way, it would appear that there can be no arbitrary upper limit on the good of the extension of life to a person. There is no point at which being a person does not involve the future-directed elements and their involvement in the process of interchange with the present and past elements. An attempt to set or discover such a general limit would appear to involve a misunderstanding of the nature of the process itself. That we may know some facts about human biology which suggest that we indeed have an end in store, and even how far in the future that end is likely to be, in no way impinges upon the intrinsic nature of the elements of hoping, desiring, planning, which are fundamental to the process of being a person. These all point towards the ever-distant horizon of the possible.

If no general limit can reasonably be set or discovered, could one be set by a person upon themselves? That my desires, hopes, and plans may fix upon particular objectives does not in itself seem to suggest that I can easily, or at all, fix these elements of myself purely upon and continent within some set of particular objectives,[d] so that they end with the completion of this set. No matter what I specifically plan for, desire, hope for, it seems that these aspects of my psychology overflow the limits of their particular objects without any particular act of will on my part. A person whose self-professed sole hope, desire, and plan in life was to stand atop Mount Everest is nonetheless likely to find himself filled with some other such goal by the time he has reached the bottom again, or indeed to discover that he already had many in store, which had merely been obscured by this overriding one. Furthermore, willing these aspects of ourselves to be contained within a fixed, time-limited framework would seem to be a very difficult task indeed, if possible at all. I may seek to direct or curtail my first-order desires (those which simply "I desire") with my second-order desires (those by which "I desire that I do or do not desire"), but that a second-order desire to have *no* desires should be effective would seem a tall order, to say the least. And as to a self-imposed limit to the temporal extension of these elements of oneself, try to imagine a person setting a particular date beyond which she will be free of all plans, hopes, desires etc. Such a picture strikes one as ridiculous. So it does not seem very reasonable that a person may even set a limit to the good of their own future extension in time. Nonetheless, it will be useful to consider

[d]It may seem that plans should clearly be fixable by persons upon and continent within specific objectives. While this is true at one level, there are two observations to be made here. Firstly that while we may fix the goal of a plan, we may not also fix, with any assurance, the temporal end-point or time-frame for that plan's fruition. A plan which we may think will take us thirty years to execute, may in fact never come to fruition in our lifetimes, but may actually have done had we lived two hundred and thirty. Secondly, plans ride in on the coat-tails of dreams, hopes, and desires.

the effect that this might have, should it be possible, or should it happen as a matter of brute fact.

Without the constant interchange between the future, present and past elements of the process of being a person we should be fixed, and frozen, ourselves objectified and unable to fulfil our autonomous will[e] or formulate rational designs and desires, let alone actualize them. Our rationality, should we still possess it, would become purely analytic of past objects,[f] stripped of instrumental potency, our autonomy stripped of meaning. The process of being a person would cease, and the *continuance* of being itself would thereby be stripped of its value. Should we lose these future-directed elements of ourselves, then, we would no longer be persons, and living would have no value.

I do not suggest, however, that this is impossible. Indeed, I think it is both possible and perhaps does happen, albeit probably rarely in the extreme, owing to the difficulties outlined above. Perhaps the case of Elina Makropulos discussed in a later section of this essay is an example of just such a person, though in this case fictional. When such a case does occur, I would argue that for that person death is not bad in any way, since they have lost their ability to value instrumentally their own futures, have stopped being a person, and so are to that extent dead in any case. More, or future life holds no further benefit, since even a desire for pleasure taken in contemplation of the past, involves a desire that this pleasure should extend into the future.

For such a being, Epicurus' conjecture concerning the badness of death becomes the only consideration, and death is not bad. Death *at a particular time* takes nothing more from them than it would at a later time. They are constituted to be only present and past directed. Their future is meaningless to them.

Death is bad, then, because of what it takes from us, and what it takes from us is our future-directed elements, and their objectives. It shears from us our possible futures. This is surely what is referred to by the

---

[e]If this assertion is accepted, it would appear even more clear that willing ourselves to have no future-directed elements is indeed impossible, since the *effective* component of willing, on this account, appears itself intrinsically future-directed.

[f]And cannot be said to involve a desire in the normal sense, as indicated in the subsequent paragraph.

common intuition that death for the young is worse than it is for the old. For a person of the biologically present human form, death at a hundred years of age cuts him off from fewer possible future goods than does death for a normal seventeen-year-old. But this should not lead one to think that death for a person of unlimited future extension in time is *far* worse than it is for an ordinary person, since while the future-directed elements make life's continuance always a good, this simply implies *some* continuance. The degree to which this may be seen as beneficial at any one time has much to do with the objectives of these future-directed elements, though it is certainly true that with unlimited scope, some such objectives would doubtless be more distantly located.

Some of you might at this point be wondering why Epicurus' model should not also annul and make irrelevant to us the value of the future goods as well. Bernard Williams' elucidation of the basic class of forward-directed desire is useful here:

> ... a man might consider what lay before him, and decide whether he did or did not want to undergo it. If he does decide to undergo it, then some desire propels him on into the future, and *that* desire at least is not one that operates conditionally on being alive, since it itself resolves the question of whether he is going to be alive. He has an unconditional or (as I shall say) a *categorical* desire. ... It is not necessarily the prospect of pleasant times that create the motive against dying, but the existence of a categorical desire and categorical desire can drive through both the existence and the prospect of unpleasant times.[7]

So even if death is nothing to us when we are dead, death most certainly is something in relation to the categorical desire for future goods. Death is the desire's frustration, and its denial, and its tragedy is the loss of the desire and the goods which are that desire's objective. This is what is bad about death. Life is instrumentally valuable to us as persons, and so long as we are persons, and possess future-directed elements in the form of desires, hopes, plans and the like, death is bad insofar as it deprives us of these and their objectives. The value of life consists in the value of living, and living as a person is an intrinsically time-extended process with indivisible forward-projected elements. As long as these elements exist for us, we are persons, and life-extension will be valuable to us. And for persons, it is a value without limitation.

Arguments Against Biogerontological Intervention

Now that we have established a basic framework in which to ground the value of life-extension, I will very briefly outline and address some standard and one or two unusual objections to the aims of aging intervention and life-extension.

Natural Aging and Human Intervention

One common claim encountered in both lay and professional philosophic conjecture is that any intervention which seeks to alter the arrangements of how things have always been, may be seen as being an attempt to change what is "natural", and is therefore to be condemned.

As it applies to biogerontology, the premises-conclusion structure of this argument may be summarized as follows:

P1: There is a category of objects and powers in the world describable as "natural".

P2: The category of "nature" is in some way intrinsically "right" and "good".

P3: Such a category may be infringed by human intervention in a way which harms the fact of its inherent "naturalness", and thus harms its good.

P4: The ordinary trajectory of and other basic facts about human lifespan and aging fall within the category of the "natural".

C: Aging interventions are wrong and bad since such intervention harms something "natural" and good.

There is often a further premise relating to the proper province of medicine, which attempts to separate "natural" aging from disease.[8] I consider this to be a variety of the same argument, and so vulnerable to the same objections I will make to the basic argument. I suggest that the above outlined argument and its cousins are unsound and should be rejected.

Apart from anything else, this argument falls foul of a standard objection in philosophy known as "the naturalistic fallacy".[9] This objection states that it is always illegitimate to make any moral statements based purely upon empirical statements about things in the world. So then, just because such and such a state of affairs happens to be the case, (it "is" so) never means, on the strength of this evidence alone, that one may then conclude that it is morally good or bad (it "ought" or "ought not" to be so). This is because moral statements and factual statements constitute entirely different conceptual categories. In and of itself, each tells you nothing about the other. Humans may age and die in a fairly standard manner and time. This appears to be a fact, though a hazy one to be sure. But there is nothing about this fact which tells me that this "ought" to be the case.

Aside from this, the argument from "nature" fails since it either does not refer to an intelligible distinction, or if it does, makes unsound or absurd assertions concerning that distinction, as we shall see.

There are two possible types of claim that can be made about nature in this context. The first is that "nature" may be defined as a set comprising any object or power that is within time and space. If this definition is accepted, neither humans, nor anything they can ever do, nor any arrangement of themselves or things in the universe they can ever make, will be anything but natural. The second type of claim that may be made is that "nature" is that set of things with which humans have not yet interfered, and that human interference creates "unnaturalness", which is usually characterized as bad. Aside from its arbitrary nature, and the oddness of the fact that a possible universe in which there never were humans would on this account have no defining condition by which it could be called "natural", there are several awkward, or unreasonable consequences of this. As regards moral claims about "unnaturalness", if we allow the general assumption to stand that "nature" is good, and any interference in it is presumed to be bad, then it would appear that *anything that humans can ever do is bad.* This is so because on this view the "natural" course of events, or whatever we don't interfere with, becomes an absolute moral standard, and all human action is by this definition "unnatural" or counter-natural. We may not take refuge in the notion that it is "okay" to interfere with things we have already interfered with, since if we cease our interference, "nature" takes its course once more. Such a conclusion is absurd.

Most devastatingly, it is impossible for a defense of this argument to be mounted once one begins to enquire just where this boundary between the natural and the unnatural lies. Are humans really unnatural in some deep sense, or is it just their actions which are? If humans are entirely so, how did they arise from a "natural" universe? If only their actions, how can a being which is wholly natural, itself act "unnaturally"? If certain facts about humans, such as their aging and life trajectory, are "natural" then what can be "unnatural" about *our own* interference in them? Such a claim relies upon *something* in humans being unnatural. But what? The problem becomes crystal clear when the argument is applied to manipulation or alteration of the human genome. For such intervention to be "unnatural", the genome itself would need to be defined as "natural". But the genome in question is *our genome*, and the source of this purported unnaturalness, *ourselves*. Considering that our abilities, insofar as they are different from the "wholly natural" animals, are different only because of differences in *our genome*, the absurdity of this line of reasoning becomes inescapable.

Overpopulation and Sociobiological Reproductive Pressures

Another common conjecture is that increased average longevity will clearly lead to overpopulation. Aside from the fact that this is mostly a practical, rather than an explicitly moral concern, it doesn't actually seem to be borne out in practice.[10] It would appear that there is ordinarily an inverse relationship between life expectancy and population growth. In poorer countries where life expectancy is low, and in other historical situations where life expectancy has been low (though less directly related to relative wealth) the birth rate has, and is still seen to be high or very high. In modern societies where the average life expectancy is longer than at any time previously, the birth rate is below replacement level in many cases. The baby boom occurred in response to (or at least simultaneous with) a perceived sudden and dramatic lowering of average life expectancy in the groups who subsequently boomed. If greatly enhanced longevity would go hand in hand with increased reproductive life span, there still appears no definite reason for alarm. One may well ask what the effect of the ability to safely reproduce anytime, say, in the next 150 years might have on the average educated and career-minded woman. I would suggest that the effect would be both positive and welcome, and would be unlikely to lead to a flood of babies. When taken to its extreme this worry can be linked to the common misconception that radically successful biogerontological intervention will lead to true immortality. Even when endogenously unlimited longevity is disentangled from its mythological cousin, I suspect that most readers of this essay will ruefully recognize what a distant prospect that is. Even if the various Gordian conundrums facing biogerontology are solved, there remains the ever present specter of cancer, which appears, at least to this commentator, more like a category of cell-state than a single disease, and is not likely to be tractable by a simple or singular "cure". This is not to mention the scourges of accident, infectious disease, war, famine, etc. Given these sorts of considerations, overpopulation as a result of life-extension seems a luxurious kind of worry.

It is worthy of note, however, that it would appear that if these conjectures are valid and sound, then the consequence of generally and greatly enhanced longevity would be a decline in the frequency of children in the population. I account this to be both a serious, sad, and perhaps morally significant consequence.

Mythic Immortality vs. Life-Extension: Problems of Finitude, Striving, Boredom and Personal Identity

It would seem a trivial observation to notice that no aim of biogerontology can be to make persons immortal in the mythic sense of being both eternal and invulnerable. Despite this, such a confusion is very common, and is apparently made even by prominent commentators in the bioethical field. For example, consider these lines from a recent and influential essay by Leon Kass on the value of a limited life, and the reason we shouldn't seek to extend it too much:

> Homer's immortals — Zeus and Hera, Apollo and Athena — for all their eternal beauty and youthfulness, live shallow and rather frivolous lives, their passions only transiently engaged, in first this and then that. They live as spectators of the mortals, who by comparison have depth, aspiration, genuine feeling, and hence a real center in their lives. Mortality makes life matter.[11]

The confusion here is obvious, and while Kass clearly acknowledges elsewhere that life will never be unlimited and that whatever biogerontologists achieve, biological persons will remain mortal, he appears not to notice the importance of this distinction. The point I would like to make here, which is relevant to many of the critiques of life-extension to be found in current literature, is that Kass and his fellow antagonists of biogerontology's life-extending potential do not take this distinction seriously enough. It is not a trivial distinction. It is vital, and categoric. Simply put, the distinction is one between an infinite set, and a finite set of presently undetermined or uncertain extent. To see the importance of this, we need to examine one or two aspects of these critiques more explicitly.

It is often held that finitude is vital to our aspiration, engagement, commitment, and striving in our lives:

> ... the remoteness of the midnight hour might influence negatively how we spend our days. For although the gift of extra time is a boon, the perception of time ahead as less limited or as indefinite may not be. All our activities are, in one way or another, informed by the knowledge that our time is limited, and ultimately that we have only a certain portion of years to use up. The more keenly we are aware of that fact, the more likely we are to aspire to spend our lives in the ways we deem most important and vital. ... Many of our greatest accomplishments are pushed along, if only subtly and implicitly, by the spur of our finitude and the sense of having only a limited time. A far more distant horizon, a sense of essentially limitless time, might leave us less inclined to act with urgency. Why not leave for tomorrow what you might do today, if there are endless tomorrows before you?[12]

A careful reading of these passages reveals the sleight-of-hand. We move from a mere sense of indefiniteness, to a definite sense of endlessness. But what about indefiniteness? Does the fact that I have a limited lifespan constitute the only condition for my doing anything in life? It seems an odd argument to assert that I go out to engage in a game of football today, only because I am aware that I cannot do it three centuries hence. Such an argument appears to miss the point of the *process* of living: it is the movement, shepherded by our autonomous will, of the forward-planned objectives into the objects of present experience and past recollection that we value. Just because I *could* put my game of football off indefinitely, does not seem to matter. I won't, because I want to experience it. After all, it is true that I could put it off until next week, next month, next year, perhaps next decade or further. But then, of course, if I continue to do so, even if I had unlimited time, *I shall never play my game of football.* This applies equally to prioritization of important, over trivial activities. There are many who exist with our presently limited span who do little or nothing with their lives. Seen in this way, such arguments may easily be re-characterized as the arguments from *laziness.* Such an argument has no real force.

But more importantly, the point is just this: it is the very indeterminacy and uncertainty which is vital to remember here. Even if we were functionally immortal-biological creatures with no endogenous limit — we would still be *mortal* and *vulnerable.* We will never know, just as we do not now, with any degree of certainty, when the "midnight hour" will strike! The war poet Keith Douglas says it best, describing the death of a young man by his hand:

> How easy it is to make a ghost.
> The weightless mosquito touches
> Her tiny shadow on the stone,
> and with how like, how infinite
> a lightness, man and shadow meet.
> They fuse. A shadow is a man
> when the mosquito death approaches.[13]

Nothing that biogerontology can achieve will change this fact. We are biological entities, and whether we are old and decrepit or young and hale, we will remain frail, mortal and vulnerable.

A related kind of worry is the concern about boredom, which involves a further concern about personal identity. The worry here generally runs like this: if we lead greatly extended lives, either we will become bored, since we will not be able to find enough variety in either life or in our own approach to it to sustain us indefinitely, or else we will, by varying our personalities and experiences so much, lose touch so completely with who we were originally that the further life we gain cannot any longer be said to benefit the person who we were. The problem is thus presented as a dilemma. One horn of the dilemma is that life will become so repetitive and our boredom will be so extreme that we will no longer have any forward-directed desires, hopes, plans, etc. and life will become valueless to us. Such a case is depicted in Karel Capek's 1922 play *The Makropulos Case*, where the character Elina Makropulos possesses an elixir of life. In the words of Bernard Williams, in his essay of the same name:

> At the time of the action she is aged 342. Her unending life has come to a state of boredom, indifference and coldness. Everything is joyless: 'in the end it is the same', she says, 'singing and silence'. She refuses to take the elixir again; she dies; and the formula is deliberately destroyed by a young woman among the protests of some older men.[14]

Such a case is importantly different from the worries expressed in the legends of Tithonius, and the story of the Struldbrugs in Swift's Gulliver's Travels, who attain immortal life, but not eternal youth, and so suffer the untold horrors of an unendingly increased decrepitude. It is quite clear that it cannot be among the aims of life-extension or more obviously of the aging intervention aspect of biogerontology to achieve a Struldbrugian style of existence. Such an outcome is explicitly contrary to biogerontological aims. Elina is depicted as rather being both physically vigorous and healthy, and frozen in the state of character she was in when she originally took the elixir.

The other horn of the dilemma is that if we do in fact find sufficient variety in our involvement in the world, in doing so we will, over time, become so different from the way we had been, that it is no longer possible to assert that life-extension into the indefinite future will be useful or good *to ourselves* as particular persons. If this is the case, then why not leave things the way they are, since if I will not be me in future, but someone else, then I may as well be dead, and someone else, of a future generation in the ordinary sense, may just as well exist in my future-self's place?[g]

Must indefinite life-extension necessarily fall foul of this dilemma? In order to examine this effectively, it is necessary to analyze the two horns more closely. In the case of the first horn, given the enormous presently existing variety of human endeavor and avenues of interest in the vast universe, and the fact that new avenues are perpetually being generated, while the old (especially in empirical or intellectual disciplines) are rarely or never exhausted (and indeed are often self-regenerating), it does not

[g]Though best set out by Williams himself, for a presentation of this case in context of biogerontology see Glannon.[15]

seem reasonable that *the world* itself should lose its interest. Surely what is taking place in such a scenario, then, is that the person *themself* fails in some way to meet or find exciting these endless challenges, intrigues, and possibilities. So the problem may be redefined as one of individual personal *character*. It may simply be that Elina Makropulos was bored essentially because she was *boring*. Seen in these terms, is it necessarily the case that all persons would be so troubled? I submit that it is not. While it may be true that some styles of character are bored with life almost from the get-go, and find no particular continuing interest in life even in the full bloom of youth, others very clearly do not fall into this category. Did Newton lose interest in his studies or activities as he aged chronologically? Did Einstein? Plato? Da Vinci? Churchill? Peter the Great? Did any such polymathic characters, at all? It seems that some characters, at least, are well suited to lives that would extend very far indeed beyond the normal lifespan limits. I, for one, feel that had I a dozen times my presently projected span, I should hardly have time adequately to pursue all my avenues of interest. I, and no doubt many others, feel deeply cramped by the shortness of span, forced rather arbitrarily to prioritize certain few among very many possible interests, to the near or total exclusion of many others. I do not account this a benefit, as the long quotation above might suggest I should. Evolutionarily speaking, in this context, the human brain (unique in the known universe) within the human frame (a typically aging mammalian model) may be seen to be maladapted to each other: a bit like a jet engine mounted on a bicycle. There is a dreadful mismatch between power/potential and physical restriction. Why ought we to accept the confused dictates of our random biological heritage? Perhaps part of the aim of biogerontology may be seen as an attempt to mitigate the bad effects of this mismatch. The fact that some characters, such as Elina, decline into a *psychological old age* early in life or when still vigorous may simply be part of the gerontological conundrum. Biogerontology, in combating physical aging, may very well combat psychological aging as well. To deny this outright is to suggest a kind of Cartesian mind/body dualism.

Another common suggestion that psychological aging is beneficial as a palliative to the prospect of death is neither comforting nor convincing. Prior to execution, I might be administered a drug that lessens my concern about my own impending doom, but in such a case would I not have as much reason to fear and despise such a drug as the execution itself? Such an intervention simply co-opts and thus reduces my autonomous will. As regards physical aging, if I do not *will* it to happen to me, the analogy holds. Given what we have established about the value of living, it would seem that this is a primary evil. As Dylan Thomas put it, lamenting his previously fierce father's decline into meekness prior to death:

> Do not go gentle into that good night,
> Old age should burn and rage at close of day;
> Rage, rage against the dying of the light.[16]

It is therefore not within the remit of biogerontology to preserve aging as a palliative to death. Nor can it be within the remit of the discipline to attempt to alter the conceptual facts about the badness of death. These are not tractable within the scope of biology.

As regards whether the fulfillment of such hardly-bounded interests of a *psychologically youthful* character-type, as described above, would necessarily involve their becoming someone else, it appears self-evident that they would not, since such interests are their own, present, and immediate interests. The situation should be examined in the light of an assertion I will now make: *no normal person who has ever lived has died having fulfilled their potential.* That some characters have little or no interest in fulfilling that potential seems irrelevant to the question, aside from the further question of whether they may be suffering a prematurely or unwarrantedly aged psychological state; or else are just plain lazy.

Furthermore, it may be seen that the situation at present itself limits the scope of our interests. There are many projects that I or others do not find interesting or will not undertake simply because we are aware that we can never finish or even significantly further them. A greatly expanded lifespan-prognosis would, I submit, expand, rather than diminish both the range and scope of our interests.

But what of the problem of personal identity? A full rehearsal of the philosophical ground is beyond the scope of this essay. But I feel that I have already addressed one issue regarding this. It is clearly the case that with some characters, *who they presently are* furnishes sufficient largesse of variety of possible engagement in the world to obviate the worry that they need become some other person in order to continue to be so engaged for a greatly extended period. As regards the deeper questions of personal identity in a lifespan, I suggest that since who I am now is very nearly in no way at all who I was when I was three, twelve, fifteen, etc. or who I will be when I am eighty, the problem is not exclusively one of indefinite life-extension, but rather both fully pertains to our present situation, and, what is more, does not appear to trouble us overmuch.

Finally, it is useful to point out what Elina Makropulos, we presently, and any future life-extended person have in common: an exit strategy. We will never be mythically immortal, and should we find that life ceases to be meaningful and fulfilling to us, we may always end it, as does Elina in the play. The opposite is, to put it mildly, not so easy, hence the protestations of the "old men". Seen in this light, the action of the young woman in destroying the elixir is both foolish and wanton, and also curiously *paternalistic*, since it presumes to make Elina's personal decision, for all. We may conjecture all we like, but we will never *know* until we have the opportunity to see for ourselves, and should we not like what we find, a remedy is always to hand. Providing the opportunity to make that choice *voluntarily* may be at least part of the aim of biogerontology.

The Problem of Incumbency and the Social Value of Life-Extension

Briefly put, the worry is that, assuming aging-intervention and life-extension are both effective and widespread in their uptake, then those who are chronologically precedent, or older in this way, will have no incentive to make way for the young, and indeed, given the above considerations, may be positively driven by their faculties and abilities to remain incumbent in positions of power and authority indefinitely. The problem is most stark when one adds to the scenario the conjecture that at least some of these persons will be of bad character, as described by David Gems:

> This is why I fear research into aging. If treatments had been available in the twentieth century that halved the rate of aging and doubled lifespan — as some mutations do in *C. elegans* — then Mao Tse Tung might still be alive. He would be the equivalent of fifty years of age, and might not be expected to die a natural death until 2059. Worse still, Joseph Stalin would be "sixty-three" and would live until 2027. Do we really want anti-aging therapies in

> the hands of Robert Mugabe, Fidel Castro, or Kim Jong Il? Historically, a great benefit of aging has been deliverance from tyranny.[17]

While one cannot but have sympathy with the view that such tyrants' demise is beneficial, is the question really one about aging or lifespan intervention? Such an argument might suggest that in order to achieve political turnover, one ought in general to *shorten* the average span of human life! In addition, it may be noticed that it was the demise of others, such as Lenin, which brought these tyrants to power in the first place. What has it fundamentally got to do with aging intervention that one good or bad character may be replaced by a better or worse? We value the longevity of the good as much as the demise of the bad. Perhaps more. Surely these questions are more about political structure than lifespan *per se*, and it would be a very strange argument indeed to suggest that we ought not to seek a cure for cancer simply because it may also benefit a tyrant, and thereby keep him in power longer.

The worry has been expressed more generally elsewhere, as here in the previously quoted "Ageless Bodies" section of the Report of the President's Council on Bioethics:

> The mature generation would have no obvious reason to make way for the next as the years passed, if its peak became a plateau. The succession of generations could be obstructed by a glut of the able. The old might think less of preparing their replacements, and the young could see before them only layers of their elders blocking the path, and no great reason to hurry in building families or careers... Families and generational institutions would surely reshape themselves to suit the new demographic form of society, but would that new shape be good for the young, the old, the familial ties that bind them, the society as a whole, or the cause of well-lived human lives?[18]

This does seem to be a concern more applicable to the ordinary structure of modern societies. But once again, it is a structural, rather than an essential concern. Its nature is political, and it is not in any case a new problem. After all, for the majority of the history of humanity, power was concentrated into the hands of a few families — the aristocracy — who handed power from one generation to the next precisely *as though* the son of one *was* his father of the previous generation. The problem is the same, at one remove, that of inter- as opposed to intrafamilial incumbency. Though as to the latter, I think it is an unfortunate phrase to speak of my parents continued living as merely "blocking [my] path"! In the intrafamilial case, simple kindness, consideration and love would appear to provide the answer. Also, it might simply mean that offspring have to shift for themselves a little more than they presently do. Would this be antagonistic to "the cause of well-lived human lives"? As to the interfamilial case, could such a problem not be tractable in like manner as with the problem of the aristocracy, with incumbency in many or all positions of hierarchical authority being limited from below, by some political constraint similar to democracy? The suggestion is simply that the process, begun in politics itself, be spread to the organizations of *all* hierarchies. Not, in principle, an unattainable end, and one which might be both desirable and more urgently demanded, and thus more likely, if intergenerational interchange is indeed slowed. As to that last point, it has already been accepted that should life-extension become generalized the interchange of generations would be slowed, but it surely would not be *stopped*. For it to be so, one must revert to the model of "mythic immortality", which has already been shown to be irrelevant.

The problem might be seen to be more acute where land ownership is a question. And yet, once again would the situation be much different than the presently existing one, with the vast majority of the land in a minority of hands? Once again, the problem is pragmatic, not essential, and the answer likely lies in land reform, perhaps in quantitative limiting of private access to ownership of land. But in the interfamilial case, this is the same kind of question we already face, where land is heritable from one generation to the next. If the above conjecture regarding the birth-rate holds, it may be that there are fewer overall people, and the problem will then be less acute than it is presently in any case.

In addition to this, returning to hierarchies of authority, it should be noted that while I may have various interests which will be benefited by my continuation into more lifespan, I will inevitably remain a unitary individual, and will not be able effectively to wear very many hats of practicing authority at once. That I may have more time overall, does not suggest that I have more days in a week, or more hours in a day. So as the complexity of human endeavor and knowledge grows, as disciplines become more and more numerous and fragmentary, so will the need for more and more individuals to devote their *whole* attention to each discipline in any one period. This is of course not to disdain the enormous *integrative* benefit which will likely be gained by many more very experienced persons being present per capita in society, but rather that the various disciplines will thereby be better able to communicate. If this picture is correct, then what we may end up with would be a far *more* integrated society, with more highly experienced subalterns, and fewer generals, with less power. In short, society may become a more cooperative and more egalitarian whole. Such a picture may appear overly rosy, but I submit that it is at least plausible, and I paint it in order to set it against the arguably overly gloomy picture painted in the above quotation.

The closely related worry that a slower interchange of generations will lead to a slower rate of change of ideas, is predicated upon the notion that the old have fewer and fewer "fresh" ideas, becoming more and more set in their ways, and that the young are needed to inject freshness and novelty into the world. The distinction between freshness and novelty will be seen to be important. But first, this worry seems directly related to the Makropulos concern about the ossification of character, and may equally be irrelevant to some, and tractable in others, if psychological aging is likewise biologically predicated. The fact that lambs become sheep in a year, while for humans it takes many more, and some humans never really become "sheep" at all, suggests that the last stated possibility is, indeed, likely true.

Beyond this, the notion that new ideas are always better than old seems an odd sort of approach from the conservative point of view, which seeks to keep the aging picture the way it presently is. Are "new" ideas necessarily always better? Should we not think that there may be some merit worth keeping in the counsel of those who have been alive longer? And are "new" ideas in any case always really "fresh"? The chronologically young have a bad track record of thinking that they have made some new discovery of approach, while they are in fact merely repeating the same bad courses of action their elders or past societies rejected long ago. Many *prima facie* "new" ideas turn out on closer examination simply to be variations on a well-worn theme. In this way, it may be harder than the proponents of this argument suggest, to have truly "new" ideas at all, and it may be easier to have genuinely "fresh" ideas only when one has been around long enough to recognize them as such. Thus it may be that by the innovation of intervening in aging, and expanding human lifespan, we may make the world both more stable, and more *profitably* conservative, in the sense of being wise, while retaining the benefits of youthfulness and energy for the discovery of truly "fresh" ideas, facilitated by both expanded knowledge and the banishment of the ossification of physical aging.

Problems of Unity: Distributive Justice, Parallel Populations, Parallel Species

I have reserved until last the set of problems which I consider to be most serious, and perhaps least tractable. Briefly put, assuming age-retarding and life-extending treatments are effective and safe, the problem is one of uptake. The less egregious version of this problem is that not all will wish to be treated, and so there will be biological disunity. The more egregious version stems from the assumption that either the treatments will be expensive, or at the very least, even if they aren't, they will not be available to all, given world poverty and population mass. It may be seen, then, that the rich or those in richer areas of the world, will begin to use their economic advantage to buy biological advantage.

The basic problem has three aspects: an economic, a moral and a political.

The economic problem is clear. It is one of both distributive justice in a straightforward sense, and also of a new kind of problem of this sort. The basic distributive problem has been stated, is obvious, classic, and needs no further treatment here. But I would suggest that since the kinds of changes likely to be necessary involve not just supplements, but changes to the biological structure of individuals, which may very well then be heritable by reproduction, the situation is quite different from anything heretofore encountered in human history. If part of the biological advantage that the wealthy buy confers advantages in terms of endogenous capability and potential, and thereby potential for both wealth, knowledge and skill acquisition at a level that is simply beyond the *physical* capabilities of the non-enhanced, then competition in an ordinary sense will no longer be possible. Consider for example, the relative advantages of a family who, by ordinary biological facts, needs to breed three to five times a century, as compared to a family who need to breed only once a century, or less. The very advantages that have been suggested in the above sections, and that are so tempting, may cause the enhanced population to be in a situation of advantage which is unreachably beyond the physical means, and indeed the *potential*, of the lives of the unenhanced. The gap would no longer simply be between rich and poor, but would rather become a categorical gulf. The poor world would be a world which is not only exogenously, but *endogenously* disadvantaged. Interventions such as the provision of medicines, food aid, and a stable local economic climate would no longer be sufficient to give even the basic circumstances of those in poorer situations parity with those with enhanced biologies. The case is similar in ordinary conditions of voluntary lack of uptake. In the rich/poor divide, one may of course suggest that there would be a trickle-down. But this is far from clear, and it appears quite clear that in the medium term at least, the gap would widen dramatically, and unprecedentedly. So there would appear parallel populations in a sense that has never yet been encountered.[h]

There is also a special kind of moral problem here, one which I shall call the problem of harm by contextual devaluation. The fact is that despite the differences in life *expectancy* between rich and poor nations, every human population on earth enjoys a roughly statistically identical *potential* lifespan. There is a unity in this sense. If some begin to enjoy a potential lifespan that is either greatly enhanced, or unlimited, the picture may be significantly morally different than it presently is. To understand this, we must return to the idea that death is bad relative to the loss of potential futures. If, as things presently stand, a ninety-year-old and a seventeen-year-old lay unconscious in a burning building, and only one could be saved in time, the instinct of a firefighter would most likely be to save the younger. The intuition upon which such a decision is based is the one just stated. But consider an alternative case, one in which two seventeen-year-olds are in the burning building, one with a presently normal lifespan prognosis, and the other with a greatly enhanced lifespan potential. Which one should the firefighter save? By the ordinary intuition, he should clearly save the *enhanced*. Could it be that by altering the background conditions of human life, what I shall call the *absolute space* of lifespan, at present unified, we will be uncoupling human populations and creating an unprecedented moral disunity? Could it be that we will alter the relative *value placed upon the lives of the parallel populations*?

[h]Harris deals at length with this problem.[19]

I consider this problem to be serious. One mitigating factor has been discussed in the section on the *Value of Living* but the question is too complex for a full treatment here. I will suggest one possible general solution below.

The moral problem thus described has, of course, a political dimension. Once again this is too complex for a full treatment here, but I will attempt a brief analysis. If we become biologically disunified, will we have either incentive, or more importantly justification, to remain politically unified? The problem may be most acute in cases such as the United States' model of democracy, which is founded upon principles of natural law. The Declaration of Independence begins:

> When, in the course of human events, it becomes necessary for one people to dissolve the political bonds which have connected them with another, and to assume among the powers of the earth, the separate and equal station to which the laws of nature and of nature's God entitle them, a decent respect to the opinions of mankind requires that they should declare the causes which impel them to the separation. We hold these truths to be self-evident, that all men are created equal ...[20]

The sense of these words is clear. It is the "natural" or in other words, biological equality, the biological *unity* of humanity, which underpins political unity. The dissolution of political bonds spoken of in the document does not reach to the more profound level which is suggested. If this does not appear clear, some research concerning the philosophical bases of this document should clarify this assertion.[21] Of course, there are different bases upon which political unity may be founded, but we must have an eye to the possible consequences within at least this one fairly dominant framework, and possibly others as well. Should the biological unity of humanity be seen to be fragmented, serious questions may arise concerning the validity of aspirations to political unity.

As I have said, a full treatment of these issues is very broad in scope, and far beyond that of this paper. But I would like to suggest one possible route out of at least the moral and political, if not the economic problem of disunity. It is, of course, suggested by the very notion of value which I have outlined in this paper: the value of personhood. No matter what biological changes may occur, we all, old biological forms and new, will be *persons*. It is this intuition above all that, for example, allows us rightly to treat those who are biologically different or lifespan-disadvantaged presently, as in the cases of Down's syndrome persons, or those suffering from progeria, as entities worthy of full moral and political respect. We are, and will remain, all of us, persons. This is the unity upon which we must focus. And it is a unity which cannot be broken.

Conclusion

I have attempted to lay out both the groundwork for a moral basis for aging intervention and life-extension. I have also outlined and addressed some commonly raised issues, and have attempted to show that while some may be serious, others are illusory or unreasonable. There are of course issues that have not been addressed, such as the delimitation of disease, but an exhaustive treatment is not possible in an essay of this length. I hope that this paper has both demonstrated the importance of ethics to the practice of biomedical gerontology, and also clarified the situation to some degree.

References

1. Harris J (2002) Intimations of Immortality — The Ethics and Justice of Life Extending Therapies. In: Freeman M (ed.) *Current Legal Problems.* OUP.
2. Milton J (1667) *Paradise Lost.* Book One.
3. Shakespeare W (1604) Hamlet III; i.
4. Epicurus. Letter to Menoeceus (124–5). Furley D (trans.) (1986) Nothing to us? In Schofield M, Sriker G (eds.) *The Norms of Nature*, Cambridge.
5. Nagel T (1970) Death. In Nagel T (1979) *Mortal Questions*, p. 2. Cambridge University Press, Cambridge.
6. Nagel T (1970) Death. In Nagel T (1979) *Mortal Questions*, pp. 9–10. Cambridge University Press, Cambridge.
7. Williams B (1972) The Makropulos case: reflections on the tedium of immortality. In: Williams B (1973) *Problems of the Self*, pp. 86, 100. Cambridge University Press, Cambridge.
8. Caplan A (1992) *Is Aging a Disease? If I were a Rich Man could I Buy a Pancreas? And Other Essays on the Ethics of Health Care*, pp. 195–209. Indiana University Press, Bloomington.
9. Moore GE (1993) Naturalistic Ethics. In: *Principia Ethica*, revised edition, pp. 89–110. Cambridge University Press.
10. Gems D (2003) Is More Life Always Better? The new biology of aging and the meaning of life. *Hastings Center Report* **33**(4): 31–39.
11. Kass LR (2001) L'Chaim and its Limits: Why Not Immortality? *First Things* **113**: 17–24.

12. Kass LR (Chair) (2003) Ageless Bodies. In: *Beyond Therapy — A report of the President's Council on Bioethics*, pp. 185–186. PCBE, Washington D.C.
13. Douglas K *How to Kill.*
14. Williams B (1972) The Makropulos case: reflections on the tedium of immortality. In: Williams B (1973) *Problems of the Self*, p. 82. Cambridge University Press, Cambridge.
15. Glannon W (2002) Identity, prudential concern, and extended lives. *Bioethics* **16**(3): 266–283.
16. Thomas D *Do not go gentle into that good night.*
17. Gems D (2003) Is More Life Always Better? The new biology of aging and the meaning of life. *Hastings Center Report* **33**(4): 31–39.
18. Kass LR (Chair) (2003) *Beyond Therapy — A report of the President's Council on Bioethics*, p. 192. PCBE, Washington D.C.
19. Harris J (2002) Intimations of Immortality — The Ethics and Justice of Life Extending Therapies. In: Freeman M (ed.) *Current Legal Problems.* OUP.
20. The Declaration of Independence. National Archives, Washington D.C.
21. Locke J (1986) *The Second Treatise of Government.* Prometheus Books, Buffalo, NY.

2

Prevention and Treatment of Skin Aging

Jerry L. McCullough† and Kristen M. Kelly
Department of Dermatology
University of California
Irvine, California

†*Irvine, California 92697-2400*
(949) 824-5515
Fax: (949) 824-5407

INTRODUCTION

Youth is a valued attribute in our culture. Yet, in Western societies, life expectancy is increasing and populations are aging.[1] These conditions create a disparity between societal values and the reality of many individuals. As such, aging concerns, especially skin manifestations of aging, have become a common reason for medical consultations.[2]

Two processes result in the skin changes associated with aging. Chronological aging has strong genetic influences and includes the effects of gravity, sleep lines, expression lines, hormonal changes and atrophy.[2] Clinically, one appreciates fine wrinkling that can be resolved with stretching; dryness, which can be uncomfortable and unpleasant to the touch; and prolonged healing times. Histologically there is mild epidermal atrophy and a moderate decrease in Langerhans cells.[3] The dermal-epidermal junction is weakened and a decrease in celluarity, elastic fibers and glycosaminoglycans is observed in the dermis. Capillaries are fragile. Diminished fat and a diminished capacity for water retention are frequently found.

The second process is extrinsic aging, the accumulation of a lifetime of effects from environmental influences including ultraviolet and chemical exposure (for example, smoking). Ultraviolet radiation, especially ultraviolet B (UVB, λ 290–320 nm), is the most significant extrinsic factor in skin aging of most individuals, and results in deep wrinkles not alleviated by stretching, irregular pigmentation, and benign and malignant neoformations.[3] Histologically there is inflammation with increased cellularity and enlarged fibroblasts. Replacement of the dermis by basophilic fibrous material, elastotic degeneration, is generally prominent.

The combination of intrinsic programmed aging and extrinsic environmental injury results in clinically observed skin changes, which are seen as "old" and even "ugly". Significant intrapsychic and interpersonal benefits can result from diminution or elimination of some of these skin changes of aging.[1] A wide range of chemical agents and surgical procedures are available for treatment of aged skin (Table 1). Proper use and selection of the available modalities can significantly improve patient appearance and emotional well-being. Unfortunately many patients have been misled or have the misconception that all aspects of aging can be "treated". It is important for physicians and patients to have an understanding of likely benefits and limitations of available treatments as well as realistic expectations and common goals for implementation of these manipulations. Scientific evaluation of and education about the modalities available for treatment of aged skin can help to achieve this goal.

Prevention of Skin Aging

Sunscreen

Any physician discussing treatments for skin aging, should begin by stressing the importance of using sunscreens, as the beneficial effects of any other skin rejuvenation steps will be minimized or even cancelled if unprotected sun exposure continues to induce damage. Broad-spectrum sunscreens help prevent further ultraviolet (UV)-induced damage and may foster skin repair.[2] Al Mahroos *et al.*[4] demonstrated that daily use of a broad-spectrum sun protection factor (SPF) 15 sunscreen effectively prevented thymidine dimer formation, a marker of UV-induced DNA damage.

Table 1 Treatments for Skin Aging.

Treatment	Recovery/ *(Discomfort)	Onset of Improvement	Effect Duration	**Expected Improvement	Estimated Cost
Topical Treatment					
Retinoids	None/(0–1)	2–3 months	All topical	0–1	$40–120/month
Hydroxy acids	None/(0)	2–3 months	treatments	0–1	$10–100/month
Kinetin	None/(0–1)	2–3 months	require	0–1	$75–120/month
Antioxidants	None/(0)	2–3 months	continuous use	0–1	$10–120/month
Copper Peptides	None/(0–1)	2–3 months		0–1	$30–120/month
Microdermabrasion	0–1 day/(0–1)	0–1 day	Requires repeat treatments	0–1	$150–300/treatment
Chemical Peels					
Superficial	0–4 days/(1–2)	2–3 peels	2–4 months	1–2	$100–200/peel
Medium	7–12 days/(2–3)	2–4 weeks	1–2 years	1–3	$500–1500
Deep	2–4 weeks/(3)	4–8 weeks	2–5 years	2–3	$3000–5000
Botulinum Toxin	0–3 days/(1–2)	1–3 days	3–6 months	1–3	$300–800/injection
Dermal Fillers	0–1 week/(1–2)	Immediate	Variable	1–3	$500–4000/injection
Lasers/Light Sources					
Pigment/Telangiectasia Removal	1–3 weeks/(1–2)	1–6 weeks	1–5 years	1–3	$300–1800
Ablative Skin Resurfacing	1–4 weeks/(2–3)	1– 4 weeks	3–7 years	2–3	$2000–7000
Nonablative Skin Resurfacing	1–4 hours/(1–2)	2–9 months	Variable	1–2	$750–3000
Dermabrasion	1–4 weeks/(3)	1–4 weeks	3–7 years	2–3	$2000–7000
Radiofrequency Devices	1–24 hours/(1–3)	1–3 months	Unknown	1–2	$2000–6000
Cosmetic surgery	1–6 weeks/(2–3)	1–6 weeks	5–7 years	2–3	$5,000–20,000

*Discomfort (0 = none; 1 = mild; 2 = moderate; 3 = severe)

**Expected Improvement (0 = none; 1 = subtle; 2 = moderate; 3 = major)

It is further important to counsel patients in the proper use of sunscreen as people often apply less than the recommended amount and fail to reapply during long outdoor exposures. Of course, sunscreens are most effective as part of a general sun protection program, which should also include good sun protection practices such as: 1) limiting midday sun; 2) judicious use of shade; 3) protective clothing; and 4) avoidance of sunlamps and tanning beds.[5]

Hormone Therapy

In developed countries, most women will spend one-third or more of their life in the post-menopausal period.[6] As such, hormone replacement therapy (HRT) as a means of maintaining "youthful skin" has been a topic of great interest.[7] Oral HRT has been demonstrated to improve skin moisture, elasticity and thickness.[7] However, the recent revelation of the adverse effects associated with oral HRT[8] make this treatment an unacceptable option for most women.

Topical estrogen may provide an alternative in some, although the potential for systemic adverse effects requires further evaluation.[9] A recent study by Fuchs *et al.*[10] asked patients to apply a cream containing 0.01% estradiol to half of the face for 6 months. Vehicle was applied to the other half. Two-millimeter punch biopsies revealed a 23% increase in epidermal thickness with estradiol.

Treatment of Skin Aging

Topical Pharmaceuticals

Topical retinoids

Vitamin A is important in epidermal growth and differentiation, as well as maintenance of dermal connective tissue integrity. There are several available topical vitamin A derivatives, termed retinoids. Numerous clinical studies have documented the effectiveness of all-trans retinoic acid (tretinoin) in repairing photodamaged skin at the clinical, histological and molecular level. Topical tretinoin (Retin-A®, Renova®) produces significant reductions in fine wrinkling, mottled hyperpigmentation, roughness and laxity in mildly to moderately photodamaged skin.[11,12] The structural changes that underlie these cosmetic improvements include epidermal thickening, increased granular layer thickness, stratum corneum compaction, decreased melanin content, deposition of new collagen and generation of new blood vessels.[13] Tazarotene (Tazorac®) is a newer topical retinoid whose effects are similar to those of tretinoin.[14] Retin-A®, Renova® and Tazorac® are all prescription products. Non-prescription retinoids are also available and include retinol and retinaldehyde. There have been no double-blind controlled studies to compare the anti-aging effectiveness of non-prescription versus prescription retinoids.

Improvements with topical retinoids may continue for up to a year or longer with continued use. Retinoid treatment must be continued indefinitely to maintain the cosmetic benefits. The most common side effect of retinoids is skin irritation, which is dose-related. New emollient creams and "microsponges", which allow slow-metered release, help reduce irritation. Topical retinoids cause the skin to be more sensitive to sunburn, although this effect normalizes with continued use. Therefore, sun-protection measures should be used during the course of treatment. Retinoids elicit their effects at the molecular level by regulating gene transcription and affecting cellular activities of proliferation and differentiation.[15–17]

Alpha and beta hydroxy acids

Alpha hydroxyacids (AHA) and beta hydroxy acids (BHA) are naturally occurring substances found in fruits, wine, milk and sugar cane. In 1989, cosmetic manufacturers began to market products with hydroxy acids (HA) for salon and home use.[18] HA products sold in department or drug stores have a HA concentration of 10% or less while cosmetologists have products with 20–30% HA concentration and doctors may distribute products with 50–70% HA concentration.

These products are applied once or twice daily to even skin pigmentation and improve texture. Efficacy depends on the pH level of the product, the HA concentration, the vehicle, frequency and area of use. These products are generally well-tolerated, although in some individuals, their use has been associated with adverse effects including erythema, swelling, dermatitis and sun sensitivity.

Ditre *et al.*[19] had 17 white subjects with moderate to severe photodamage apply a 25% concentration AHA lotion, pH 3.5, to their forearm twice daily. After an average of 6 months, skin analysis revealed increased thickness, increased acid mucopolysaccharides, improved quality of elastic fibers and increased density of collagen. Other studies demonstrated increased collagen gene expression[20] and cell turnover in HA-treated skin.

The above-described histologic changes correlate with observed clinical improvements. Stiller *et al.*[21] had 74 women with moderately severe photodamage apply topical 8% glycolic or l-lactic acid to the face and forearms twice a day for 22 weeks. A statistically significant greater percentage of patients achieved improvement in overall severity of photodamage, mottled hyperpigmentation, sallowness and roughness with HA application as compared to vehicle ($p < 0.05$).

Kinetin

Kinetin (N6-furfuryladenine) is an essential plant growth factor that retards senescence of plants[22] and delays age-related changes in human skin fibroblasts in culture.[23] In a 52-week study of 96 subjects, twice daily application of kinetin improved skin roughness (63%), mottled hyperpigmentation (32%) and fine wrinkles (17%).[24] Treatments also improved skin-barrier function as measured by a decrease in transepidermal water loss.[24] Extended treatment with kinetin was well-tolerated and did not cause clinical signs or subjective symptoms of irritation.[25] Kinetin products offer the patient an alternative to more irritating anti-aging products for improving some of the signs and symptoms of mildly to moderately photodamaged skin.

Antioxidants

Ultraviolet radiation and air pollutants result in the generation of reactive oxygen species and other free radicals in the skin.[26] These are known to induce many of the clinical and histologic changes associated with photodamage.[27] To counteract these harmful effects, the skin has an antioxidant system. Two important components of this system are vitamin C and vitamin E. Antioxidant effects of vitamin C include neutralization of free radicals and regeneration of vitamin E, while vitamin E protects cell membranes from peroxidation.[28]

Studies, mostly animal and a few human, have demonstrated photoprotective, and in some cases, photorejuvenation benefits of topical application of these antioxidants and as such, vitamin C and vitamin E have become popular additives in cosmeceuticals. Fitzpatrick and Rostan[29] had 10 patients apply a vitamin C complex with 10% ascorbic acid and 7% tetrahexyldecyl ascorbate in an anhydrous polysilicone gel base to half of the face and an inactive polysilicone gel base to the opposite site. After 12 weeks, 4 of 10 patients showed improvement in peri-orbital, cheek and peri-oral areas. Average improvement on the treatment side was 25% compared to 7.7% on the gel-base control side. Equal improvement was noted for hydration and pigmentation. No inflammation or adverse effects were reported for either side. Biopsies showed increased Grenz zone collagen and increased type I collagen mRNA.

Copper peptides

Copper has become another popular ingredient for facial creams.[30] Copper is applied topically as a peptide because it is not well-absorbed through the gastrointestinal tract. In serum, copper combines with a tripeptide complex glycyl-L-histidyl L lysine to form GHK-copper. GHK-copper may promote wound healing and functions as a catalyst for antioxidants and a cofactor for free radical scavengers.[30,31] One study demonstrated that a copper peptide complex could increase the release of newly synthesized collagen.[32]

Leyden *et al.*[33] performed a study in which 67 female subjects with mild to moderate photodamage applied a GHK-copper face cream twice a day for 12 weeks. After one week, improvements were noted in skin laxity, clarity and overall appearance as compared to placebo. Improvement was noted in fine lines at week two, and in wrinkles at week eight. Ultrasound demonstrated an increase in overall skin density. No adverse effects were reported.[33]

Surgical Interventions

Topical pharmaceuticals offer effective skin aging prevention and one method of rejuvenation. However, the improvement associated with these products occurs slowly and in many cases is subtle. Some patients seek faster and more dramatic options, which in some cases, can be provided by the myriad of surgical interventions currently available for rejuvenation. We will briefly discuss the rejuvenation potential of some of the most commonly utilized modalities including microdermabrasion, chemical peels, botulinum toxin, dermal fillers, lasers and light sources, dermabrasion, radiofrequency devices and cosmetic surgery.

Microdermabrasion

Microdermabrasion directs a stream of fine abrasives (generally aluminum oxide crystals although other salt crystals may be used) at the skin using a compressed air system.[34] Superficial cellular debris and the particles are then aspirated into a separate container and discarded. With two passes, skin exfoliation to a depth of 15–25 micrometers has been reported.[35] Multiple treatments are generally recommended. Microdermabrasion is commonly used to improve skin texture and even skin tone. This procedure has become popular as it is low-risk, rapid, available to patients of all skin types, and results in minimal or no down time.

Tan *et al.*[36] evaluated 10 subjects who received up to 6-weekly microdermabrasion treatments. Immediately following the final treatment, a temporary increase in skin roughness consistent with mild abrasion was noted and mild flattening of wrinkles thought to be secondary to temporary edema was documented. One week later, 2 patients demonstrated a decrease in skin roughness, but wrinkle improvement had resolved. Histologic evaluation 1 week after the final treatment revealed orthokeratosis, a perivascular mononuclear infiltrate, slight edema and vascular ectasia.

Chemical peels

Chemical peels have a long history of use for skin rejuvenation. Chemical peels can be divided into superficial, medium depth and deep, and these distinctions are important as the magnitude of effect as well as associated risks and recovery time, generally increase with the depth of the peel.

1) Superficial Chemical Peels
   Superficial peeling agents injure the epidermis and include glycolic acids (concentrations of 10–70%), trichloracetic acid (concentrations

of 10–25%) and Jessner's solution (resorcinol 14 g, salicylic acid 14 g, 85% lactic acid 14 g and 95% ethanol quantity sufficient to add up to 100 cc).[37] Multiple peels performed at weekly, bi-weekly or monthly intervals are generally required to achieve the desired effect of more even skin tone and a "refreshed" appearance. Healing generally involves mild to moderate peeling and erythema that lasts 1–4 days. Complications are rare except for occasional transient hyperpigmentation.[37]

Newman *et al.*[38] applied glycolic acid (50%) gel to one side of the face, forearms, and hands of 41 subjects, once weekly for 4 weeks. One week post the final treatment, a dermatologist not previously involved in the study evaluated subjects. Statistically significant improvements included a decrease in rough texture, improvement of fine wrinkles, decrease in the number of solar keratoses and slight lightening of solar lentigines. Histologic analysis revealed thinning of the stratum corneum and an increase in the granular layer. Some specimens demonstrated increased dermal collagen thickness.

2) Medium Depth Chemical Peels

Medium depth chemical peels damage through the epidermis into the lower papillary or upper reticular dermis (a depth of approximately 0.6 mm).[34] Jessner's solution + trichloracetic acid or 35–50% trichloracetic acid alone are commonly used. A single procedure is generally performed to remove actinic keratoses, resolve dyschromias and perhaps, improve mild rhytides. Post-treatment patients develop erythema, edema and crust, which resolve by 7–12 days.[37] Erythema may last longer, in some cases 30 days or more. Repeat peels can be performed at six-month intervals but may not be required. Adverse effects include skin discoloration and occasionally, scarring.

Tse *et al.*[39] performed medium depth peels (70% glycolic acid plus 35% TCA or Jessner's solution plus 35% TCA) in 13 patients. Both peels effectively removed actinic keratoses and lightened solar lentigines, but neither peel reduced wrinkles. Swelling, crusting and erythema were noted within 24 hours. Complete re-epithelialization occurred within 7–10 days. Erythema lasted up to 60 days in some patients, but resolved more quickly in most. Post-inflammatory hyper or hypopigmentation and scarring were not seen. Histologic assessment revealed a hyperplastic epidermis, a Grenz zone of new collagen and neoelastogenesis.

3) Deep Chemical Peels
The Baker/Gordon formula (3 ml Phenol, USP, 88%, 2 ml tap or distilled water, 8 drops septisol liquid soap, 3 drops croton oil) is used for the classic deep chemical peel.[37] This peel causes injury to the reticular dermis, which increases procedure-associated risks, but it also results in new collagen formation, leading to wrinkle reduction. Sedation and/or analgesia are generally used during the peel and in the post-operative period. Phenol is cardiotoxic and intra-procedure monitoring is required. Liver and kidney toxicity may also occur. An open wound is created, which requires considerable post-operative care and requires 2–4 weeks to heal. Erythema may last longer. Adverse effects include skin discoloration (temporary hyperpigmentation or permanent hypopigmentation) and scarring. Both laser skin resurfacing and dermabrasion can achieve results equal to these peels. The common use of these alternatives and safety concerns has made performance of the Baker/Gordon peel less common. However, some physicians believe they get optimal results with this modality and consider it an important tool in their armamentarium for treating aged skin.

Botulinum toxin

Since the 1970's, *Clostridium botulinum* has been used as a therapeutic agent.[40] Since its introduction by Drs. Alistair and Jean Carruthers for treatment of frown lines at the 1991 meeting of the American Society of Dermatologic Surgeons, botulinum toxin has been used safely, with excellent cosmetic results, for facial rejuvenation.[40] By blocking the release of the neurotransmitter acetylcholine, and inducing a temporary chemical paresis, botulinum toxin eliminates muscle contraction, which results in dynamic wrinkles and furrows.[40]

Multiple preparations of type A and type B botulinum toxin are available clinically. These preparations are not identical and require different dosing. Botulinum toxin injections are used to improve forehead lines, glabellar furrows, crow's feet, and certain peri-oral rhytides. A brow lift can also be achieved. Effects are noted 1–3 days after injection[41] and last for 3–6 months. Side effects are generally mild and transient and include bruising, headaches and dry mouth. Systemic neuromuscular disease can be potentiated by botulinum toxin and thus this agent is contraindicated in individuals with such a history. Lid or brow ptosis may occur and can be treated with iopidine and neosynephrine.

Dermal fillers

Filler substances can be injected into the skin to soften or eliminate wrinkles or soft-tissue defects such as deep creases or depressed scars. Xenogenic (bovine) collagen is a commonly used filler with proven safety and reproducibility. However, clinical effects are relatively brief (3–6 months) and approximately 2% of the population is allergic to bovine collagen.[42] As such, while bovine collagen continues to be utilized frequently, a wide range of additional filler substances have become available in the United States and/or Europe,[42,43] including human collagen (autologous or allogenic), autologous fat, hyaluronic acid, calcium hydroxylapatite, polymethyl methacrylate, and silicone. These substances are in different stages of investigation and/or regulatory body approval and each has advantages and disadvantages. Evaluation of safety, biocompatibility, immunogenicity, duration and acceptability of the clinical response, and stability at the injection site must be carefully considered in selecting the optimal agent for each patient. At the current time, clinicians and patients must choose between the well-known and relatively safe but temporary fillers (such as bovine collagen) or more newly available, potentially longer acting fillers (including hyaluronic acid products), and permanent fillers (polymethylmethacrylate) with less available information and potentially more risk.

Narins *et al.*[44] compared the efficacy and safety of nonanimal stabilized hyaluronic acid gel (Restylane; Q-Med, Uppsala, Sweden) to bovine collagen (Zyplast, McGhan Medical Inc., Santa Barbara, CA). One hundred and eight patients with prominent nasolabial folds were treated with injected hyaluronic acid gel on one side of the face and collagen on the contra lateral side. The two fillers had similar immediate effects but the hyaluronic acid filler demonstrated superior maintenance of improvement in 62% of patients at 6 months. The intensity, frequency, and duration of local injection-site reactions were similar for the two products and no long-term adverse effects were noted for either product.

The use of botulinum toxin A in combination with fillers may improve and prolong the cosmetic effect and is considered to be the gold standard for wrinkle improvement by some clinicians. A recent study by Carruthers and Carruthers[45] evaluated 38 patients with moderate to severe glabellar rhytides, half of whom received a non-animal sourced hyaluronic acid filler (NASHA) alone (Restylane, Q-Medical Corporation, Uppsala, Sweden and Toronto, Canada) while the other half received NASHA and botulinum toxin A (BOTOX, Allergan Inc., Irvine, CA). The NASHA + botulinum toxin A group achieved a better response on rest and maximum response with a median time to return to pre-injection furrow status of 32 weeks as compared to the NASHA alone group with a median time for return to pre-injection furrow status of 18 weeks.

Lasers/Light sources

Thermal injury during laser or light treatment results from heat generated by absorption of photon energy. For treatment of skin aging, this thermal energy can be used to selectively remove a pigmented target (unwanted lentiginous solar damage or telangiectasia) or to less specifically remove (laser skin resurfacing) or heat (non-ablative resurfacing) tissue to achieve wrinkle reduction.

Pigment/Telangiectasia removal

Laser or light source emitted radiant energy can be used to target superficial melanin for removal of lentiginous sun damage or oxyhemoglobin for creation of vessel thrombosis and elimination of unwanted telangiectatic vessels.[46] A wide range of lasers and non-coherent high intensity flashlamps (intense pulsed light sources) can be used to effectively remove unwanted benign pigmented lesions (lentigos) or telangiectasia, which may be associated with aging. Recovery time, onset of improvement and number of treatments required to achieve a desired result varies with several factors including the device utilized, the extent of lesions to be removed and the willingness of the patient to experience temporary crusting or purpura in order to achieve the desired result. A complete discussion of all available laser devices and treatment options with these instruments is beyond the scope of this manuscript. However, in general, light-based devices are highly effective for removal of lentigines and telangiectasia. As an example, Kilmer *et al.*[47] treated 37 subjects with lentigines with a frequency-doubled Q-switched Nd:YAG laser (532 nm, 2 mm spot size, 10 nanosecond pulse duration) and achieved 75% pigment removal in 60% of lesions after 1 treatment. Mild, transient erythema, hypopigmentation and hyperpigmentation were noted in several patients but resolved without treatment in 3 months. No scarring or other adverse effects were noted.

These devices are less effective for removal of other dyschromias, such as post-inflammatory hyperpigmentation and melasma, and cannot be relied upon to safely and completely remove pre-malignant or malignant lesions such as lentigo maligna or melanoma. As such, evaluation and diagnosis by a clinician trained in skin pathology is important prior to initiation of light-based therapies.

Ablative laser skin rejuvenation

Ablative laser skin rejuvenation (LSR) uses collimated light absorbed by tissue water and converted to heat to precisely remove the epidermis and superficial dermis,[48] achieving reduction in epidermal dyschromias and surface irregularities as well as wrinkles. Two lasers are commonly utilized: 1) the carbon dioxide (CO_2) laser with a wavelength of 10,600 nm; and 2) the erbium:yttrium aluminum garnet (Er:YAG) laser with a wavelength of 2940 nm. Combination devices are also utilized.

Some of the rejuvenation seen after LSR is a result of removal and consequent regeneration of the epidermis, which generally results in improved skin tone and more even coloration. The mechanism for wrinkle reduction is less completely understood. One proposed mechanism is collagen denaturation and subsequent formation of new healthier collagen, 100–400 μm below the skin surface.[49] Heat-induced collagen contracture is a second probable mechanism of rhytid improvement proposed for CO_2 LSR, which creates a zone of thermal injury as well as ablation.[50] This mechanism is not likely a factor in Er:YAG LSR in which the injury is almost exclusively ablative.

LSR is generally performed as an outpatient surgical procedure using local anesthesia and sedation or general anesthesia. The procedure lasts from 30 minutes to 2 hours, depending on the extent of the areas treated, and patients generally experience mild to moderate discomfort during and after the procedure.

LSR is an ablative technique, which depending on the aggressiveness of the procedure, can achieve results comparable to medium or deep chemical peels or dermabrasion. Some surgeons feel lasers offer more precise control as compared to these other modalities. LSR is the most effective laser procedure for wrinkle reduction and as described above, has the added benefit of simultaneously improving surface irregularities and uneven skin color. Periorbital and perioral wrinkles show the most improvement, while areas of strong muscle activity such as the glabella exhibit the poorest response.[48] However, epidermal disruption or removal occurs and the open wound requires extensive care during the first 1–2 weeks during which time, there is a risk of bacterial, viral and fungal infections. Medications to prevent these infectious complications are generally utilized. Abnormal or delayed wound healing may result in skin dyspigmentation and scarring. In addition, the wound resolves with significant erythema, which often lasts for weeks or months. As such, while the potential for benefit is great, patients should be well-informed about the post-procedure healing phase and complication risks.

Rostan *et al.*[51] treated 16 patients with a 950 microsecond pulsed CO_2 laser or variable pulsed Er:YAG laser followed by short pulsed Er:YAG ablation. Three months post-treatment, overall photoaging assessment improved by an average of 57%. At a 4-week post-operative visit, 14 patients had mild or trace erythema while 2 patients had moderate to severe erythema. Two patients still had mild residual erythema at the 3-month post-operative visit. All patients also experienced edema, which was still present to a mild degree in 5 patients at a 4-week post-operative visit.

Non-ablative photorejuvenation

In order to diminish the risks associated with traditional LSR, while still attempting to achieve comparable rhytid reduction, non-ablative methods of laser or light source skin rejuvenation were developed. Variable success has been achieved, but as of yet no method has proven completely satisfactory.

Non-ablative photorejuvenation is designed to confine selectively, without any epidermal damage, thermal injury to the papillary and upper reticular dermis leading to fibroblast activation and synthesis of new collagen and extracellular matrix material.[52] The skin surface is not removed or modified. What really occurs may be more accurately referred to as dermal "remodeling" or "toning" as a wound healing response is initiated by laser-induced thermal injury, resulting in subsurface collagen regeneration.

Laser heating should be confined to a zone 100–500 μm below the skin surface, where the majority of the aforementioned histologic changes associated with photoaging occur.[53] More superficial injury may be ineffective for rhytid reduction; deeper injury may result in scarring. Wavelengths from 1.3–1.8 μm best fit this need. Alternatively, light sources, which target subsurface structures (e.g. blood vessels), which reside at the desired dermal depth, can be targeted.

In general, non-ablative photorejuvenation is a relatively low-risk procedure, which can be performed with little or no lifestyle interference (no obvious evidence that a "procedure" has been done). Multiple treatment sessions (3–6) are required and improvement occurs slowly over months. Many patients and physicians are willing to accept the subtle, incremental improvements in trade for significantly reduced discomfort and fewer complications. As an example, a multicenter study evaluated peri-orbital rhytid improvement in 35 adults after 3 treatments performed at 2-week intervals with the CoolTouch®, a 1.32 μm Nd:YAG laser used in combination with CSC.[54] Small but statistically significant clinical improvements were noted in the mild, moderate and severe rhytid groups 12 weeks after the final laser treatment. A final assessment performed 24 weeks after the last treatment showed statistically significant clinical improvement only in the severe rhytid group. The procedure was found to be safe, although 4 sites (5.6%) developed transient hyperpigmentation and 2 sites (2.8%) developed barely perceptible pinpoint-pitted scars. Subsequent device improvements improved safety and decreased the incidence of adverse effects to even lower levels.

It should be noted that traditional CO_2 and Er:YAG LSR as well as chemical peels and dermabrasion offer the patient not only rhytid reduction but also improvement in epidermal atypia and melanocyte maldistribution, which present clinically as skin roughness and dyspigmentation.

As non-ablative photorejuvenation does not affect the epidermis, the cosmetic benefits of epidermal rejuvenation are lost. However, relatively low-risk adjuvant treatments including tretinoin, glycolic acids, microdermabrasion or light chemical peels can be used to address pigmentary and textural changes in patients undergoing non-ablative photorejuvenation.

Dermabrasion

Dermabrasion is a method of ablative resurfacing, which uses wire brushes, diamond fraises and serrated wheels attached to a dermabrader to remove the skin surface.[55] This procedure is completely different from microdermabrasion (see above) in terms of depth of injury, cosmetic efficacy, healing time and associated risks, although we have found that patients frequently confuse the two terms/techniques.

Prior to the procedure, the skin is frozen with refrigerant topical anesthesia.[55] The skin is then held taut while the procedure is performed, one anatomic unit at a time. Small, portable hand-held devices are most commonly used today. Dermabrasion creates a wound and subsequent healing response similar to that seen with LSR and the resultant epidermal regeneration removes or diminishes surface irregularities while new collagen formation achieves wrinkle reduction. Again, like LSR, the epidermal disruption or removal associated with dermabrasion creates an open wound, which requires extensive care and puts the patient at risk for development of viral, bacterial or fungal infections, skin dyspigmentation and scarring. Re-epithelialization occurs over 5–7 days but residual erythema commonly lasts 4 weeks.[55] Results and associated risks for dermabrasion procedures are technique/operator-dependent (perhaps more so than chemical peels and laser resurfacing because of lack of intraoperative hemostasis and more cumbersome instrumentation), and thus, it is important that the surgeon is experienced and skilled.

Holmkvist *et al.*[56] treated half of the peri-oral area of 15 patients with a pulsed CO_2 laser and the other half of the area with dermabrasion using a hand engine-drive diamond fraise or a medium-grade drywall sanding screen. Dermabrasion resulted in more bleeding during the immediate post-operative period. Significantly more crusting and initial erythema (up to 1 month post-treatment) was noted on the CO_2 laser-treated side. All patients demonstrated wrinkle improvement and there was no significant difference between the treatment sides. Fine wrinkles were more responsive than deep wrinkles with both treatments.

Radiofrequency devices

The first radiofrequency device designed for skin rejuvenation was the ThermaCool TC™ System (Thermage, Inc., Hayward, CA), which received 510 K clearance for non-invasive treatment of facial wrinkles and rhytids. The ThermaCool™ System utilizes two electrodes on the skin to produce an electric field.[52] Ions and charged molecules in the tissue within the electric field move and or rotate. The inherent resistance to the movement of these ions and molecules in tissue causes heat. A cryogen spray inside a cooling tip creates a cooling device to protect the epidermis.

The ThermaCool System® has been used for cheek,[57] neck,[58] brow[59] and breast[60] lifting. Improvement has been noted for all of these indications. Patients generally undergo 1 treatment, but in some cases, a second treatment has added benefit. Side effects are generally limited to transient erythema and edema. Small, depressed scars have been noted in $< 1\%$ of cases.

A recently published multicenter study,[60] evaluated 86 subjects after a single peri-orbital treatment with the ThermaCool TC™ System. Three independent reviewers compared pre-treatment and up to 6 months post-treatment photographs. Of the evaluable treatment sites, 83.2% demonstrated improvement of at least 1 Fitzpatrick wrinkle score (a 9-point scale), and 61.5% of eyebrows were lifted by at least 0.5 mm. Fifty percent of subjects were satisfied or very satisfied with wrinkle improvement. Three patients had small areas of scarring at the six-month follow-up, but system and technique improvements have made this complication less common in subsequent studies.

Syneron, Inc. (Richmond Hill, Ontario, CA) developed two combined light source and radiofrequency devices for skin rejuvenation: the Aurora combines bipolar radiofrequency and a pulsed light source and is used for overall skin renewal while the Polaris WR combines bipolar radiofrequency and a 900 nm diode laser for enhanced wrinkle reduction. The radiofrequency component is a bipolar unit as opposed to the monopolar unit of the Thermage device. It is important to note that the mechanism of this device is quite different from that of the ThermaCool system, and as such, clinical effects are likely to be different.

Doshi and Alster[61] recently published an abstract in which they treated 20 patients with mild to moderate facial rhytides with the Polaris WR System. Patients received three passes per treatment (radiofrequency energy 40–100 J/cm^3, optical energy 15–50 J/cm^2). Additional passes were performed over some rhytides. Up to 3 treatments were performed at 3-week intervals. The authors reported significant improvement in facial rhytides and modest improvement in skin laxity.

Cosmetic surgery

Cosmetic surgery is the most aggressive treatment approach for wrinkles and skin laxity. Cosmetic surgery generally "pulls up" or removes skin and sub-dermal tissue producing tightening and smoothing. A vast array of procedures can be performed depending on the area and type of desired improvement. Cosmetic surgery generally does not improve skin roughness and dyspigmentation, and so these procedures are frequently combined at the same or different treatment sessions with other rejuvenation options including topical preparations, laser removal of pigment/telangiectasia and superficial or medium depth chemical peels in order to achieve optimal results.

The greatest rewards in terms of degree and duration of wrinkle reduction can be achieved with cosmetic surgery, but risks and recovery time are also increased. General anesthesia is often required and because deeper tissue is manipulated, adverse effects including bleeding, infection and scarring may occur more frequently as compared with other rejuvenation options. Recovery varies with the extent of the procedure but swelling and bruising may last 1–6 weeks. Newer endoscopic procedures may minimize recovery times.

Conclusion

A wide range of options is available to clinicians performing and patients seeking treatment for changes associated with skin aging. Clinicians should have knowledge of available options and counsel patients appropriately in order to help select treatments, which will meet realistic improvement expectations. Pre-treatment counseling will also help to minimize associated risks by allowing for proper follow-up and adverse effect management. It is important that patients followed for skin aging are also monitored for the development of skin cancers and that any pre-malignancies or malignancies be treated prior to addressing cosmetic concerns. A clinician who wisely and judiciously uses the many modalities available for treatment of skin aging, can provide their patients a great service, a more attractive appearance whose psychological benefits can result in greater self-esteem and improved physical well-being.[1]

References

1. Koblenzer CS (1996) Psychologic aspects of aging and the skin. *Clinics in Dermatol* **14**: 171–177.
2. Gendler EC (1997) Topical treatment of the aging face. *Dermatol Clin* **15**: 561–567.
3. Goihman–Yahr M (1996) Skin aging and photoaging: an outlook. *Clinics in Dermatol* **14**: 153–160.
4. Al Mahroos M, Yaar M, Phillips TJ, Bhawan J, Gilchrest BA (2002) Effect of sunscreen application on UV-induced thymine dimmers. *Arch Dermatol* **138**: 1480–1485.
5. Ramirez R, Schneider J (2003) Practical guide to sun protection. *Surg Clin North Am* **83**: 97–107.
6. Kligman AM, Koblenzer C (1997) Demographics and psychological implication for the aging population. *Dermatol Clin* **15**: 549–553.
7. Sator P-G, Schmidt JB, Sator MO, Humber JC, Honigsmann H (2001) The influence of hormone replacement therapy on skin ageing: a pilot study. *Maturitas* **39**: 43–55.
8. Beral V, Million Women Study collaborators (2003) Breast cancer and hormone-replacement therapy in the million women study. *Lancet* **362**: 414–415.
9. Felner EE, White PC (2000) Prepubertal gynecomastia; indirect exposure to estrogen cream. *Pediatrics* **105**: E55.
10. Fuchs KO, Solis O, Tapawan R, Parranjpe J (2003) The effects of an estrogen and glycolic acid cream on the facial skin of postmenopausal women: a randomized histologic study. *Cutis* **71**: 481–488.
11. Weinstein GD, Nigra TP, Pochi PE, Savin RC *et al.* (1991) Topical tretinoin for treatment of photodamaged skin. A Multicenter Study. *Arch Dermatol* **127**: 659–665.
12. Olsen EA, Katz HI, Levine N, Nigra TP *et al.* (1997) Tretinoin emollient cream for photodamaged skin: results of 48-week, multicenter, double-blind studies. *J Am Acad Dermatol* **37**: 217–226.
13. Gilchrest BA (1999) Treatment of photodamage with topical tretinoin: an overview. *J Am Acad Dermatol* S27–S36.

14. Phillips TJ, Gottlieb AB, Leyden JJ *et al.* (2002) Efficacy of 0.1% tazarotene cream for the treatment of photodamage: a 12-month multicenter, randomized trial. *Arch Dermatol* **138**: 1486–1493.
15. Giguere V, Ong ES, Segui P *et al.* (1987) Identification of a receptor for the morphogen retinoic acid. *Nature* **330**: 624–629.
16. Fisher GJ, Talwar HS, Lin J, Voorhees JJ (1999) Molecular mechanisms of photoaging in human skin in vivo and their prevention by all-trans retinoic acid. *Photochem Photobiol* **69**: 154–157.
17. Varani J, Warner RL, Gharaee–Kermani M *et al.* (2003) Vitamin A antagonizes decreased cell growth and elevated collagen-degrading matrix metalloproteinases and stimulates collagen accumulation in naturally aged human skin. *J Invest Dermatol* **114**: 480–486.
18. Kurtzweil P (1998) Alpha hydroxy acids for skin care. *FDA Consum Mag* **32**: 30–35.
19. Ditre CM, Griffin TD, Murphy GF, Sueki H, Telegan B, Johnson WC, Yu RJ, Van Scott EJ (1996) Effects of α-hydroxy acids of photoaged skin: a pilot clinical, histologic, and ultrastructural study. *J Amer Acad Dermatol* **34**: 187–195.
20. Bernstein EF, Lee J, Brown DB, Yu R, Van Scott E (2001) Glycolic acid treatment increases type I collagen mRNA and hyaluronic acid content of human skin. *Derm Surg* **27**: 429–433.
21. Stiller MJ, Bartolone J, Stern R, Smith S, Kollias N, Gillies R, Drake LA (1996). Topical 8% glycolic acid and 8% l-Lactic acid creams for the treatment of photodamaged skin. *Arch Dermatol* **132**: 631–636.
22. Van Staden J, Cook EL, Nooden LD (1988) Cytokinins and senescence. In: Nooden LD, Leopold AC (eds.) *Senescence and Aging in Plants* pp. 281–328, Academic Press, New York, NY.
23. Rattan SI, Clark BF (1994) Kinetin delays the onset of aging characteristics in human fibroblasts. *Biochem Biophys Res Commun* **201**: 665–672.
24. McCullough JL (1999) Furfuryladenine-A new antiaging topical: research and clinical experience. In: *Skin & Allergy News: Developments in Topical Skin Treatments: an Update* pp. 3–5, Skin Disease Education Foundation Symposium.
25. McCullough JL, Weinstein GD (2002) Clinical study of safety and efficacy of using topical kinetin 0.10% (Kinerase®) to treat photodamaged skin. *Cosmetic Dermatol* **15**: 29–32.
26. Dreher F, Maibach H (2001) Protective effects of topical antioxidants in humans. *Curr Probl Dermatol* **29**: 157–164.
27. Black HS (1987) Potential involvement of free radical reactions in ultraviolet light-mediated cutaneous damage. *Photochem Photobio* **46**: 213–221.
28. Lawrence N (2000) New and emerging treatments for photoaging. *Derm Clinics* **18**: 99–112.
29. Fitzpatrick RE, Rostan EF (2002) Double-blind, half-face study comparing topical vitamin C and vehicle for rejuvenation of photodamage. *Derm Surg* **28**: 231–236.
30. Baumann LS (2001) Cosmeceutical critique: copper. In: *Skin & Allergy News*, 30.
31. Contato C, Gavioli R, Guerrini R, Kozlowski H, Mlynarz P, Pasti C, Puli F, Remelli M (2001) Copper complexes of glycyl-histidyl-lysine and two of its synthetic analogues: chemical behaviour and biological activity. *Biochim Biophys Acta* **1526**: 199–210.
32. Oddos T, Aurelie J-L, Ries G (2002) Requirement of copper tripeptide glycyl-l-histidyl-l-lysine (GHK) complex formation for collagen synthesis activity in normal human dermal fibroblasts. Poster presentation Annual Meeting of the American Academy of Dermatology.

33. Leyden JJ, Grove G, Stephens TJ, Finkey MB, Barkovic BA, Appa Y (2002) Skin benefits of copper peptide containing facial cream. Poster presentation Annual Meeting of the American Academy of Dermatology.
34. Holck DEE, Ng JD (2003) Facial skin rejuvenation. *Curr Op Ophthalmol* **14**: 246–252.
35. Rubin MG, Greenbaum SS (2000) Histologic effects of aluminum oxide microdermabrasion on facial skin. *J Aesthetic Dermatol* **1**: 237–239.
36. Tan MH, Spencer JM, Pires LM, Ajmeri J, Skover G (2001) The evaluation of aluminum oxide crystal microdermabrasion for photodamage. *Dermatol Surg* **27**: 943–949.
37. Brody HJ (1992) *Chemical Peeling*. Mosby-Year book, Inc., St. Louis, Mo.
38. Newman N, Newman A, Moy LS, Babapour R, Harris AG, Moy RL (1996) Clinical improvement of photoaged skin with 50% glycolic acid: double-blind vehicle-controlled study. *Derm Surg* **22**: 455–460.
39. Tse Y, Ostad A, Lee H-S, Levine VJ, Koenig K, Kamino H, Ashinoff R (1996). A clinical and histologic evaluation of two medium-depth peels. *Derm Surg* **22**: 781–786.
40. Cather JC, Cather JC, Menter A (2002) Update on botulinum toxin for facial aesthetics. *Derm Clinics* **20**: 749–761.
41. Blitzer A, Binder WJ, Aviv JE, Keen MS, Brian MF (1997) The management of hyperfunctional facial lines with botulinum toxin. *Arch Otolaryngol Head Neck Surg* **123**: 389–392.
42. Alster TS, West TB (2000) Human-derived and new synthetic injectable materials for soft-tissue augmentation: current status and role in cosmetic surgery. *Plastic Reconstruc Surg* **105**: 2515–2525.
43. Lemperle G, Morhenn V, Charrier U (2004) Human histology and persistence of various injectable filler substances for soft tissue augmentation. *Aesth Plast Surg*. On-Line First edition.
44. Narins RS, Brandt F, Leyden J, Lorenc ZP, Rubin M, Smith S (2003) A randomized, double-blind, multicenter comparison of the efficacy and tolerability of restylane versus zyplast for the correction of nasolabial folds. *Derm Surg* **29**: 588–595.
45. Carruthers J, Carruthers A (2003) A prospective, randomized, parallel group study analyzing the effect of BTX-A (Botox) and nonanimal sourced hyaluronic acid (NASHA, Restylane) in combination compared with NASHA (Restylane) alone in severe glabellar rhytides in adult female subjects: treatment of severe glabellar rhytides with a hyaluronic acid derivative compared with the derivative and BTX-A. *Derm Surg* **29**: 802–809.
46. Anderson, RR, Parrish, JA (1981) The optics of human skin. *J Invest Derm* **77**: 13–19.
47. Kilmer S, Wheeland RG, Goldberg DJ, Anderson RR (1994) Treatment of epidermal pigmented lesions with the frequency-doubled Q-switched Nd:YAG laser. *Arch Dermatol* **130**: 1515–1519.
48. Kelly KM, Nelson JS (1998) Carbon Dioxide Laser Resurfacing of Rhytides and Photodamaged Skin. *Lasers Med Sci* **13**: 232–241.
49. Kuo T, Speyer MT, Ries WR, Reinisch L (1998) Collagen thermal damage and collagen synthesis after cutaneous laser resurfacing. *Lasers Surg Med* **23**: 66–71.
50. Fitzpatrick RE, Rostan EF, Marchell N (2000) Collagen tightening induced by carbon dioxide laser versus erbium:YAG laser. *Lasers Surg Med* **27**: 395–403.
51. Rostan EF, Fizpatrick RE, Goldman MP (2001) Laser resurfacing with a long pulse erbium:YAG laser compared to the 950 millisecond pulsed CO_2 laser. *Lasers Surg Med* **29**: 136–141.

52. Nelson JS, Majaron B, Kelly KM (2002) What is Non-ablative Photorejuvenation of Human Skin? *Sem Cutan Med Surg* **21**: 238–250.

53. Kelly KM, Majaron B, Nelson JS (2001) Nonablative laser and light rejuvenation. *Arch Facial Plast Surg* **3**: 230–235.

54. Kelly KM, Nelson JS, Lask GP, Geronemus RG, Bernstein LJ (1999) Cryogen spray cooling in combination with nonablative laser treatment of facial rhytides. *Arch Dermatol* **135**: 691–694.

55. Gold MH (2003) Dermabrasion in dermatology. *Am J Clin Dermatol* **4**: 467–471.

56. Holmkvist KA, Rogers GS (2000) Treatment of perioral rhytides a comparison of dermabrasion and superpulsed carbon dioxide laser. *Arch Dermatol* **136**: 725–730.

57. Alster TS, Tanzi EL (2003) Treatment of prominent nasolabial folds and cheek laxity with a nonablative radiofrequency device. *Lasers Surg Med* **15S**: 34.

58. Tanzi EL, Alster TS (2003) Improvement of neck laxity with a nonablataive radiofrequency device: a lifting experience. *Lasers Surg Med* **15S**: 34.

59. Fitzpatrick RE, Gernoemus R, Goldberg D, Kaminar M, Kilmer S, Ruiz-Esparza J (2003) First multicenter study of a new non-ablative radiofrequeny device to tighten facial tissue. *Lasers Surg Med* **15S**: 35.

60. Fitzpatrick R, Geronemus R, Goldberg D, Kaminer M, Kilmer S, Ruiz–Esparza J (2003) Multicenter study of noninvasive radiofrequency for periorbital tissue tightening. *Lasers Surg Med* **33**: 232–242.

61. Doshi SN, Alster TS (2004) A novel device combining diode laser and radiofrequency energy for rhytides and skin laxity. *Lasers Surg Med* **S16**: 31.

3

Industrial Approaches Towards Developing Effective Skin Care Products

Thomas Blatt, Horst Wenck and Franz Stäb
Beiersdorf AG, Hamburg, Germany
Email: Franz.Staeb@beiersdorf.com

INTRODUCTION

The skin is the only organ whose aging can be detected by the naked eye already out an early stage. The current expansion of knowledge in modern biogerontology has facilitated both a detailed and in-depth insight into the molecular mechanisms of age-dependent changes in organs and their cells. Skin aging is experienced as an amplified wrinkle formation, a loss of elasticity, uneven pigmentation, loss of moisture and increased roughening and itching of skin. What consumers of cosmetics increasingly expect from modern cosmetic products is a deceleration, cessation or even a reversal of the physiological processes leading to these signs of premature skin aging.

These continuously advancing consumer demands require intensive technological endeavours in the cosmetic industry, which apply particularly to the field of modern skin research on skin aging. The scientifically based research activities on skin aging are facilitated by the skin's accessibility to examination of aging mechanisms in the form of slightly invasive procedures and non-invasive biophysical measurements. In addition to these *in vivo* or *ex vivo* methods, many experiments can also be performed successfully on cultured cells or on three-dimensional cultured skin models. Besides the development and optimization of new formulation technologies, the focus in cosmetic industry is especially on the search for new active ingredients, which should ideally lead to perceivable improvements of the state of aging skin or, at the very least, to exert objective, biophysically measurable effects.

Every seriously undertaken search for a new topically applied molecular "fountain of youth", however, requires an detailed understanding of the mechanism of aging processes in the skin. Aging is understood as the result of a complex interaction of biological processes that are caused by both, genetic (chronological or intrinsic aging) and environmental or behavioral processes (premature or extrinsic aging).

Prominent Aging Theories

For a long time, there have been efforts to explain scientifically the fundamental causes and mechanisms of aging, resulting in different theories and concepts. Each of these theories focuses on one facet of the complex mechanisms involved in skin aging processes. The following are a few of the most prominent theories, specifically mentioned with reference to skin aging.

The "program theory" should be understood as the functional decline of the skin as a result of genetic and stochastic processes (intrinsic aging). Today more than 7000 genetic regulators are in discussion; their exact regulatory mechanisms are, however, not yet fully understood. One of the oldest and most well-known theories for the possible causes of skin aging is the "free radical theory".[1] Underlying the cellular and molecular composition the skin is a complex chemical redox system, in which — in healthy youthful skin — oxidative and reductive processes are kept in balance by the endogenous antioxidant system. With increasing age and influenced by negative environmental factors, this balance is shifted, and the oxidative process finally causes the formation of "Reactive Oxygen Species" (ROS).[2,3] Some of these ROS belong to the species of free radicals, especially to oxygen radicals. It is estimated, that about 10 billion ROS produced daily in every cell in the body, particularly during cellular respiration in mitochondria. The number and function of mitochondria, which act as small power stations responsible for energy supply in all human cells containing a nucleus, decreases with increasing age.[4] The dysfunction of the respiratory chain in mitochondria and the decline of endogenous antioxidant systems can lead to an increased production of ROS, both in the cells and in the surrounding tissue. The formation of ROS is also facilitated by exogenous noxes such as ultraviolet light, ozone, cigarette smoke, smog and alcohol, therefore, contributes via similar mechanisms to extrinsic skin aging.[5,6] An incomplete elimination of these ROS by the endogenous anti-oxidative defence system in skin fuels the aging process. According to the latest scientific findings, ROS regulate certain gene segments by affecting activation and deactivation of integral constituents of so-called redox-sensitive signalling cascades.[7–9] Enzymatic and non-enzymatic protective mechanisms are involved to avoid such oxidative signalling pathways, minimizing oxidative damage.[10] Ultimately the kinetics of the formation and neutralization of oxidative stressors is determined by the redox-status of the skin, which seems to be genetically and environmentally influenced.[11]

Accumulation of age-dependent functional flaws and specifically oxidative damage in the mitochondria of skin cells cause an insufficient energy production and supply[12] that can favor cutaneous aging (the theory of "age-dependent energy deficit"). It has been shown that the levels of ROS formed in the mitochondria of the skin are directly proportional to maximal lifespan of skin cells.

Numerous *in vitro* studies of cultivated cells originating from humans and animals have proven that normal cells have a limited doubling potential, whereas cancer cells are unrestricted in their ability to divide (the Hayflick model).[13] A prominent difference in these cells is the presence of a special enzyme in tumor cells, telomerase, which is inactive in normal cells ("telomere theory"). Because of this difference, it is not possible for a non-tumor cell to repair the DNA shortening on the telomere region (at the end of linear chromosomes) that occurs as a consequence of cell division. After countless cell divisions, the DNA in the normal cells reaches a critical length, which terminates cell division and results in cell death.[13] In addition to all these cellular processes, an accumulation of molecular damage (damage theory) as well as an age-dependent reduction in the ability for cellular repair (repair theory) is observed.[15]

The "neuroendocrine theory" of aging assumes that hormones strongly influence or determine the aging process, thereby controlling the biological clock of the cells. According to this theory, hormones regulate the activity of all enzymes that are essential to cellular repair and protection from damage. At this stage it should be accepted that human, and especially cutaneous aging, is less a single-caused than a complex, multi-faceted phenomenon.

Physiological Targets in Anti-Age Prophylaxis and Treatment

In the cosmetic industry, the postulated theories and explanations on mechanisms of aging are the basis for scientific approaches in search for new concepts in anti-aging treatment of skin. In contrast to the internal organs, the skin as the outermost protective barrier covering the complete area of body is particularly exposed to external influences. As the result of external challenges, the aging process in skin is not only influenced by genetic, intrinsic factors, but influenced to a far greater extent (80%) by extrinsic factors, especially by sun exposure.[16] Intrinsic mechanisms of skin aging seem to be only basically involved in formation of fine lines and shallow wrinkles in advanced age. Therefore the research in cosmetic industry is focused on the identification and qualification of new active principles mainly to fight against these extrinsic factors found to be harmful for the skin.

While UVB-light is known to induce directly DNA-damages, UVA-light induces damages more indirectly via the generation of ROS. One of the most important fields of research is therefore the prevention and the repair of sun-induced skin damages. During sun exposure endogenous absorbers of UV-light and photosensitizers in skin such as riboflavine, porphyrine, tryptophan, urocanic acid etc. can be involved in generation of ROS. The UV-induced ROS are thought to be the biggest players in generation of age-related damages in skin that abound as deep wrinkles and furrows, and mechanistically defined as photo-aging. This can be ascribed to a lack of regeneration of the dermal connective tissue, i.e. the structural reorganization of collagen, glycosaminoglycans and elastin in the dermis. Additionally, the loss of glycosaminoglycans in the dermal skin tissue causes a reduction in the tissue fluid content and water binding capacity in the epidermal and dermal layers, which seems to play an important role in the formation of wrinkles. In the skin the balance between assembly and breakdown of collagen shifts with increasing age in direction of favoring collagen breakdown, which is accomplished by collagen degrading enzymes such as the collagenase MMP-1.[16]

Even in young skin, the regeneration of collagen (over a period of months) is a relatively slow moving process. UV-radiation augments the process of degeneration by stimulation of both, the activation of collagenases and the oxidation, and enhanced glycation of collagen and other connective tissue components. Aside from this, UV-light causes the accumulation of elastotic material, a non-functional mass of elastic fibres. Thus, the tensile characteristics of the skin are altered, so that it becomes generally thinner, less elastic, and less resistant against stress. Phenotypically, this extrinsically caused and accelerated premature skin aging is manifested in an advanced state as the formation of coarse, deep furrows and folds, as well as aggravated elastoses.

Furthermore, the process of premature skin aging leads to an impairment of the denticulation of the epidermal/dermal junction and to a reduction of the number of so-called papillae, each of which harbors a blood capillary growing out of the dermis. These structural changes are considered as histological hallmarks of aging skin morphology. It is associated with the reduction of the capillary diameter, as well as capillary density in aged skin. A well-functioning blood capillary system contributes to the adequate nutrient supply of the upper skin layers, and to the structural integrity and complexion. Even though the mechanisms of both extrinsic and intrinsic skin aging cause a fundamental change in the appearance of the skin, the contribution of the extrinsic portion, however, seems to be predominant.

A characteristic feature of aging skin is the declining ability to regenerate. This is particularly striking in the longer time span needed for renewal of the epidermal layer. This so-called "epidermal turnover" takes about 28 days in young adult skin and the turnover slows down with age, taking about 40–60 days in the very old. For example, the conservation of a holiday tan in older skin is prolonged, due to the slower epidermal turnover, and the produced melanin is not released so fast by scaling of the horny layer. Furthermore, as skin gets older, the UV-induced tanning intensity becomes more irregular. However, the scientific knowledge about the pathophysiology of age spots or melasma is still poor. The appearance of age spots can be a result of a decreasing ability of melanocytes to equally distribute melanin packets (melanosomes) to the surrounding keratinocytes or of a localized overproduction of melanin. One can only speculate whether this is primarily caused by an uneven distribution of melanocytes in skin or rather a dysregulation of physiological processes in melanocytes.

Older skin is more sensitive to UV-exposure due to the age dependent atrophy of the skin,[17] UV-radiation penetrates deeper into the skin and the damages escalate. The immunological defence system is also significantly reduced with increasing age. Thus intensive sun exposure can promote neoplastic cell transformation (for example, melanoma). The incidence of skin tumors increases with age. Therefore, the consequent protection of skin of all ages against the sun is the most important task in keeping skin young and healthy. The cosmetic industry makes great efforts to optimize *in vivo* functional UVB- and UVA-filter combinations and antioxidant systems, especially for sunscreens and face care products.

Often described and subjectively felt, dry aged skin cannot only be attributed to the distinctive defect of the epidermal water barrier of the horny layer. Rather it also reflects regenerative processes, as well as a worsening in the water storage capacity, due to a diminished production of cutaneous moisturization factors (for example, amino acids, hyaluronic acid, pyrrolidone carboxylic acid and glycerine) that bind water in the horny layer. Besides the reduced water retention capacity, the age-dependent reduction in sebum secretion of the sebaceous glands also plays a role in the formation of dry, aged skin. As the sebaceous glands seem to be predominantly hormonally controlled, the age-dependent disruption in the hormonal system contributes to the condition and function of aged skin. An overt example of the endocrine influence on skin aging is the exacerbation of dry skin and increased wrinkle formation that occur with menopausal hormonal changes. Accordingly, specialized hormone treatments can lead to an improvement in the condition of the skin. Lately, new scientific insights have led to a merging of the fields of dermatology and endocrinology.

ACTIVE COSMETIC INGREDIENTS AND THEIR POTENTIAL

Moisturizers

Because of their dry skin, aging consumers tend to favor rich skin care formulations that include moisturizers with high water-binding properties, e.g. glycerine. Increasingly, modern cosmetics attempt to satisfy the more increasing demands of consumers for products with preventative or even regenerative performance. Besides preventing early skin aging, products must also smooth or improve the appearance or condition of existing wrinkles as well as the weakened regenerative performance of the skin. In the meantime, skin research in the cosmetic industry has already revealed ways to target improvements to aging skin.

UV-FILTERS

In addition to avoiding sun exposure, protection of the skin against UV-damage can be provided by the use of products that utilize highly efficient UV filter technologies. For high photo-protection, product formulations should employ an efficient UVA/B filter combination. Besides UVB protection, which delivers anti-erythemal screening UVA filter performance plays a decisive role in the prevention of photo-aging. Because UVA radiation is a principle contributor to skin aging by the production of ROS and the subsequent activation of collagen degrading enzymes, broad adsorptive performance of the employed filters offers protection against light-dependent aging.

ANTIOXIDANTS

Physiological events in the skin are based on physical and chemical processes, including redox-cascades. The redox-system of the skin is balanced by antioxidative and oxidative reactions. Skin has an anti-oxidative defence system as a direct protective barrier against endogenous and exogenous/environmental oxidative stress (UV-light) in the form of enzymatic and non-enzymatic antioxidant-systems.[18] These anti-oxidative protectants are higher concentrated in the epidermis than in the dermis, thus accounting for the higher environmental stress in the epidermis. Substances such as flavonoids, vitamins A, C and E, coenzyme Q10, as well as carotinoids are components of a healthy diet and can support the cellular systems in their protective function.

Topical application of substances such as vitamin E and vitamin C, as well as in particular the plant derived flavonoid derivative alpha-glycosylrutin show a profound protective potential against premature UV light-induced skin aging.[19] This positive activity, however, requires that topically applied antioxidants will adequately interact with the natural endogenous redox system of the skin. Therefore, not all of the established antioxidants achieve the desired protective effect when applied to the skin. The water soluble, bioavailable flavonoid alpha-glcosylrutin (AGR) can build up a skin protective depot in the living layer of the skin, in which the inherent glutathione redox system protects against oxidative damage and reduces UV-induced inflammation.[19–21] Similarly, the water-soluble antioxidant vitamin C can, amongst other activities, function as a cofactor in the collagen synthesis, thereby supporting skin regeneration in deeper layers. Knowing the causative involvement of UV-induced oxidative stress reactions in the cutaneous aging process,[16] the best prevention and radical modulation can be reached, and to some extent improved, by providing focused and customized topical treatment strategies.

Exfoliation

For the treatment of age-damaged skin, particularly for anti-scaleness and anti-wrinkle efficacy, countless principles are available on the market, whose effects are based on the removal of the outer horny layers of skin (exfoliation or peelings). Commonly used agents are so-called AHAs (alpha hydroxy acids) which are often endogenous metabolites (lactate) or other naturally occurring substances such as fruit acids. Depending on the substance used, the respective depth of treatment in skin can be determined by adjusting the pH, the topical concentration and treatment length employed. The activity of these agents is in generally based on the induction of skin regeneration by exfoliation and subclinical inflammation, which appears to be comparative to a superficial wound healing process. Thus the skin looks younger after the application.

Vitamin A and its Derivatives

Vitamin A and its derivatives have been used as active ingredients in the cosmetic industry for many years. Their activity is essentially based on the interaction with specific nuclear receptors, whose activation regulates collagen synthesis, improving the structure of the skin. Regrettably, besides concentration-dependent skin irritant properties, these vitamin A ingredients are also highly sensitive to light-dependent and oxidative processes, which substantially reduces their activities. New cyclodextrine-based formulation technologies enable the efficient stabilization of these active ingredients without limiting the activity.[22]

Coenzyme Q10 and Creatine

All cells, and thus also skin cells, need energy to grow, for protection and repair and above all else, for regeneration and cell division. To maintain this capacity for cellular life the mitochondria, small intracellular organelles operating as small power-plants in the cells, are imperative. Beside the mitochondrial energy supply, the cells also have a further possibility for energy storage in the creatine/phospho-creatine system. According to latest insights, this occurs in the human skin and is responsible for a demand dependent, and if required, extremely fast energy supply. In so doing, energy in the form of phospho-creatine, stored in large amounts in the cell, can be supplied by the cleavage and transfer of the phosphate. Creatine, as well as another energy metabolite, coenzyme Q10, can both be synthesized in human cells. From the age of about 30 onwards, a reduction in the cellular concentration of these compounds in the skin can be determined.[23] As a fat-soluble oxidative substance, coenzyme Q10 protects the cell membrane and organelles.[24] It especially plays a role in the electron transport system during the energy production (ATP) in cellular respiration of the mitochondria, preventing a chronic energy deficiency in aging cells.[25] The topical application of the skin's own coenzyme Q10 and creatine, identical to that found in humans, can activate a wide spectrum of synthetic processes, ultimately resulting in a reduction of wrinkle depth by balancing energy deficits. Besides this, the regeneration activity of aged skin can be stimulated by the external application of these active ingredients.

Anti-inflammatory Actives

There are different reasons why skin can become irritated. Independent of age, in cold, dry winter time, the skin is more sensitive against irritation than in summer time. People with so-called sensitive skin especially have to protect their skin against dryness, intensive sun exposure, mechanical stress and environmental noxes. More than 15% of the human population in Central Europe and Japan suffers from atopy or sun allergy (polymorphic light eruption), with increasing tendency. Special skin care regimens adopted to the specific needs of sensitive skin are developed and provided from the cosmetic industry. For prophylaxes against sun allergy and sunburn, sunscreens with high SPF in combination with an effective antioxidant system (e.g. alpha-glycosylrutin) are highly effective.[20] Actives isolated from herbal extracts (e.g. Licochalcone A) proved to be effective against mechanical stress like razor burn, but can be also effective in skin care for dry atopic skin or rosacea. Old skin is reported to be more sensitive against skin irritation due to restricted defence and repair mechanisms. Therefore, a prophylactic anti-inflammatory treatment appears to be recommended for elderly skin as well.

Future Perspectives

The physiology of skin aging is a complex, multi-faceted phenomenon. Despite the fact that the different molecular causes of the aging process are still not understood in detail, there is a general consensus that age alone is not the only mechanism in this process. In the future, the application of modern molecular and biological methods in skin research such as the DNA chip technology (micro-arrays) and the proteomic technology will allow new insight into the aging process — the involved genes and gene products, and their genetic control mechanisms. The DNA chip allows, through the alignment of many single stranded DNA sequences on a small planar surface, a systematic and above all parallel expression analysis of activated genes. The proteomic technology allows an analogous quantitative and qualitative detection of proteins. These new technologies, the accessibility of the skin, and the improved possibilities to culture *in vivo* resembling skin models, will increasingly contribute to a better understanding of the regulation of the intrinsic and extrinsic aging process, and the positive effects of topically applied age-specific agents. These technologies can, in the long term, provide fundamental knowledge about the control mechanisms of the human aging process as a whole.

REFERENCES

1. Harman D (1956) Aging: a theory based on free radical and radiation chemistry. *J Gerontol* **11**: 298–300.
2. Shigenaga MK (1994) Oxidative damage and mitochondrial decay in aging. *Proc Natl Acad Sci USA* **91**: 10771–10778.
3. Sohal RS (1991) Hydrogen peroxide production by mitochondria may be a biomarker of aging. *Mech Ageing Dev* **60**(2): 189–198.
4. Richter C (1995) Oxidative damage to mitochondrial DNA and its relationship to aging. In: Esser K, Martin GM (eds.) *Molecular Aspects of Aging*, pp. 99–108. J. Wiley & Sons, Chichester.
5. Blatt T (2001) Aging of human skin cells *ex vivo* and *in vitro*. In: Bertoni–Freddari C, Niedermüller H (eds.) *Current Concepts in Experimental Gerontology*, pp. 93–101, Vienna Aging series. Facultas–Verlag, Wien.
6. Swartz HM (1995) Free radicals in aging: theories, facts and artefacts. In: Esser K, Martin GM (eds.) *Molecular Aspects of Aging*, pp. 77–97. J Wiley & Sons, Chichester.
7. Bender K (1997) UV-induced signal transduction. *Photochem Photobiol* **37**: 1–17.
8. Knebel A (1996) Dephosphorylation of receptor tyrosine kinases as target of regulation by radiation, oxidants or alkylating agents. *EMBO J* **15**(19): 5314–5325.
9. Suzuki YJ (1997) Oxidants as stimulators of signal transduction. *Free Rad Biol Med* **22**: 269–285.
10. Sauermann G (1999) Ultraweak Photon Emission of humans skin *in vivo*: influence of topically applied antioxidants on human skin. In: Packer L (ed.) *Methods in Enzymology*, pp. 419–428, volume 300, Academic Press.
11. Podda M (1997) UV-irradiation depletes antioxidants and causes oxidative damage in a model of human skin. *Free Rad Biol Med* **24**: 55–65.
12. Nohl H (1997) Imbalance of oxygen activation and energy metabolism as a consequence or mediator of aging. *Exp Gerontol* **32**: 485–500.
13. Hayflick L (1961) The serial cultivation of human diploid cell strains. *Exp Cell Res* **25**: 585–621.
14. Zglinicki T von (1995) Mild hyperoxia shortens telomeres and inhibits proliferation of fibroblasts: A model for senescence? *Exp Cell Res* **220**: 186–193.
15. Lithgow GJ (1996) Mechanisms and evolution of aging. *Science* **273**: 80.
16. Scharffetter–Kochanek K (1997) UV-induced reactive oxygen species in photocarcinogenesis and photoaging. *Biol Chem* **378**: 1247–1257.

17. Yuan H (1996) Increased susceptibility of late passage human diploid fibroblasts to oxidative stress. *Exp Gerontol* **31**(4): 465–474.
18. Vessey DA (1993) The cutaneous antioxidant system. In: Fuchs J, Packer L (eds.) *Oxidative Stress in Dermatology*, pp. 81–103, Marcel Dekker Inc., New York.
19. Stäb F (2000) Alpha-Glucosylrutin — an innovative antioxidant in skin protection. *SÖFW-Journal* **127**: 2–8.
20. Hadshiew I (1997) Effects of topical applied antioxidants in experimentally provoked polymorphous light eruption (PLE). *Dermatology* **195**: 362–368.
21. Stäb F (2000) Topically applied antioxidants in skin protection. In: Packer L, Sies H (eds.) *Methods in Enzymology*, pp. 465–478, Academic Press.
22. Raschke T (2003) Encapsulation technologies in cosmetics. *SÖFW-Journal* **129**: 73–78.
23. Kalen A (1989) Age-realted changes in the lipid composition of rat and human tissues. *Lipids* **24**: 579–584.
24. Frei B (1990) Ubiquinol-10 is an effective lipid-soluble antioxidant at physiological concentrations. *Proc Natl Acad Sci USA* **87**: 4879–4883.
25. Hoppe U (1999) Coenzyme Q10, a cutaneous antioxidant and energizer. *Biofactors* **9**(2–4): 371–378.

4

Strategies to Reduce Age-Related Skeletal Muscle Wasting

Gordon S. Lynch

Department of Physiology, The University of Melbourne, Australia

Thea Shavlakadze and Miranda D. Grounds

School of Anatomy and Human Biology, The University of Western Australia

Crawley, Perth, Western Australia, 6009. Australia

INTRODUCTION

Some of the most serious consequences of aging are its effects on skeletal muscle. The term *sarcopenia* is widely used to describe the progressive loss of muscle mass and quality with increasing age.[1] Sarcopenia is characterized by a gradual decline in strength and a slowing of movement that increases the risk of injury from sudden falls and the need for assistance for the frail elderly to accomplish even the most basic tasks required for independent living. Although sarcopenia affects the elderly regardless of ethnicity, gender, or wealth, it should be recognized that health status, physical activity, and nutrition, can play an important role in slowing the rate of physical decline and preserving functional independence and quality of life.

As the proportion of older persons in the population continues to escalate, sarcopenia will have widespread implications and place increasing demands on public health. Even though it is generally agreed that the effects of aging on skeletal muscle are inevitable, there is debate as to whether these deleterious changes can be stopped or reversed. Therapeutic strategies are needed to slow the effects of aging on skeletal muscle, and to restore and preserve muscle size and strength so that quality of life can be maintained or improved. In this review, we discuss the importance of physical activity in slowing the effects of aging, but emphasize the fact that exercise alone will not prevent the sarcopenia. We also describe some of the other contributing factors that affect muscle quantity and quality, such as age-related changes in circulating levels of muscle anabolic hormones and growth factors. Neural influences in the preservation of muscle size and strength are also described as well as some of the interventions that have been proposed for the preservation of motor units and neurotransmission during aging. In addition, the importance for continued research into the development and testing of safe and efficacious strategies that can be applied clinically for combating sarcopenia are be reiterated.

Underlying Mechanisms of Age-Related Changes in Skeletal Muscle

With advancing age, there is a slow but progressive loss of skeletal muscle mass that results in gradual decline in muscle function. The underlying mechanisms responsible for these age-related changes in muscle quantity and quality involve a complex interaction of many factors that affect neuromuscular transmission, muscle architecture, fiber composition, excitation-contraction (E-C) coupling, and metabolism. There continues to be considerable research investigating strategies that can reduce or reverse these age-related changes in skeletal muscle structure and function, especially muscle atrophy and weakness, and a number of these interventions will be described. Sarcopenia in men and women is associated with significant changes in body composition, which involve a decrease in body mass and skeletal muscle mass with increased fat and reduction of the muscle force.[2] Sarcopenia is not limited to humans, since most mammals are affected by the progressive loss of muscle mass with advancing age including, but not limited to, rats, mice, cats, rabbits, dogs, and horses. One of the major issues is whether the decline in muscle force producing capacity is due to a loss of contractile protein resulting from loss of individual muscle fibers, or to a decrease in the ability of the remaining myofibers to produce force, i.e. an age-related decrease in specific force (sP_o) or the force per cross-sectional area.[3–6] Differences in basal muscle protein turnover may also explain the age-related loss in muscle mass in humans.[7,8] Several studies have investigated these age-related effects at the single muscle fiber (cellular) level and some of them reported decreases in the maximum force producing capacity (P_o), specific force (sP_o) or force per cross-sectional area, and maximum velocity of shortening.[9–11] Alterations in the mechanisms of excitation-contraction (E-C) coupling are thought to contribute to the changes in muscle contractility with age.[12,13] These age-related alterations include: a reduction in the amount of Ca^{2+} available for triggering contraction and a reduction in Ca^{2+} release due to dihydropyridine (DHPR)-ryanodine receptor (RyR) uncoupling,[14] impairment of sarcoplasmic reticulum (SR) Ca^{2+}-pump function,[15] abnormalities in the regulation of RyRs[16] and decreased turnover of Ca^{2+}-ATPase and RyR protein.[17]

Other general age-related systemic changes such as alterations in levels of anabolic hormones and neuronal function also deleteriously affect skeletal muscle and other tissues. The decline in overall body function further impacts upon the loss of muscle mass and much of this is due to a decreased capacity for exercise.[18] While there is compelling evidence that problems with reinnervation can lead to impaired function of regenerated muscles, the proposal that the capacity of muscle precursor cells for new muscle formation (myogenesis) *in vivo* is not a limiting factor in healthy individuals, even in very old age, is controversial.[18,19] There is consistent evidence that the rate of myoblast activation may be a little slower in older muscles, but cross-transplantation experiments of whole muscle grafts between young (2 months) and old (24 months) mice, show no marked differences in the overall capacity of these cells to become activated and fuse to form new muscle (Grounds *et al.*, manuscript in preparation). Under tissue culture conditions, many studies show a reduced replicative capacity of myoblasts derived from aged compared with adult muscles of mice and humans. However, the extent to which this behavior under *in vitro* conditions accurately reflects the more complex *in vivo* situation is unclear. For example, extraction of myoblasts may be less efficient from aged muscles that contain more interstitial connective tissue, and myoblasts conditioned by the aged environment may have a sub-optimal response to tissue culture conditions that have been optimized for cells extracted from younger muscles. Elegant *in vivo* parabioisis experiments where the circulatory systems of young and aged mice are joined, reveal that systemic factors from young mice can enhance the rate at which myoblasts in old muscle respond to muscle injury, supporting the idea of no intrinsic difference of the response of satellite cells with age, although clearly the aged environment modifies their capacity to respond (T. Rando, personal communication). These observations raise the possibility that administration of (yet to be identified) systemic factors could be used to boost the regenerative response of aged muscle under certain conditions.

Undoubtedly, factors that contribute to the age-related decrease in muscle mass and function include changes in endocrine and cytokine levels, a decline in food intake and energy balance, age-related changes in skeletal muscle metabolism and biochemistry,[20] and especially decreased physical activity and impaired neuronal function. Before describing some of the hormonal and growth factor approaches that have been proposed for treating sarcopenia, we must first discuss the most effective intervention for slowing the rate of loss of muscle function with advancing age, namely exercise. We will outline the role of exercise in preserving muscle mass and strength, and identify the best exercises that can slow the rate of sarcopenia and improve overall muscle function in older adults.

Exercise and Aging

Maintenance of skeletal muscle mass is highly dependent on muscle activity (muscle contraction), which initiates signal transduction pathways responsible for protein synthesis and muscle growth.[21] It is well-recognized that reduced physical activity contributes to the age-related loss of muscle mass, accelerates osteoporosis and sets up a vicious cycle of increasing disability and impaired mobility.[22] The loss of functional independence can lead to depression and is painful for the individual and their families and carers.[23] Thus, physical fitness, especially muscle strength, is critical for the preservation of functional independence, the promotion of good health, and improved overall quality of life for the elderly. Home-based exercise programs have proved successful for improving functional independence and have been advocated in the elderly with benefits for reducing nursing home and health care costs;

e.g. a six month physical activity program of balance exercises and strength training reduced the level of disabilities in muscle strength, balance, and mobility by 45% in frail elders living at home.[24] In addition to the widely recognized effects on physiological functions, regular exercise has been shown to have a protective effect against depression.[25] Resistance exercise (or strength training) has been shown to be the most effective way to actively avoid losing muscle strength with age and improve the ability to perform the tasks of daily living, even in nonagenarians.[26] Lack of strength, especially around the ankle, predisposes elderly individuals to falls[27] and to the fear of falls, which results in the decrease of their physical activities.[28,29] Improved strength due to resistance training may lead to reduced fears about the likelihood of falling, increased confidence about posture and response time and improved balance.[30] Although improvements in strength and power can be demonstrated in the elderly following training, the greatest benefits of exercise are likely to be seen when it is incorporated early in life (well before the age of 50 years) and maintained for as long as possible in order to produce the desired preventive effect.[22] Exercise intensity is also an important factor, with the greatest benefits being observed when resistance training involves more intense, higher velocity lifting of heavier rather than lighter loads,[26] although such intense exercise may not be possible for a considerable proportion of older people especially the frail elderly. Furthermore, flexibility training such as yoga, balance training such as tai chi, and swimming, which are milder forms of exercise can be beneficial.[2] An exercise prescription involving resistance training for muscle strength, aerobic exercise for maintaining cardiovascular fitness, as well as flexibility and balance exercises, has been proposed as the best combination for maintaining independent living.[31]

Use it or Lose it? Exercise Alone Will Not Prevent Sarcopenia

Despite the obvious benefits bestowed by exercise training, it must be made clear that exercise alone cannot *prevent* age-related changes in skeletal muscle function. Even elite Master's level athletes who train and compete at the highest levels, for a significant portion of their lifespan, simply do not perform as well as they did when they were younger, and they still exhibit obvious age-related losses in muscle mass and strength.[32] An active lifestyle, while likely to slow the relentless effects of aging on muscle structure and function, will not prevent sarcopenia completely. Thus, the popular adage "use it or lose it" when applied to aging, is only partly true, and other factors such as age-related changes in circulating levels of muscle anabolic hormones and growth factors must also be considered when developing effective strategies to combat sarcopenia.[23]

Innervation and Neurotrophic Factors

Contraction of skeletal muscle occurs in response to stimulation at the neuromuscular junction by signals from nerves. Many age-related changes in the central nervous system are deleterious for innervation of skeletal muscle and much muscle wasting may be the result of functional denervation.[2] Neuromuscular function declines with age and while involvement of many neurotrophic and/or myotrophic candidate molecules has been proposed, studies show no clear association between the age-related changes in muscle mass, neurotransmission and levels of neurotrophic factors such as nerve growth factor, neurotrophin-3 (NT-3), neurotrophin-4 (NT-4), brain-derived neurotrophic factor (BDNF), or glial-derived neurotrophic factor.[33,34] Local blockade of neuromuscular transmission decreases levels of NT-4 mRNA in skeletal muscle, whereas increased levels of mRNA of NT-4,[35] NT-3 and BDNF[36] occur in response to increased activity of skeletal muscle. Ciliary neurotrophic factor (CNTF) and insulin-like growth factor-1 (IGF-I) both have myotrophic and neurotrophic effects[36] and in mammals, both show a decline in synthesis with age.[37,38] IGF-1 is clearly very important for maintaining muscle mass (discussed later) but may also directly affect motor neurons, since administration of low doses of IGF-1 in combination with glycosaminoglycans dramatically prevent the motor neuron loss in wobbler mice, that are a model for motor neuron disease.[39]

Although CNTF has myotrophic effects,[40] it is most well-known for its neurotrophic properties. CNTF stimulates axonal sprouting and reinnervation of denervated myofibers.[41] As CNTF is the main neurotrophic factor that shows decreased levels with advancing age, it is one of the most likely candidates contributing to age-related impairment of neurotransmission and muscle sprouting in response to transient denervation[41] that result in the deterioration of muscle performance. CNTF participates in the survival of motor neurons and reduces denervation atrophy of skeletal muscles.[42] In preliminary experiments in old rats, CNTF levels were restored by exogenous CNTF administration, and a strong correlation was found between CNTF production and muscle performance.[37] Follow-up studies in rats[43] and humans[38] support the notion of a strong relationship between CNTF and age-related changes in muscle mass and function. Of related interest to sarcopenia is the clinical administration of neurotrophic factors to maintain neuronal function and for treating neuromuscular diseases.[44,45] Reports of success with respect to rescue of skeletal muscle function have been claimed, although many of these are unsubstantiated.[46] It has been shown that administration of CNTF retards the adverse progressive motor neuron dysfunction and improves muscle strength in wobbler mice[44] and also accelerates regeneration of transected sciatic nerve and muscle re-innervation in rats.[47] While innervation is clearly essential for skeletal muscle contraction (required for maintenance of muscle mass), relatively little is known about the muscle-derived factors that maintain neuromuscular contact and function and their decline with age.

Systemic Changes in Circulating Hormones and Other Factors

Age-related changes in body composition are commonly attributed to alterations of systemic endocrine function in aging individuals leading towards hormonal imbalances.[2] Studies examining steroid (testosterone, oestradiol/noretisterone acetate, dehydroepiandrosterone) treatment in healthy individuals have not provided major health benefits and several undesirable side effects have been identified, including a greater risk for cardiovascular disease and some cancers.[23,46] In March 2004, the United States Department of Health and Human Services (HHS) issued warnings to companies that manufacture, market and distribute products containing androstenedione, which was stated to act like a steroid once it was metabolized and could therefore pose similar health risks as anabolic steroids. Androstenedione (also called "anabolic steroid precursor" or "andro") is produced naturally in humans and can be converted in the body to testosterone. Generally, androstenedione is advertised as a dietary supplement that can enhance athletic performance based on its claimed anabolic and androgenic properties, which stimulate muscle growth and increase production of testosterone.[48,49] A press release from the United States Food and Drug Administration (FDA) reported that androstenedione when taken over time and in sufficient quantities, may increase the risk of serious and life-threatening diseases, and subsequently a number of athletic organizations, including the International Olympic Committee, banned use of androstenedione.

Growth Hormone/IGF-I

Of particular interest in relation to sarcopenia is the decreased activity of the growth hormone (GH)/IGF-I axis.[50–52] Although it was a widely held notion that sarcopenia was directly related to an age-related decline in GH secretion, this view has been contested and numerous studies in humans do not support a benefit of GH administration on muscle protein synthesis.[2,46] Synthetic peptides that cause the release of GH (GH secretagogues, e.g. benzoazepines and their analogs) have been used clinically but their efficacy is unclear, as are the benefits of commercial hormone replacement therapies.[53]

The growth factor IGF-I, seems to be particularly important for maintaining muscle mass in the elderly.[53–55] The mechanisms by which IGF-I maintains muscle mass and might prevent age-related muscle atrophy have been reviewed elsewhere.[18,21,46,56] It is now recognized that there are several isoforms of IGF-I, some present in the circulation, and others are expressed locally in skeletal muscle and other tissues.[57,58] Of particular importance for maintaining muscle mass are the two IGF-I isoforms produced locally in myofibers, which arise from differently spliced mRNA precursors.[58] Aging is associated with decreased serum levels of IGF-I[59] and also down-regulation of the local IGF-I in muscle.[60,61] Muscles of old rats[60] and humans[61] have less robust upregulation of one of the muscle specific IGF-I isoforms (mechano growth factor or MGF) in response to physical overload, compared to young individuals. Since MGF is involved in myoblast proliferation,[62] this might help to partly explain muscle wasting in elderly. IGF-1 produced locally within muscles in response to exercise is responsible for hypertrophy and maintaining muscle mass and transgenic over-expression of the other muscle specific

IGF-I isoform (Class 1 Ea isoform) in muscles of mice prevents or reduces the age-related loss of muscle mass[63] and the extent of damage in dystrophic myofibers.[64,105] The challenge is to translate these experimental benefits of muscle restricted elevated IGFs to the clinical situation. It is noted that systemic administration of IGFs is not recommended as this results in hypertrophy of cardiac muscle and heart failure and also prostate cancer.[2]

Treating Sarcopenia is not Simply "Hormone Replacement Therapy"

On the basis of the evidence presented, simple hormone "top up" strategies to restore hormone levels in the elderly have not been successful, especially if they have not been performed in conjunction with a resistance exercise program.[65–68] Developing supplementation strategies with different combinations of low-dose anabolic compounds has merit for attenuating the loss of muscle mass and improving muscle function in the elderly, and this has received some attention.[68] In 2002, a report was convened by the Therapeutic Goods Administration (TGA) to assess the findings regarding the safety of the US Women's Health Initiative (combined Hormone Replacement Therapy, HRT) trial. The TGA Report made recommendations regarding the relative benefits and risks of combined HRT (estrogen and progestin) for post-menopausal women, and concluded that the benefits did not outweigh the risks for side effects, including coronary heart disease. Furthermore, recent evidence also indicates that there may be increased risks of combined therapy in accelerating other age-related pathologies such as Alzheimer's disease.[69]

β_2-Adrenoceptor Agonists

It has long been known that some β_2-adrenoceptor agonists (β_2-agonists), agents that stimulate the β_2 receptors of the sympathetic nervous system and which are widely used to treat asthma, have powerful anabolic effects on skeletal muscle. The application of β_2-agonists (particularly clenbuterol) for increasing muscle mass and performance and their adverse effects has been reviewed elsewhere.[70] Clenbuterol mediates hypertrophy of skeletal muscle[71,72] and inhibits denervation-induced atrophy through its action on β_2-adrenergic receptors.[71] However, the use of clenbuterol is limited by numerous undesirable side effects, including sweating, tachycardia, tremors, effects on the central nervous system and serious heart complications that may be lethal.[70]

At an equimolar dose to clenbuterol, another β_2-agonist, fenoterol, has a 10–15% greater anabolic effect on rat fast-twitch (EDL) and slow-twitch (soleus) muscles.[72] In a follow-up study, 4 weeks of daily administration of fenoterol completely ameliorated the age-related loss of muscle mass and strength in aged (28 month old) F344 rats and this was attributed to hypertrophy of existing fibers and not by increased myofiber number.[73] This was one of the first studies to demonstrate complete restoration of both muscle mass and strength to (adult) control levels following β_2-agonist administration.[73] Aging is also characterized by a slowing of movements, due to factors such as muscle remodeling, where denervated fast fibers become reinnervated by slow motoneurons. One advantage of treating sarcopenia using β_2-agonists over other anabolic agents, relates to the ability for some of these β_2-agonist to cause fiber type transitions within skeletal muscles, typically from slow- to fast-twitch.[70,72] Thus powerful β_2-agonists such as fenoterol may not only prevent the loss of muscle mass and strength, but may also help retain a higher proportion of fast-twitch fibers that will attenuate the characteristic slowing of contraction in aged mammals.[74]

Despite the positive attributes of β_2-agonists for treating sarcopenia, there are several deleterious side effects especially when administered in high doses. Given that β_2-agonist act via the β_2-adrenoceptors, and there exists a population of these adrenoceptors in the heart, it is difficult (perhaps impossible) to separate the hypertrophic effects on skeletal muscle from those on the heart. Not surprisingly, cardiac hypertrophy has been observed in nearly all studies that have examined the effects of β_2-agonist administration on skeletal muscle.[74] Most studies have employed high doses in order to produce skeletal muscle hypertrophy and therefore have also resulted in significant (and potentially deleterious) increases in cardiac mass. This has so far limited the clinical potential of β_2-agonists for sarcopenia. To prevent or reduce these detrimental effects on tissues other than skeletal muscle, the challenge is to devise treatments that utilise different β_2-agonists that are capable of eliciting skeletal muscle hypertrophy at low doses, and following short duration treatments.

Risks and Benefits of Muscle Anabolic Therapies for Sarcopenia

For anabolic therapies, it could be argued that concerns regarding potential pharmaceutical toxicity and safety issues are usually only related to high doses. Although the use of powerful muscle anabolic agents will likely have dose-related side effects, developing low-dose, short-term treatment strategies are likely to have less toxic effects and their clinical merit is worthy of testing. This will require extensive pre-clinical and clinical studies to determine the optimum dosage and regimen of administration that will produce significant improvements in muscle mass and strength without causing deleterious side effects such as cardiovascular complications or tumor formation.

As society faces an increasing burden of aging-related health issues, the need to develop safe and effective therapies for sarcopenia to promote independent living is deserving of immediate attention. Unlike HRT for post-menopausal women where, under some circumstances, risks may outweigh benefits, the potential benefits for successful treatment of sarcopenia with short-term low-dose muscle anabolic agent may be effective in restoring muscle mass and far outweigh the possible risks.[23]

Myostatin

Myostatin, a member of the transforming growth factor-beta superfamily, is another circulating growth factor that has attracted much attention since the absence of myostatin results in increased muscle size, known as double muscling, in mice and cattle.[75] Myostatin is a negative regulator of myogenesis and suppresses myoblast proliferation[76] and myogenic differentiation.[77] High serum and muscular levels of myostatin are associated with the cachexia (wasting) associated with HIV-infected men.[78] It has been postulated that age-related deficits in both GH and testosterone may lead to an increase in myostatin expression and a disassociation in autocrine IGF-I effects on muscle protein synthesis, both of which could contribute to sarcopenia.[79] Experiments in mice show that blocking myostatin with antibodies[80] and silencing of the myostatin gene results in decreased fat content.[81] Whether strategies to decrease myostatin activity might reduce muscle sarcopenia in humans remains speculative.

Vitamin D

The importance of circulating vitamin D in maintaining muscle mass and strength is often overlooked and vitamin D deficiency is extremely prevalent in the elderly. Vitamin D is formed from calciferol that is mainly produced in the skin in response to sunlight. Levels of vitamin D and muscle strength decline with age. Indeed, low levels of vitamin D are associated with reduced physical performance, muscle strength and physical function, and increased risk of falls in older people.[82] With age there is a decreased ability to manufacture vitamin D, resulting in marked deficiency, leading to muscle atrophy with selective loss of type IIB muscle fibers and other disorders. Fifty per cent of subjects with vitamin D deficiency display hip fractures.[82] Since pain, weakness, and fear of falling are all factors that may cause older people to restrict their outdoor activities, it is quite possible that low vitamin D levels result from illness and disuse that limits outdoor activities and exposure to sunlight. While studies have found that vitamin D supplementation is a simple, safe, and low cost intervention that may reduce the number of bone fractures,[83] a recent review concluded that there was insufficient evidence that vitamin D supplementation alone could improve physical performance in older people,[84] and that confirmation of a benefit from vitamin D combined with calcium supplementation would only come from larger, well-designed clinical trials. It is suggested that vitamin D may act by stimulating IGF-I signaling.[18] Clearly vitamin D deficiency should be carefully monitored in the elderly; it is readily diagnosed and appears to be reversed by administration of calcium and vitamin D supplements.[82]

Inflammatory Cytokines

Another important circulating cytokine that severely affects skeletal muscle is the pro-inflammatory cytokine tumor necrosis factor-alpha (TNF-α). Increased inflammation is associated with many age-related problems (e.g. arthritis, inflammatory myopathies, coronary heart disease, cancers) and the associated elevated of TNF-α and interleukin-6 (IL-6) can directly contribute to loss of skeletal muscle tissue in aging humans. The role of TNF-α and other inflammatory cytokines in extensive muscle wasting (cachexia) associated with severe inflammatory situations like cancer and acquired immunodeficiency syndrome has attracted considerable attention.[85] Increased plasma TNF-α and IL-6 are often observed in healthy elderly people and higher levels of these cytokines are associated with lower muscle mass and strength in well-functioning older men and women.[86] TNF-α expression is also elevated locally in skeletal muscles of the elderly[87] and inflammation, measured as high levels of inflammatory markers IL-6, *C*-reactive protein and IL-1RA, and is significantly associated with poor physical performance and muscle strength in older persons.[88] A relationship between IL-6 and levels of growth factors, such as IGF-I, and the preservation of muscle mass in the elderly has recently been proposed[89] and sarcopenia in men was found to reflect a withdrawal of anabolic stimuli (such as growth hormone/IGF-I), but in women was due to increased catabolic stimuli (such as cellular IL-6). Increasing evidence supports the idea that the effects of TNF-α on muscle atrophy may also be mediated in part via interference with IGF-I signaling.[90] Attempts to minimize cachexia have focused on anti-inflammatory drugs to block TNF-α action. This might be done by blocking the function of TNF-α with either specific antibodies such as Remicade (Infliximab®) or by soluble receptors marketed as Etanercept (Enbrel®).[91] Remicade is a highly specific anti-inflammatory intervention and it clearly delays and reduces the breakdown of muscles from dystrophic *mdx* mice.[92] Remicade is very successful clinically in the treatment of inflammatory diseases like rheumatoid arthritis; while it can have some side effects[93] it might be useful for amelioration of inflammatory cytokine induced muscle loss with aging. Reduction of serum TNF-α by administration of *L*-carnitine has been proposed to prevent muscle loss secondary to heart failure,[94] since *L*-carnitine reduces TNF-α serum levels in rats with experimentally induced chronic heart failure: however, such *L*-carnitine treatment did not significantly increase muscle size. Since high levels of TNF-α result in cachexia and act by increasing protein degradation as well as directly interfering with IGF-1 signaling,[18] reduced TNF-α combined with either elevated IGF to increase overall protein synthesis, or a high protein diet, may have additive benefits.

Metabolism, Energy Balance and Nutrition

Imbalance between energy intake and energy expenditure is considered to be one of the main reasons for decreased lean body mass. This aspect includes altered food intake, combined with changes in basic metabolic rate and biochemistry. Aging is associated with dramatic changes in the biochemistry and enzymatic activity of skeletal muscle resulting (among other effects) in reduced capacity to synthesize new proteins, up-regulation of pathways leading to increased protein breakdown (catabolism), and increased oxidative cell damage.[20] There are strong advocates that suggest that an optimum intake of micronutrients and metabolites, which varies with age and genetic constitution, would tune up metabolism and give a marked increase in health, particularly for the poor, young, obese and elderly, at little cost.[95]

The use of specific amino acid supplements and their derivatives has recently attracted much attention to reduce muscle protein breakdown and increase muscle mass. Supplementation with a nutrient mixture of β-hydroxy-β-methylbuturate (HMB), a metabolite of leucine, has beneficial effects for increasing muscle strength, and administration of HMB to elderly individuals involved in strength-training, resulted in increased muscle strength as well as a significant decrease in fat mass.[96]

Another nutritional supplement which has attracted attention for increasing muscle mass and strength is creatine. Supplementation with creatine has been advocated for older individuals because creatine enhances muscle strength and explosive power after only 5–7 days in young adults.[97] Creatine supplementation in normally active older men (59–72 years of age) increased several indices of muscle performance, including functional tests, without adverse side effects and it was concluded that creatine may be a useful therapeutic strategy for older adults to attenuate loss in muscle strength and performance.[98]

Caloric restriction, which refers to a nutritious dietary regimen low in calories, is one of the very few interventions that extends longevity in animal models.[99] This extends life expectancy by 30–40% if initiated early in the animal's life, and by about 20% if started in middle age.[100] In a recent study of aged (26–28 month old) F344 rats, long-term (life-long) calorie restriction was an effective intervention against the loss of muscle function with age.[101] It is proposed that the main mechanism responsible for such longevity is reduced reactive oxygen species formation and thus reduced oxidative damage to cellular structures.[102] The other mechanism, which underlies life-extending benefits of caloric restriction, is its ability to reduce fat deposition and maintain insulin sensitivity.[103] These observation must be balanced against the requirement for a healthy high protein diet to maintain muscle mass, and a progressive loss of appetite that can lead to "anorexia of aging" and weight loss is a major cause of muscle wasting in the elderly.[104]

Conclusion

There is clearly a profound need for therapeutic strategies that can slow the effects of aging on skeletal muscle structure and function; namely to restore muscle size and strength in the frail elderly so that their quality of life can be maintained or improved. Physical activity, particularly resistance (strength) training, can play an important role in slowing the effects of aging, but exercise alone will not prevent the gradual decline in muscle quantity and quality. Other factors, such as nutrition and age-related changes in circulating levels of muscle anabolic hormones and growth factors, must also be considered when developing strategies to combat sarcopenia. Neurotrophic strategies may also be an area of focus since many of these factors have myotrophic as well as neurotrophic effects and healthy innervation is clearly crucial for skeletal muscle structure and function. Although we have identified many promising candidates and strategies that have potential for ameliorating sarcopenia and improving muscle function in the elderly, few pharmacological interventions seem justified at this stage. Much research is needed to test the safety and efficacy of various exciting experimental strategies before any of them could be recommended for potential clinical application.

Acknowledgments

GSL is grateful for research grant support from the Muscular Dystrophy Association (USA) and Pfizer Pharmaceuticals (USA) and the research of TS is made possible by an International Postgraduate Research Scholarship from the University of Western Australia.

References

1. Roubenoff R, Castaneda C (2001) Sarcopenia-understanding the dynamics of aging muscle. *JAMA* **286**: 1230–1231.
2. Shavlakadze T, Grounds MD (2003) In: Rattan S (ed.) *Modulating Aging and Longevity*, pp. 139–166. Kluwer Academic Publishers, Britain.
3. Larsson L (1995) Motor units: remodeling in aged animals. *J Gerontol A Biol Sci Med Sci* **50**: 91–95.
4. Brown M, Hasser EM (1996) Complexity of age-related change in skeletal muscle. *J Gerontol A Biol Sci Med Sci* **51**: B117–123.
5. Larsson L, Ramamurthy B (2000) Aging-related changes in skeletal muscle. Mechanisms and interventions. *Drugs Aging* **17**: 303–316.
6. Lynch GS, Hinkle RT, Chamberlain JS, Brooks SV, Faulkner JA (2001) Force and power output of fast and slow skeletal muscles from mdx mice 6–28 months old. *J Physiol* **535**: 591–600.
7. Volpi E, Kobayashi H, Sheffield–Moore M, Mittendorfer B, Wolfe RR (2003) Essential amino acids are primarily responsible for the amino acid stimulation of muscle protein anabolism in healthy elderly adults. *Am J Clin Nutr* **78**: 250–258.
8. Yarasheski KE (2003) Exercise, aging, and muscle protein metabolism. *J Gerontol A Biol Sci Med Sci* **58**: M918–922.
9. Frontera WR, Suh D, Krivickas LS, Hughes VA, Goldstein R, Roubenoff R (2000) Skeletal muscle fiber quality in older men and women. *Am J Physiol Cell Physiol* **279**: C611–618.
10. Trappe S, Williamson D, Godard M, Porter D, Rowden G, Costill D (2000) Effect of resistance training on single muscle fiber contractile function in older men. *J Appl Physiol* **89**: 143–152.
11. Lowe DA, Warren GL, Snow LM, Thompson LV, Thomas DD (2004) Muscle activity and aging affect myosin structural distribution and force generation in rat fibers. *J Appl Physiol* **96**: 498–506.
12. Wang ZM, Messi ML, Delbono O (2000) *L*-Type Ca^{2+} channel charge movement and intracellular Ca^{2+} in skeletal muscle fibers from aging mice. *Biophys J* **78**: 1947–1954.
13. Plant DR, Lynch GS (2002) Excitation-contraction coupling and sarcoplasmic reticulum function in mechanically skinned fibers from fast skeletal muscles of aged mice. *J Physiol* **543**: 169–176.
14. Delbono O, O'Rourke KS, Ettinger WH (1995) Excitation-calcium release uncoupling in aged single human skeletal muscle fibers. *J Membr Biol* **148**: 211–222.
15. Narayanan N, Jones DL, Xu A, Yu JC (1996) Effects of aging on sarcoplasmic reticulum function and contraction duration in skeletal muscles of the rat. *Am J Physiol* **271**: C1032–1040.
16. Damiani E, Larsson L, Margreth A (1996) Age-related abnormalities in regulation of the ryanodine receptor in rat fast-twitch muscle. *Cell Calcium* **19**: 15–27.
17. Ferrington DA, Krainev AG, Bigelow DJ (1998) Altered turnover of calcium regulatory proteins of the sarcoplasmic reticulum in aged skeletal muscle. *J Biol Chem* **273**: 5885–5891.
18. Grounds MD (2002) Reasons for the degeneration of ageing skeletal muscle: a central role for IGF-1 signaling. *Biogerontology* **3**: 19–24.
19. Grounds MD (1998) Age-associated changes in the response of skeletal muscle cells to exercise and regeneration. *Ann N Y Acad Sci* **854**: 78–91.

20. Carmeli E, Coleman R, Reznick AZ (2002) The biochemistry of aging muscle. *Experimental Gerontology* **37**: 477–489.
21. Rennie MJ, Wackerhage H, Spangenburg EE, Booth FW (2004) Control of the size of the human muscle mass. *Annu Rev Physiol* **66**: 799–828.
22. Tseng BS, Marsh DR, Hamilton MT, Booth FW (1995) Strength and aerobic training attenuate muscle wasting and improve resistance to the development of disability with aging. *J Gerontol A Biol Sci Med Sci* **50**: 113–119.
23. Lynch GS (2004) Tackling Australia's future health problems: developing strategies to combat sarcopenia — age-related muscle wasting and weakness. *Intern Med J* **34**: 294–296.
24. Gill TM, Baker DI, Gottschalk M, Peduzzi PN, Allore H, Byers A (2002) A program to prevent functional decline in physically frail, elderly persons who live at home. *N Engl J Med* **347**: 1068–1074.
25. Strawbridge WJ, Deleger S, Roberts RE, Kaplan GA (2002) Physical activity reduces the risk of subsequent depression for older adults. *Am J Epidemiol* **156**: 328–334.
26. Fielding RA, LeBrasseur NK, Cuoco A, Bean J, Mizer K, Fiatarone Singh MA (2002) High-velocity resistance training increases skeletal muscle peak power in older women. *J Am Geriatr Soc* **50**: 655–662.
27. Buchner DM, Cress ME, de Lateur BJ, Esselman PC, Margherita AJ, Price R, Wagner EH (1997) The effect of strength and endurance training on gait, balance, fall risk, and health services use in community-living older adults. *J Gerontol A Biol Sci Med Sci* **52**: M218–224.
28. Vellas B, Cayla F, Bocquet H, de Pemille F, Albarede JL (1987) Prospective study of restriction of activity in old people after falls. *Age Ageing* **16**: 189–193.
29. Tinetti ME, Speechley M (1989) Prevention of falls among the elderly. *N Engl J Med* **320**: 1055–1059.
30. Brown M, Sinacore DR, Host HH (1995) The relationship of strength to function in the older adult. *J Gerontol A Biol Sci Med Sci* **50**: 55–59.
31. Singh MA (2002) Exercise comes of age: rationale and recommendations for a geriatric exercise prescription. *J Gerontol A Biol Sci Med Sci* **57**: M262–282.
32. Hawkins SA, Wiswell RA, Marcell TJ (2003) Exercise and the master athlete — a model of successful aging? *J Gerontol A Biol Sci Med Sci* **58**: 1009–1011.
33. Frostick S, Yin Q, Kemp G (1998) Schwann cells, neurotrophic factors, and peripheral nerve regeneration. *Microsurgery* **18**: 397–405.
34. Keller–Peck CR, Feng G, Sanes JR, Yan Q, Lichtman JW, Snider WD (2001) Glial cell line-derived neurotrophic factor administration in postnatal life results in motor unit enlargement and continuous synaptic remodeling at the neuromuscular junction. *J Neurosci* **21**: 6136–6146.
35. Funakoshi H, Belluardo N, Arenas E, Yamamoto Y, Casabona A, Persson H, Ibanez CF (1995) Muscle-derived neurotropin-4 as an activity-dependent trophic signal for adult motor neurons. *Science* **268** (5216): 1495–1499.
36. Gomez–Pinilla F, Ying Z, Opazo P, Roy RR, Edgerton VR (2001) Differential regulation by exercise of BDNF and NT-3 in rat spinal cord and skeletal muscle. *Eur J Neurosci* **13**: 1078–1084.
37. Guillet C, Auguste P, Mayo W, Kreher P, Gascan H (1999) Ciliary neurotrophic factor is a regulator of muscular strength in aging. *J Neurosci* **19**: 1257–1262.

38. Roth SM, Schrager MA, Ferrell RE, Riechman SE, Metter EJ, Lynch NA, Lindle RS, Hurley BF (2001) CNTF genotype is associated with muscular strength and quality in humans across the adult age span. *J Appl Physiol* **90**: 1205–1210.
39. Vergani L, Losa M, Lesma E, Di Giulio AM, Torsello A, Muller EE, Gorio A (1999) Glycosaminoglycans boost insulin-like growth factor-I-promoted neuroprotection: blockade of motor neuron death in the *wobbler* mouse. *Neuroscience* **93**: 565–572.
40. Peroulakis ME, Forger NG (2000) Ciliary neurotrophic factor increases muscle fiber number in the developing levator ani muscle of female rats. *Neurosci Lett* **296**: 73–76.
41. English AW (2003) Cytokines, growth factors and sprouting at the neuromuscular junction. *J Neurocytol* **32**: 943–960.
42. Mousavi K, Miranda W, Parry DJ (2002) Neurotrophic factors enhance the survival of muscle fibers in EDL, but not SOL, after neonatal nerve injury. *Am J Physiol Cell Physiol* **283**: C950–959.
43. Fraysse B, Guillet C, Huchet–Cadiou C, Camerino DC, Gascan H, Leoty C (2000) Ciliary neurotrophic factor prevents unweighting-induced functional changes in rat soleus muscle. *J Appl Physiol* **88**: 1623–1630.
44. Mitsumoto H, Ikeda K, Holmlund T, Greene T, Cedarbaum JM, Wong V, Lindsay RM (1994) The effects of ciliary neurotrophic factor on motor dysfunction in wobbler mouse motor neuron disease. *Ann Neurol* **36**: 142–148.
45. Elliott JLM , Snider WDM (1996) Motor neuron growth factors. *Neurology* **47**(4)Supp 2: 47S–53S.
46. Lynch GS (2002) Novel therapies for sarcopenia: ameliorating age-related changes in skeletal muscle. *Exp Opin Ther Patents* **12**: 11–27.
47. Zhang J, Lineaweaver WC, Oswald T, Chen Z, Zhang F (2004) Ciliary neurotrophic factor for acceleration of peripheral nerve regeneration: an experimental study. *J Reconstr Microsurg* **20**: 323–327.
48. King DS, Sharp RL, Vukovich MD, Brown GA, Reifenrath TA, Uhl NL, Parsons KA (1999) Effect of oral androstenedione on serum testosterone and adaptations to resistance training in young men: a randomized controlled trial. *JAMA* **281**: 2020–2028.
49. Broeder CE, Quindry J, Brittingham K, Panton L, Thomson J, Appakondu S, Breuel K, Byrd R, Douglas J, Earnest C, Mitchell C, Olson M, Roy T, Yarlagadda C (2000) The Andro Project: physiological and hormonal influences of androstenedione supplementation in men 35 to 65 years old participating in a high-intensity resistance training program. *Arch Intern Med* **160**: 3093–3104.
50. Yarasheski KE (1994) Growth hormone effects on metabolism, body composition, muscle mass, and strength. *Exerc Sport Sci Rev* **22**: 285–312.
51. Taaffe DR, Jin IH, Vu TH, Hoffman AR, Marcus R (1996) Lack of effect of recombinant human growth hormone (GH) on muscle morphology and GH-insulin-like growth factor expression in resistance-trained elderly men. *J Clin Endocrinol Metab* **81**: 421–425.
52. Lange KH, Andersen JL, Beyer N, Isaksson F, Larsson B, Rasmussen MH, Juul A, Bulow J, Kjaer M (2002) GH administration changes myosin heavy chain isoforms in skeletal muscle but does not augment muscle strength or hypertrophy, either alone or combined with resistance exercise training in healthy elderly men. *J Clin Endocrinol Metab* **87**: 513–523.

53. Schwartz RS (1995) Trophic factor supplementation: effect on the age-associated changes in body composition. *J Gerontol A Biol Sci Med Sci* **50**: 151–156.

54. Borst S, Lowenthal D (1997) Role of IGF-I in muscular atrophy of aging. *Endocrine Rev* 7: 61–63.

55. Roubenoff R (2003) Sarcopenia: effects on body composition and function. *J Gerontol A Biol Sci Med Sci* **58**: 1012–1017.

56. Musaro A, Rosenthal N (2002) The role of local insulin-like growth factor-1 isoforms in the pathophysiology of skeletal muscle. *Curr Genomics* **3**: 149–162.

57. Winn N, Paul A, Musaro A, Rosenthal N (2002) Insulin-like growth factor isoforms in skeletal muscle aging, regeneration and disease. In *Cold Springs Harbour Symposium on Quantitative Biology*, LXVII pp. 507–518.

58. Shavlakadze T, Winn N, Rosenthal N, Grounds MD Reconciling data from transgenic mice that over-express IGF-1 specifically in skeletal muscle. Submitted.

59. Ravaglia G, Forti P, Maioli F, Nesi B, Pratelli L, Cucinotta D, Bastagli L, Cavalli G (2000) Body composition, sex steroids, IGF-1, and bone mineral status in aging men. *J Gerontol A Biol Sci Med Sci* **55**: M516–521.

60. Owino V, Yang SY, Goldspink G (2001) Age-related loss of skeletal muscle function and the inability to express the autocrine form of insulin-like growth factor-1 (MGF) in response to mechanical overload. *FEBS Lett* **505**: 259–263.

61. Hameed M, Orrell RW, Cobbold M, Goldspink G, Harridge SD (2003) Expression of IGF-I splice variants in young and old human skeletal muscle after high resistance exercise. *J Physiol* **547**: 247–254.

62. Yang SY, Goldspink G (2002) Different roles of the IGF-I Ec peptide (MGF) and mature IGF-I in myoblast proliferation and differentiation. *FEBS Lett* **522**: 156–160.

63. Barton–Davis ER, Shoturma DI, Musaro A, Rosenthal N, Sweeney HL (1998) Viral mediated expression of insulin-like growth factor I blocks the aging-related loss of skeletal muscle function. *Proc Natl Acad Sci USA* **95**: 15603–15607.

64. Barton ER, Morris L, Musaro A, Rosenthal N, Sweeney HL (2002) Muscle-specific expression of insulin-like growth factor I counters muscle decline in mdx mice. *J Cell Biol* **157**: 137–148.

65. Balagopal P, Proctor D, Nair KS (1997) Sarcopenia and hormonal changes. *Endocrine* 7: 57–60.

66. Proctor DN, Balagopal P, Nair KS (1998) Age-related sarcopenia in humans is associated with reduced synthetic rates of specific muscle proteins. *J Nutr* **128**: 351S–355S.

67. Bross R, Storer T, Bhasin S (1999) Aging and Muscle Loss. *Trends Endocrinol Metab* **10**: 194–198.

68. Greenlund LJ, Nair KS (2003) Sarcopenia — consequences, mechanisms, and potential therapies. *Mech Ageing Dev* **124**: 287–299.

69. Shumaker SA, Legault C, Rapp SR, Thal L, Wallace RB, Ockene JK, Hendrix SL, Jones BN, 3rd, Assaf AR, Jackson RD, Kotchen JM, Wassertheil–Smoller S, Wactawski–Wende J (2003) Estrogen plus progestin and the incidence of dementia and mild cognitive impairment in postmenopausal women: the Women's Health Initiative Memory Study: a randomized controlled trial. *JAMA* **289**: 2651–2662.

70. Lynch GS (2002) Beta-2 Agonists. In Bahrke MS, Yesalis CE (eds.) *Performance-enhancing Substances in Sport and Exercise. Human Kinetics*, pp. 47–64. USA.

71. Hinkle RT, Hodge KM, Cody DB, Sheldon RJ, Kobilka BK, Isfort RJ (2002) Skeletal muscle hypertrophy and anti-atrophy effects of clenbuterol are mediated by the beta2-adrenergic receptor. *Muscle Nerve* **25**: 729–734.
72. Ryall JG, Gregorevic P, Plant DR, Sillence MN, Lynch GS (2002) beta 2-Agonist fenoterol has greater effects on contractile function of rat skeletal muscles than clenbuterol. *Am J Physiol Regul Integr Comp Physiol* **283**: R1386–1394.
73. Ryall JG, Plant DR, Gregorevic P, Sillence MN, Lynch GS (2004) Beta2-Agonist administration reverses muscle wasting and improves muscle function in aged rats. *J Physiol* **555**: 175–188.
74. Ryall JG, Plant DR, Lynch CD Making old muscles young again: a therapeutic role for beta 2-agonists. *Physiol News* **56**: 33–35.
75. McPherron AC, Lee SJ (1997) Double muscling in cattle due to mutations in the myostatin gene. *Proc Natl Acad Sci USA* **94**: 12457–12461.
76. Thomas M, Langley B, Berry C, Sharma M, Kirk S, Bass J, Kambadur R (2000) Myostatin, a negative regulator of muscle growth, functions by inhibiting myoblast proliferation. *J Biol Chem* **275**: 40235–40243.
77. Rios R, Carneiro I, Arce VM, Devesa J (2002) Myostatin is an inhibitor of myogenic differentiation. *Am J Physiol Cell Physiol* **282**: C993–999.
78. Gonzalez-Cadavid NF, Taylor WE, Yarasheski K, Sinha-Hikim I, Ma K, Ezzat S, Shen R, Lalani R, Asa S, Mamita M, Nair G, Arver S, Bhasin S (1998) Organization of the human myostatin gene and expression in healthy men and HIV-infected men with muscle wasting. *Proc Nat Acad Sci USA* **95**: 14938–14943.
79. Marcell TJ, Harman SM, Urban RJ, Metz DD, Rodgers BD, Blackman MR (2001) Comparison of GH, IGF-I, and testosterone with mRNA of receptors and myostatin in skeletal muscle in older men. *Am J Physiol Endocrinol Metab* **281**: E1159–1164.
80. Whittemore LA, Song K, Li X, Aghajanian J, Davies M, Girgenrath S, Hill JJ, Jalenak M, Kelley P, Knight A, Maylor R, O'Hara D, Pearson A, Quazi A, Ryerson S, Tan XY, Tomkinson KN, Veldman GM, Widom A, Wright JF, Wudyka S, Zhao L, Wolfman NM (2003) Inhibition of myostatin in adult mice increases skeletal muscle mass and strength. *Biochem Biophys Res Commun* **300**: 965–971.
81. Lin J, Arnold H, Della-Fera M, Azain M, Hartzell D, Baile C (2002) Myostatin knockout in mice increases myogenesis and decreases adipogenesis. *Biochem Biophys Res Commun* **291**: 701–706.
82. Eriksen EF, Glerup H (2002) Vitamin D deficiency and aging: implications for general health and osteoporosis. *Biogerontology* **3**: 73–77.
83. Trivedi DP, Doll R, Khaw KT (2003) Effect of four monthly oral vitamin D3 (cholecalciferol) supplementation on fractures and mortality in men and women living in the community: randomised double blind controlled trial. *BMJ* **326**: 469.
84. Latham NK, Anderson CS, Reid IR (2003) Effects of vitamin D supplementation on strength, physical performance, and falls in older persons: a systematic review. *J Am Geriatr Soc* **51**: 1219–1226.
85. Argiles JM, Meijsing SH, Pallares-Trujillo J, Guirao X, Lopez-Soriano FJ (2001) Cancer cachexia: a theraputic approach. *Med Res Rev* **21**: 83–101.

86. Visser M, Pahor M, Taaffe DR, Goodpaster BH, Simonsick EM, Newman AB, Nevitt M, Harris TB (2002) Relationship of interleukin-6 and tumor necrosis factor-alpha with muscle mass and muscle strength in elderly men and women: the Health ABC Study. *J Gerontol A Biol Sci Med Sci* **57**: M326–332.
87. Greiwe J, Cheng B, Rubin D, Yarasheski K, Semenkovich C (2001) Resistance exercise decreases skeletal muscle tumor necrosis factor alpha in frail elderly humans. *FASEB* **15**: 475–482.
88. Cesari M, Penninx BW, Pahor M, Lauretani F, Corsi AM, Rhys Williams G, Guralnik JM, Ferrucci L (2004) Inflammatory markers and physical performance in older persons: the InCHIANTI study. *J Gerontol A Biol Sci Med Sci* **59**: 242–248.
89. Payette H, Roubenoff R, Jacques PF, Dinarello CA, Wilson PW, Abad LW, Harris T (2003) Insulin-like growth factor-1 and interleukin 6 predict sarcopenia in very old community-living men and women: the Framingham Heart Study. *J Am Geriatr Soc* **51**: 1237–1243.
90. Fernandez–Celemin L, Pasko N, Blomart V, Thissen JP (2002) Inhibition of muscle insulin-like growth factor I expression by tumor necrosis factor-alpha. *Am J Physiol Endocrinol Metab* **283**: E1279–1290.
91. Fam AG (2001) Recent advances in the management of adult myositis. *Monthly Focus* 1265–1277.
92. Grounds MD, Torrisi J (2004) Anti-TNFalpha (Remicade) therapy protects dystrophic skeletal muscle from necrosis. *FASEB* **18**: 676–682.
93. Keating GM, Perry CM (2002) Infliximab: an updated review of its use in Crohn's disease and rheumatoid arthritis. *BioDrugs* **16**: 111–148.
94. Vescovo G, Ravara B, Gobbo V, Sandri M, Angelini A, Della Barbera M, Dona M, Peluso G, Calvani M, Mosconi L, Dalla Libera L (2002) *L*-Carnitine: a potential treatment for blocking apoptosis and preventing skeletal muscle myopathy in heart failure. *Am J Physiol Cell Physiol* **283**: C802–810.
95. Ames BN (2004) A role for supplements in optimizing health: the metabolic tune-up. *Arch Biochem Biophys* **423**: 227–234.
96. Vukovich M, Stubbs N, Bohlken R (2001) Body composition in 70-year-old adults responds to dietary beta-hydroxy-beta-methylbutyrate similarly to that of young adults. *J Nutr* **131**: 2049–2052.
97. Volek JS, Kraemer WJ, Bush JA, Boetes M, Incledon T, Clark KL, Lynch JM (1997) Creatine supplementation enhances muscular performance during high-intensity resistance exercise. *J Am Diet Assoc* **97**: 765–770.
98. Gotshalk LA, Volek JS, Staron RS, Denegar CR, Hagerman FC, Kraemer WJ (2002) Creatine supplementation improves muscular performance in older men. *Med Sci Sports Exerc* **34**: 537–543.
99. McCay C, Crowell M, Maynard L (1935) The effect of retarded growth upon the length of life span and upon the ultimate body size. *J Nutr* **10**: 63–79.
100. Butler RN, Fossel M, Harman SM, Heward CB, Olshansky SJ, Perls TT, Rothman DJ, Rothman SM, Warner HR, West MD, Wright WE (2002) Is there an antiaging medicine? *J Gerontol A Biol Sci Med Sci* **57**: B333–338.

101. Payne AM, Dodd SL, Leeuwenburgh C (2003) Life-long calorie restriction in Fischer 344 rats attenuates age-related loss in skeletal muscle-specific force and reduces extracellular space. *J Appl Physiol* **95**: 2554–2562.
102. Ramsey JJ, Colman RJ, Binkley NC, Christensen JD, Gresl TA, Kemnitz JW, Weindruch R (2000) Dietary restriction and aging in rhesus monkeys: the University of Wisconsin study. *Exp Gerontol* **35**: 1131–1149.
103. Das M, Gabriely I, Barzilai N (2004) Caloric restriction, body fat and ageing in experimental models. *Obes Rev* **5**: 13–19.
104. Chapman IM, MacIntosh CG, Morley JE, Horowitz M (2002) The anorexia of ageing. *Biogerontology* **3**: 67–71.
105. Shavlakadze T, White J, Hoh JF, Rosenthal N, Grounds MD (2004) Targeted expression of insulin-like growth factor-I reduces early myofiber necrosis in dystrophic mdx mice. *Mol Ther* **10**: 829–843.

5

Antioxidants and Aging in Human Beings

Éric Le Bourg
Centre de Recherche sur la Cognition Animale
U.M.R. C.N.R.S. n° 5169
Université Paul-Sabatier, 118 route de Narbonne
F-31062 Toulouse cedex 4, France
Fax: 33 5 61 55 61 54, e-mail: lebourg@cict.fr

INTRODUCTION

Antioxidants are fashionable. Most people know this word, even if they do not know what antioxidants are. The purpose of this article is to review the effects of antioxidant use on aging and mortality in human beings. This article is not concerned with antioxidant use in animals, which is reviewed elsewhere.[1] Before reviewing the literature, the context of the research on antioxidant use in human beings needs to be explained.

For some years, some companies have aggressively promoted antioxidant use as a means to delay aging, which creates many problems for biogerontologists and health practitioners. Indeed, there is a gap between biogerontologists, most of whom are not convinced that antioxidants are the miraculous elixir for successful aging, and people who routinely use antioxidant supplements. The increasing use of antioxidants by elderly without any medical advice is a problem in some highly developed countries, particularly the United States, and some biogerontologists have raised the alarm against the so-called anti-aging products.[2–7] More recently, 51 scientists issued a warning to the public with respect to the unscientific claims of various anti-aging therapies.[8] Since antioxidants are a big business, it is useful to explain why companies sell antioxidants as a means to fight aging.

Free Radical Theory of Aging and the Use of Antioxidants

After the nuclear bombing of the two Japanese cities in 1945 and the beginning of the nuclear weapons race, studying the effects of radiation attracted the attention of many biologists. In this context, Denham Harman[9] proposed "a theory based on free radical and radiation chemistry" to explain aging. This theory gave a biochemical explanation for an earlier theory of aging, "the rate of living theory" by Pearl,[10] which states that life-span is inversely related to the metabolic rate. Harman[9] accepted that aging "seems to be a more or less direct function of the metabolic rate" and proposed that highly reactive derivatives of oxygen, the free radicals, are produced during normal metabolism. The organism is unable to counteract all the damages caused by free radicals on macromolecules, because the balance between oxidants and antioxidants is in favor of oxidants. With time, unrepaired damages are expected to accumulate and put at risk the homeostasis of the organism, and to finally provoke aging and death. The rate of living theory[10] has been tested in different ways and today, many biogerontologists do not consider it an appropriate explanation of the aging process.[2,3,11] However, the fate of the free radical theory was different and this theory is today one of the leading theories of aging, even if the term "oxidative stress hypothesis" has supplanted the original one.[12] Yet, being a leading theory does not mean that this theory has been fully verified and accepted by the community of biogerontologists, and this is exactly the current status of the oxidative stress hypothesis of aging.

One of the expectations of the free radical theory was that adding antioxidants to the diet could change the balance between oxidants and antioxidants, which should increase longevity and delay aging. Since the early antioxidant supplementation studies carried out by Harman in rodents,[13] many other studies have been carried out, but contrary to Harman's view,[14] it does not seem that "there is now a growing consensus ... that (free radical) reactions are a major cause of aging, possibly the only one". It has to be said that Harman was not the only scientist to advocate antioxidant use since the double Nobel laureate Linus Pauling considered that mega-doses of vitamin C can be useful in fighting cancer[15] (also, see the debate between Pauling and Moertel[16]).

Despite the fact that the oxidative stress hypothesis is still a matter of debate, companies have not waited the resolution of the issue before promoting antioxidant use as a means to retard aging. American lobbies of anti-aging medicine won a strategic battle in the United States when, in 1994, the Dietary Supplement Health and Education Act (DSHEA) was passed. As indicated in a report of the US General Accounting Office,[17] the "DSHEA does not require manufacturers of dietary supplements to demonstrate either safety or efficacy to FDA (US Food and Drug Administration) prior to marketing them". The FDA "can ask a court to halt its sale ... if FDA *subsequently* (italics ours) determines that a dietary supplement is unsafe". Therefore, companies can sell any product up to the moment when FDA will be able to prove that it is dangerous: profits are for the companies, but the burden of the proof that a product is harmful lies with the FDA and, obviously, all the risk is for the consumer. This incredible 1994 act had adverse health consequences and the report of the US General accounting office[17] stressed that "some dietary supplements have been associated with potentially serious health consequences for senior citizens". It would be wrong to believe that the DSHEA act is of no concern outside the United States, because it provoked a strong growth of the anti-aging business with attempts to sell antioxidants outside United States. With the growth of internet, it has become possible at any place to purchase various products and, among them, antioxidants.

The main problem regarding antioxidant use without medical advice is obvious: most customers are not physicians and are thus totally unable to reach appropriate decisions about their own health. Ingesting these magic pills is like adopting the behavior of a sect-addict: a guru tells you that a magic pill is good for you and you ingest it. While customers may be unaware that the product could be inefficient or even unsafe, companies selling these products or organizations promoting their consumption seem to know that. In an article reporting "clinical trials using novel antioxidant supplements", Villeponteau *et al.*[18] from the HealthSpan Sciences company wrote: "we wished to identify an oral formulation that maximally reduced oxidative stress. The initial decision was whether such a formulation should be a novel drug or a nutritional supplement. An anti-aging drug … has several drawbacks in comparison to a nutritional supplement. First, the development time for a novel nutraceutical product is on the order of 6–12 months versus 6–12 years for a new pharmaceutical product, which requires FDA approval. Second, safety is a paramount concern for any anti-aging product, as it would have to be taken by healthy people for a very long period of time. In the case of an anti-aging drug, FDA approval would be particularly difficult, as even minor side effects might disqualify such a drug for long-term use by healthy adults. In contrast, *nutritional supplements are expected to have less long-term side effects* (italics ours) and, in any case, do not require FDA approval. Given these considerations, we initially set out to develop effective nutritional products that might reduce oxidative stress".

Villeponteau *et al.*[18] are perfectly right: developing a new efficient and safe drug is a costly and lengthy task while developing a supplement without studying its long-term effects is cheap and fast, with a quick and high return on investment. The basic question is to know who is the most probable winner in this game: the dealer or the customer? The answer is obvious!

Deciding whether antioxidant use may be useful to elderly is thus not an easy task. Basically, a scientist writing a review on any topics has to collect data, present them clearly, and conclude. If one is concerned with antioxidants, extra-scientific matters can pollute the task. One has to take into account that some scientists, due to their own positions about antioxidant use, can reach biased conclusions and we cannot dismiss *a priori* that the conclusions of some studies are contaminated by links between the authors and the anti-aging business.[18]

Keeping in mind the context of experimentation on antioxidants, we can now review some results. We will focus this review on the effects of antioxidants on mortality and on cognitive function, as a surrogate for the physiological state of elderly people. If antioxidants have a beneficial effect on aging and longevity, lower mortality and better cognitive status with higher antioxidant intake are expected.

Effects of Antioxidants on Aging and Mortality

According to the free radical theory, supplementing the elderly with antioxidants could be useful. However, a distinction has to be made between the effect of antioxidants on age-related diseases and in people free of diseases. Antioxidants are essential to life and it may be hypothesized that they could help elderly suffering diseases linked to free radical attacks. For instance, macular degeneration, which causes blindness in one quarter of nonagenarians,[19] has been associated with smoking, which increases markers of oxidative stress even for teenagers.[20] Taylor *et al.*[19] hypothesized that vitamin E supplementation could prevent the development or progression of early and late stages of macular degeneration, but a 4-year supplementation in 55–80 year-old people showed no significant difference (e.g. 8.6% incidence of the early stage in the supplemented group vs 8.1% in the control one). However, a positive result could not have been considered as evidence that the main cause of aging is linked to free radical attacks, because all people age but only some develop macular degeneration. Proving that antioxidants prevent or delay aging would require observing beneficial effects of antioxidants in disease-free people, for instance on mortality.

Circulating Antioxidants and Resistance to Free Radicals in Very Old People

Studying old and very old people is of interest before reviewing results on mortality. Very old people could be well-protected against free radicals because nonagenarians and centenarians have survived to what has killed most of their counterparts such as, possibly, free radical attacks.

Cristol *et al.*[21] studied the antioxidant defense mechanisms in 75–99 year-old women and 20–45 year-old ones. The activity of superoxide dismutase and the glutathione content decreased with age in red blood cells, while the plasma glutathione peroxidase activity and the vitamin E level did not change. Despite this decrease of antioxidant mechanisms, lipid peroxidation did not change with age. Garibaldi *et al.*[22] studied plasma antioxidants and protein oxidative damage in 25–89 year-old Italian people. The plasma level of lipophilic antioxidants (vitamins A and E) was similar in both genders and increased with age, and the level of protein oxidative damage did not change with age or sex. Whatever the cause of this last result could be (better protection against free radicals linked to an appropriate lifetime Mediterranean diet?), it shows that oxidative damage is not discernible in plasma proteins of elderly. Hyland *et al.*[23] observed that the non-enzymatic antioxidant capacity of plasma was higher in nonagenarians (86–94 year-old, $n = 138$) than in middle-aged controls (40–60 year-old, $n = 18$), which indicates that nonagenarians could be well-protected against free radicals and could explain their high longevity. However, the results of this study have to be considered with caution because the authors did not specify the gender of subjects. It could be that the middle-aged control group had, more or less, a 50% sex-ratio but it is probable that the oldest group was strongly women-biased. Secondly, while the old group was sampled from people living in the community, the control one "was drawn among professionals working in the laboratories" in a hospital of the same city, which probably explains its small size. Thus, it cannot be excluded that the results are partly biased, because other differences than age could differentiate the two groups.

Mecocci *et al.*[24] studied plasma antioxidants in the 50–100 year-old range. When excluding centenarians, the non-enzymatic plasma antioxidants (e.g. vitamins A, C, E) decreased with age. Thus, nonagenarians had lower plasma antioxidants than younger people, at variance with the vitamin E results of Cristol *et al.*[21] and Garibaldi *et al.*[22] The activity of glutathione peroxidase increased with age and that of superoxide dismutase increased non-significantly in plasma but significantly in red blood cells. Including centenarians in the analysis modified the results, because vitamin A and E levels were higher in this group than in all other ones and the effect of age on enzymatic activity was no longer significant. However, the vitamin C level was the lowest in centenarians, as also observed by Polidori *et al.*[25] Mecocci *et al.*[24] concluded that the antioxidant activity of the lipophilic vitamins A and E could be more important in longevity than that of the other antioxidants. However, they stressed that the biological effect of vitamins A and E could also be linked to the immune system.

Nevertheless, the previous studies of Hyland *et al.*[23] and Mecocci *et al.*[24] do not show that nonagenarians or centenarians better resist free radical attacks. Indeed, a high level of antioxidants in centenarians could mean that they are undergoing high oxidative attacks and marshalling last ditch efforts before death, but not that they are more able to resist free radicals. Measuring susceptibility to free radical attacks is a means to test whether centenarians better resist free radical attacks. Rabini *et al.*[26] showed that the lipid peroxide level and the susceptibility to peroxidation of erythrocyte membranes increased with age from the 21–40 year-old group to the 81–99 year-old one. By contrast, centenarians got values similar to those of the 41–60 year-old group. This result, showing that centenarians are protected against free radicals as young people, and better than elderly, is in contrast with that of Polidori *et al.*[25] showing that the lipid peroxide level is higher in centenarians than in 60–99 year-old people. It is thus difficult to know whether protection against free radicals is the key for a high longevity, but these results call for other studies on the link between free radicals and successful aging in centenarians.

These few studies do not show unequivocally that elderly suffer more from oxidative damage than younger counterparts or that antioxidant levels decrease with age. Similarly, Polidori *et al.*[27] concluded in a review article that "research conducted so far allows to hypothesize that plasma and serum small-molecular antioxidant vitamins do not change remarkably with healthy aging, with the possible exception of vitamin C, and at least until the very last years of human life". However, low plasma antioxidants concentrations have been reported in people suffering from Alzheimer's disease or of mild cognitive impairment.[28]

It could be that free radical attacks do not really increase with age in healthy people. The results of centenarians could be explained in two ways: centenarians have a particular antioxidant profile but it does not have an influence on their longevity, or their antioxidant profile does strongly contribute to their longevity. If this latter explanation were true, it would remain to explain why antioxidant levels are not linked to age in younger groups. Obviously, it could also be, as hypothesized above, that centenarians suffer hard free radical attacks and all remaining resources are invested before dying soon, which could explain their high antioxidant levels. A possible test of this hypothesis would be to correlate the antioxidant profile of centenarians with their longevity.

Antioxidants and Mortality

A means to increase plasma antioxidant levels could be to add antioxidants to the diet or to increase consumption of fruits and vegetables. On the one hand, an antioxidant cocktail with vitamins C and E increased the plasma level of these vitamins and decreased lipid peroxidation in the elderly[29] or decreased free radical attacks in the blood of 19–63 year-olds.[30] On the other hand, a high intake of vegetables and fruits, rich in antioxidants, increased the plasma vitamin C level but had no effect on lipid peroxidation in healthy adults,[31] but decreased free radical attacks in the blood of 19–63 year-olds.[30] A review article[32] concluded that supplementation with antioxidants does not seem to reduce oxidative damage, except in what concerns vitamin E and, maybe, vitamin C.

Therefore, it does not seem that there is a mechanistic link between antioxidant intake and oxidative damage, even if plasma antioxidant levels increase. In any case, this result does not show that antioxidants have no effect on mortality and does not preclude studying the effects of antioxidants on mortality (Table 1). Relative risks reported below are significant, except if the contrary is indicated.

Enstrom *et al.*[33] examined the relation between vitamin C intake and mortality in 25–74 year-old people followed for 10 years. Vitamin C intake was determined from food and vitamin C supplements consumed

Table 1 Effect of antioxidant level, determined from food intake, supplements use or blood samples, on all-cause mortality in human beings.*

Antioxidant	Age (years)	Follow-up duration (years)	Relative risk	Reference
Vitamin C: from food intake and supplements	25–74	10	**0.65** (men) 0.90 (women)	33
Vitamin C in blood samples	60 and older	9–12	**0.54**	34
Vitamin E in blood samples	60 and older	9–12	0.98	34
Carotenoids in blood samples	60 and older	9–12	0.78	34
Vitamin C in blood samples	75–84	4	**0.54**	35
Vitamin E in blood samples	75–84	4	0.89	35
Carotenoids in blood samples	75–84	4	0.76	35
Vitamin E in blood samples	65 and older	13	1.04 and 1.23 (vascular and non-vascular mortality)	36
Vitamin E in blood samples	65 and older	7	1.11	37
Carotenoids in blood samples	65 and older	7	1.45	37
Vitamins C and E supplements	35–60	7	0.63 (men) 1.03 (women)	39

*For all studies, the 1.00 relative risk is that of the lowest antioxidant concentration, except for de Waart *et al.*[37] for which the 1.00 relative risk is that of the highest antioxidant concentration. When antioxidants had not significant effects, the best relative risk is given, which is not necessary that of the highest antioxidant level. Positive significant effects on mortality are shown in bold characters.

by about 10% of participants. Higher vitamin C intakes were associated with lower mortality and vitamin C supplements seemed to have a positive role in that regard. However, these positive effects were mainly observed in men (relative risk for overall mortality: 0.65 in men and not significant in women). Thus, this study shows that vitamin C may have a positive effect on mortality but, like for the other studies deducing antioxidant content from the consumed food, it has to be stressed that food contains many other antioxidants than vitamin C that could explain the observed results.

Sahyoun *et al.*[34] showed that a high level of vitamin C in the plasma of people aged 60 or over was predictive of a low overall mortality 9 to 12 years later (relative risk: 0.54), particularly from heart disease, while vitamin E or carotenoids were not. In this study, about one third of participants used vitamin supplements. The benefit of a high plasma level of vitamin C was observed in people with "higher dietary plus supplemental intake since vitamin C supplements without consideration of diet were not associated with mortality".

Fletcher *et al.*[35] correlated blood antioxidant levels with mortality after 4 years of follow-up in 75–84 year-old British people. The vitamin C level was associated with low all-cause mortality (relative risk: 0.54) but vitamin C supplements, taken by about 50% of participants, had no effect on this result. However, the authors stressed that British people have low vitamin C levels and that about 60% of participants to the study would have been included in the lowest vitamin C level quintile of Sahyoun *et al.*[34] Otherwise, Fletcher *et al.*[35] reported that vitamin E and carotene were not associated with mortality.

The absence of effect of the vitamin E blood level on mortality was confirmed[36,37] in people aged 65 years and older followed, respectively, for 13 and 7 years. By contrast, these latter authors showed that a low level of carotenoids was associated with a high mortality (relative risk: 1.73).

Concerning antioxidant supplementation, Losonczy *et al.*[38] reported in a cohort of 67–105 year-old people followed for 9 years that vitamin E supplements reduced the risk of all-cause (relative risk: 0.66) or of coronary disease (relative risk: 0.53) mortality. Vitamin C supplementation had no effect on mortality. However, the authors did not record dietary intake and thus, "possible beneficial effects of nutritional habits ... could contribute to reduced risk of mortality".

Hercberg *et al.*[39] supplemented the diet of 35–60 year-old people with nutritional doses of vitamins C and E, carotene, zinc and selenium for 7 years. The all-cause mortality was similar in the placebo and antioxidant-supplemented groups (less than 2% of the subjects died during the study), but there was a not significant trend for a gender by treatment interaction because the supplemented men had a lower relative risk (0.63) than women (1.03). However, as the subjects of this study were rather young, it is impossible to conclude about possible effects in elderly.

On the one hand, these results seem to indicate that the blood vitamin C level can predict mortality, but that other antioxidants are poorly linked to mortality. On the other hand, taking vitamin supplements is not clearly associated with the reported results. In other words, the reported positive effects appear to be more linked to an appropriate diet than to supplements, as it is particularly clear from the study by Fletcher *et al.*[35] Indeed, as emphasized by Ward,[40] we may wonder whether "supplemental antioxidant vitamins are as effective as a diet that is rich in fruits and vegetables, soya beans and olive oil. There are thousands of antioxidant substances in these foods, many of which are more powerful than vitamin E, ascorbic acid and β-carotene". Thus, reviewing the studies connecting diets rich in fruits and vegetables and mortality is of concern.

Diet and Mortality

Many components of food contain antioxidants, particularly fruits and vegetables. Obviously, we can hypothesize that an adequate diet would be more efficient, and cheaper, to lower mortality than antioxidant supplements because the best antioxidant cocktail will never contain more than some antioxidants. Therefore, effects of appropriate diets on mortality can be expected to be stronger than are those of antioxidant supplements (Table 2). Relative risks reported below are significant, except if the contrary is indicated.

Greeks aged 70 years and older who adhered to a Mediterranean diet, which is rich in fruits and vegetables and low in meat, had a low overall mortality after 5 years of follow-up (relative risk: 0.83), but the various components of the diet (e.g., vegetables or fruits) were not linked to survival.[41] This lower mortality is not linked to a peculiar region, because similar results have been observed after 6 years of follow-up in 65–70

Table 2 Effect of diet and various food items on all-cause mortality in human beings.*

Food item	Age (years)	Follow-up duration (years)	Relative risk	Reference
Mediterranean diet	70 and older	5	**0.83**	41
Mediterranean diet	65–70	6	**0.79**	42
Fruits	54	16–26	**0.87** and 0.92 for 16 and 26 years of follow-up respectively	43
Fruits and vegetables rich in carotenoids	66 and older	5	**0.54** (cardiovascular mortality)	44
Fruits	65 and older	5	**0.52**	45
Yoghurt	65 and older	5	**0.38**	45
Coffee	65 and older	5	**0.35**	45
Red wine	40–60	12–18	**0.67**	47
Red wine	65 and older	3	**0.69**	48
Candy	60 and older	5	**0.73**	49

*For all studies, the 1.00 relative risk is that of the lowest food item concentration. Positive significant effects on mortality are shown in bold characters.

year-old Danish people following a Mediterranean diet[42] (relative risk: 0.79). As for the Greek study, the individual components of the diet were not linked to mortality, if one excepts that a high consumption of cereals increased mortality (relative risk: 1.10).

In Sweden,[43] a high fruit intake in 54 year-old people followed for 16 years was predictive of a low mortality (relative risk: 0.87), but it was marginally significant if participants were followed for 26 years (relative risk: 0.92, $p = 0.051$). However, a high vegetable intake was not linked to mortality. Intake of fruits and vegetables rich in carotenoids in 66 years or older people followed for 5 years was also shown (relative risk: 0.54) to decrease cardiovascular mortality.[44]

Beyond fruits and vegetables, other components of the diet contain antioxidants as well. Reviewing all food items is however beyond the scope of this review and we give below just a few examples of food items with positive effects on mortality of elderly.

Fortes *et al.*[45] correlated various components of the diet and mortality after 5 years of follow-up in 65 years and older Italian people. Decreased mortality was associated with high fruit consumption (relative risk: 0.52), but also with yogurt (relative risk: 0.38) and coffee intakes (relative risk: 0.35; obviously it was tasty Italian coffee!). Vegetable and pasta intakes were not significantly associated with a low mortality. By contrast, meat intake not significantly increased mortality (relative risk: 1.82; not significant), which seems in accordance with studies showing that a high meat consumption decreases longevity (reviewed in Singh *et al.*[46]).

In France, a moderate consumption of red wine in 40–60 year-old people followed for 12–18 years has been shown to lower overall (relative risk: 0.67) and cardiovascular mortality.[47] In another study, subjects aged 65 years and older had a lower (relative risk: 0.69) all-cause mortality after 3 years of follow-up.[48] However, one should not forget that, contrary to what happens with fruits, high wine consumption increases mortality. Other food items containing antioxidants may also have positive effects on mortality if consumed with moderation. For instance, Lee and Paffenbarger[49] reported that subjects aged 60 years or older consuming moderate amounts of candy-sweets had a lower mortality after 5 years of follow-up (relative risk: 0.73) than non-consumers. The authors linked this result to the antioxidant content of chocolate. However, as for red wine, a higher mortality was observed with high candy consumption.

All these results clearly indicate that following a diet rich in fruits and vegetables is a game with no loser: all participants have a good chance to win extra longevity. Other components of the diet, possibly due to their high antioxidant content, have positive effects but have to be consumed with moderation. Coffee, red wine, or candy, and possibly other food items, seem to decrease mortality but the limit between their positive effects, when taken at a moderate dose, and their harmful effects, can vary between individuals. In such conditions, no physician or nutritionist will ever prescribe them to elderly people without discernment while they can afford recommending fruits and vegetables. It is probable that the effects of diets rich in fruits and vegetables are due, at least partly, to their antioxidant content. However, a meal with more fruits and vegetables will contain less of other food items because, due to satiation, people reduce consumption of these items. Since it seems, for instance, that meat consumption has negative effects on mortality, increasing fruits and vegetables consumption could lessen mortality on their own, but also due to the decreased consumption of food items increasing mortality.

Finding that antioxidants or a diet containing antioxidants decrease mortality does not mean that they have positive effects on aging, as it can be inferred from e.g. cognitive status. Reviewing some studies is thus of interest.

Antioxidants and Aging

Some studies have tested the hypothesis that antioxidants have a positive effect on cognition in the elderly (Table 3). Warsama Jama *et al.*[50]

Table 3 Effect of antioxidant level, determined from food intake, supplements use or blood samples, on cognitive aging in human beings.

Antioxidant	Age (years)	Positive effect?	Reference
β-carotene from food intake	55–95	Yes	50
Vitamins C and E from food intake		No	50
Various vitamin supplements	66–90	Yes	51
β-carotene and vitamin C in blood samples	50–75	No	55
β-carotene and vitamin C from food intake	65–94	Yes	52
Vitamin E from food intake	65–94	No	52
Carotenoids in blood samples	59–71	Yes	53
Vitamin E in blood samples	59–71	No	53
Vitamin E in blood samples	50–75	Yes	55
Vitamin E in blood samples or from food intake	65–91	Yes	56
One year supplementation with a cocktail of vitamins and minerals	80–90	No	57
One year supplementation with a cocktail of vitamins and minerals	65 and older	Yes	58
Vitamins C and E from food intake	55 and older	Yes	59
Supplementation with vitamins C and/or E	71–93	Yes	60
Supplementation with vitamins C and E	70 and older	Yes	59
Supplementation with vitamin C alone	70 and older	No	59
Vitamin E from food intake and supplements	65 and older	Yes	62
Vitamin C from food intake and supplements	65 and older	No	62
Vitamin E in blood samples	65 and older	Yes	63
Vitamin A in blood samples	65 and older	No	63

correlated current dietary antioxidants intake with cognitive performance in 55–95 year-old people. A lower β-carotene intake was associated with impaired cognitive function while intake of vitamins C and E was not linked to cognitive status.

La Rue *et al.*[51] correlated cognitive performance in 66–90 year-old people with current nutritional components (e.g. vitamin C or E) and those recorded 6 years before. One of the cognitive tests was weakly correlated to current nutritional status, 50% of the coefficient correlations with the various nutritional components being significant (9/18), while for other comparisons the number of significant correlations could be probably explained by chance. However, some correlation coefficients indicated a negative association with supplementation. Clearly speaking, since all coefficients were low, the highest one being $r = 0.29$, it could be that they did reflect spurious associations. Nevertheless, people taking vitamin supplements, about 50% of participants, had a significantly, but slightly, higher cognitive performance than non-supplemented people.

Perrig *et al.*[52] correlated memory performance in 65–94 year-old people with current antioxidant plasma levels and those recorded 22 years before. Interestingly, antioxidant plasma levels were significantly correlated at the individual level, despite the long interval between measurements (e.g. for vitamin E: $r = 0.47$, $p < 0.001$). The results of some memory tests were correlated with current β-carotene and vitamin C levels, but not with the vitamin E one. Furthermore, the vitamin C level recorded 22 years before memory testing was also a predictor of performance. However, all these effects, even if significant (p being about 0.035), were very weak.

Berr *et al.*[53] correlated cognitive performance in 59–71 years people with antioxidants concentration in plasma or red blood cells. A low level of carotenoids was linked with a poor cognitive performance while plasma vitamin E, glutathione peroxidase and superoxide dismutase in red blood cells, were not associated with performance. Lipid peroxidation level was also not linked to cognitive status. However, when these people were examined four years later, a high lipid peroxidation level was found to be associated with a higher cognitive decline.[54]

Schmidt *et al.*[55] correlated plasma antioxidant levels with cognitive performance in 50–75 year-old subjects. A low vitamin E level was linked to poor cognitive performance but this significant effect was very low.

Vitamin C, β-carotene, or other antioxidants were not linked to cognitive status.

Ortega *et al.*[56] correlated cognitive performance in 65–91 year-old people and vitamin E level in serum or determined from food intake. No subject took vitamin E supplements. Low cognitive status was found to be associated with low vitamin E level in serum or in food intake. The authors observed that vitamin E intake was low when compared to the recommended dietary intake and concluded that vitamin E intake of elderly could be improved.

Regarding supplementation, Baker *et al.*[57] tried to reverse cognitive malfunction in 80–90 year-old women with a cocktail of vitamins and minerals. After one year, the supplemented group had a much higher concentration of vitamins in blood while the non-supplemented group had a lower level of vitamins than before the start of the study. However, cognitive function decreased to the same extent in both groups during this one-year study and both groups got similar cognitive scores at the start and end of the study. This study also showed that cognitive scores did not decrease if women developed a vitamin depression during the study. By contrast, Chandra[58] reported that 65 years and older people supplemented for one year with a cocktail of vitamins and minerals had an improved cognitive performance while no change was observed in the placebo group. This study also showed that people with blood-nutrient levels below the reference standard had low cognitive scores. Grodstein *et al.*[59] investigated the effect of antioxidant supplementation on cognition in 70 years and older women. Thirty three percent of women used supplements containing both vitamins C and E, while lower percentages took supplements with only vitamin C or E. Users of both vitamins had slightly better scores than non-supplemented women on some cognitive tests and these scores increased with duration of supplementation. However, when a global cognitive score was used, the relative risk for a low score was decreased in long-term vitamins C and E users but not significant. Users of only vitamin C or vitamin E had no better global cognition scores than non-users.

Concerning the risk for developing Alzheimer's disease and dementia, Masaki *et al.*[60] studied the effect of supplementation with vitamins C and E some years before cognition testing by cognitive performance and occurrence of dementia in 71–93 year-old men. These supplements protected from vascular dementia but not from Alzheimer's disease. Among non-demented people, using supplements was associated with a better cognitive function. Engelhart *et al.*[61] studied in 55 years and older people not demented at baseline the risk to develop dementia after 6 years of follow-up. A higher vitamins C and E content from dietary intake at baseline or from supplements was associated with a lower risk of Alzheimer's disease. Morris *et al.*[62] reported, in 65 years and older people without APOEε4 allele only, an association between a high vitamin E intake from food and a low risk of developing Alzheimer's disease. Helmer *et al.*[63] also reported in people 65 years and older that a low vitamin E concentration in plasma, which could not be considered as reflecting a vitamin E deficiency, was associated with a higher incidence of dementia up to 9 years after blood collection. In these two last studies, respectively, vitamins C and A had no effect on the risk of developing a dementia.

To sum up, it can be concluded that antioxidant supplementation or plasma antioxidant levels have minor effects on cognitive aging in the participants of the previous studies who were considered as healthy. Concerning the risk for developing a dementia, a low vitamin intake could be predictive; however, it is premature to conclude that antioxidant supplements protect from dementia. Otherwise, associations between low cognitive status and low vitamin plasma level, when they exist, could be due to a too low dietary intake.

Diet and Aging

Only a few studies have been done on the association between cognitive aging and diets containing antioxidants, even if some studies correlating various components of the diet and cognitive status have been done (reviewed in Solfrizzi *et al.*[64]). In this review, the authors concluded that "cognitive disorders may derive from severe vitamin deficiencies, with some improvement in cognitive performance after vitamin supplementation ... but, ... in healthy older people who consume normal amounts of vitamin, there is only a preliminary evidence of a preventive role of vitamin supplementation on cognitive decline". However, one has to keep in mind that some previously cited studies deduced antioxidant intake from food intake but did not actually measure antioxidant intake. Obviously,

Table 4 Effect of diet and various food items on cognitive aging in human beings.

Food item	Age (years)	Positive effect?	Reference
Fruits, vegetables, vitamin C from food intake	65–90	Yes	65
Red wine	65 and older	Yes (risk of dementia after 3 years of follow-up)	48
Moderate alcohol consumption	55 and older	Yes (risk of dementia after 6 years of follow-up)	66

a high antioxidant intake is linked to a high consumption of fruits, for instance. In such conditions, some of the previously cited studies could be cited here as well (Table 4).

Ortega *et al.*[65] correlated cognitive performance in 65–90 year-old people and dietary intake. Higher cognitive scores were linked to a higher consumption of total food, vegetables, fruits and vitamin C (determined from food intake). Obviously, these results could be due to the fact that subjects with better cognition have also better dietary habits but they also suggest that appropriate nutrition could help to protect from cognitive decline.

Orgogozo *et al.*[48] studied in 65 years and older persons not demented at baseline the effect of red wine drinking on incidence of dementia and Alzheimer's disease. After 3 years of follow-up, moderate drinking (250–500 mL a day) was associated with lower relative risks of incident dementia (0.19) and Alzheimer's disease (0.28) as compared to non drinking. Mild drinkers (less than 250 mL a day) had only a lower relative risk of Alzheimer's disease (0.55). The same study also reported a lower all-cause mortality in moderate (relative risk: 0.66) and mild drinkers (relative risk: 0.69). Similarly, Ruitenberg *et al.*[66] reported in 55 years and older people that moderate alcohol consumption was associated with a lower risk of dementia after 6 years of follow-up.

Conclusions

The message is rather clear: a diet rich in fruits and vegetables has positive effects on all-cause mortality, and these effects could be linked to their antioxidant content. However, antioxidant supplements themselves appear to be less efficient than an appropriate diet.

Positive effects of the diet could also be linked to other variables, because people adopting an appropriate diet often adopt a healthy lifestyle, with positive effects on mortality. For instance, Strandhagen *et al.*[43] reported that the proportion of smokers was lower in people eating more fruits. However, in most cases, the previously reported results on mortality relative risks were adjusted for many variables such as, age, sex, smoking habit, and so on.

Concluding that a diet rich in fruits and vegetables has positive effects on all-cause mortality does not mean in any way that an appropriate diet can increase longevity far beyond the current records. No one will probably equate the 122-years longevity of the current world record-woman, Jeanne Calment, by eating fruits or taking antioxidants. Simply, people adopting an appropriate diet have a higher chance to live long than those preferring a more dangerous one.

A crucial question is whether antioxidant supplements are useful to elderly, because they are available without medical advice and, "in general, people will choose a 'magic bullet' in preference to the more difficult behavioural change of adopting a well balanced diet".[40] Supplementing a bad diet with antioxidants is probably not appropriate, because there is a little chance that even the best cocktail of antioxidants will mimic the action of antioxidants present in food. It is still unknown whether the positive effects of some diets on overall mortality are due to antioxidants or to other compounds and, therefore, more research on the matter is needed. However, people wishing to improve their life when old are probably wiser in adopting an appropriate diet than to consume many pills daily for the rest of their life. Furthermore, while there is no obvious risk when ingesting fruits daily for lifetime, nobody can tell whether lifetime antioxidant supplementation is safe and high doses can be toxic.[40] To be perfectly clear, eating too many fruits can provoke diarrhea: you will be informed immediately and will adopt counter-measures. Intoxication with any chemical product, such as vitamin E, is more insidious and time can be needed before discovering it.[40] Obviously, when people have a too low intake of antioxidants from food and suffer from vitamins deficiency, using supplementation under physician supervision can be useful. However, for a long-term use, switching to a diet containing more antioxidants seems more appropriate.

Future research will probably show whether antioxidants can improve everyday life of healthy elderly or if they are useful in some age-related diseases only. If antioxidants improve tne health of elderly people, physicians will prescribe them, as they prescribe other medicines. If they are useful to the general population, cheap antioxidants will probably be added to all industrial food items, as it the case today with iodine in salt or fluorine in toothpaste. In the meantime, it seems appropriate to resist advertisements for miraculous anti-aging products and to eat generous amounts of fruits and vegetables.

A final word must be said about people selling anti-aging products. These products can be antioxidants, hormones, or anything else. Companies selling them claim that these products can help to fight aging or even to extend life span. For instance, the Life Extension Foundation claimed in an advertisement to be able to "obtain life extension drugs from around the world" and particularly therapies that "prevent and treat degenerative brain diseases including Alzheimer's disease, Parkinson's disease, etc". Unfortunately, such therapies did not exist when this advertisement was printed a few years ago and still do not exist. The duty of biogerontologists is to stress that these companies are mere charlatans and that their concern is not health but money.

References

1. Le Bourg E (2003) Antioxidants as modulators. In: Rattan SIS (ed.) *Modulating Aging and Longevity*, pp. 283–303, Kluwer Academic Publishers, Dordrecht, The Netherlands.
2. Austad SN (1997) *Why We Age*. John Wiley and sons, Inc., New York, USA.
3. Arking R (1998) *Biology of Aging*. Second edition. Sinauer Associates, Inc., Sunderland, USA.
4. Le Bourg E (1998) *Le Vieillissement en Questions*. C.N.R.S. Editions, Paris, France.
5. Le Bourg E (2000) Gerontologists and the media in a time of gerontology expansion. *Biogerontol* **1**: 89–92.
6. Kirkwood TBL (2000) *Time of our lives*. Weidenfeld and Nicolson, London, UK.
7. Olshansky SJ, Carnes BA (2001) *The quest for immortality*. W.W. Norton and Company, New York, USA.
8. Olshansky CA, Hayflick L, Carnes BA (2002) No truth to the fountain of youth. *Sci Amer* June issue, 92–95.

9. Harman D (1956) Aging: a theory based on free radical and radiation chemistry. *J Geront* **11**: 298–300.
10. Pearl R (1928) *The rate of living*. Knopf, London, UK.
11. Lints FA (1989) The rate of living theory revisited. *Gerontology* **35**: 36–57.
12. Sohal RS, Mockett RJ, Orr WC (2002) Mechanisms of aging: an appraisal of the oxidative stress hypothesis. *Free Rad Biol Med* **33**: 575–586.
13. Harman D (1968) Free radical theory of aging: effect of free radical reaction inhibitors on the mortality rate of male LAF_1 mice. *J Geront* **23**: 476–482.
14. Harman D (1998) Extending functional life span. *Exp Geront* **33**: 95–112.
15. Pauling L (1983) Vitamin C and longevity. *Agressol* **24**: 317–319.
16. Pauling L, Moertel C (1986) A proposition: megadoses of vitamin C are valuable in the treatment of cancer. *Nutr Rev* **44**: 28–32.
17. United States General Accounting Office (2001) Health products for seniors: "anti-aging" products pose potential for physical and economic harm. Report to chairman, special committee on aging, US senate 2001; http://www.gao.gov.
18. Villeponteau B, Cockrell R, Feng J (2000) Nutraceutical interventions may delay aging and the age-related diseases. *Exp Geront* **35**: 1405–1417.
19. Taylor HR, Tikellis G, Robman LD, McCarthy CA, McNeil JJ (2002) Vitamin E supplementation and macular degeneration: randomised controlled trial. *Brit Med J* **325**: 11–16.
20. Kim SH, Kim JS, Shin HS, Keen CL (2003) Influence of smoking on markers of oxidative stress and serum mineral concentrations in teenage girls in Korea. *Nutrition* **19**: 204–243.
21. Cristol JP, Abderrazick M, Favier F, Michel F, Castel J, Leger C, Descomps B (1999) Impairment of antioxidant defense mechanisms in elderly women without increase in oxidative stress markers: "a weak equilibrium". *Lipids* **34**: S289.
22. Garibaldi S, Valentini S, Aragno I, Pronzato MA, Traverso N, Odetti P (2001) Plasma protein oxidation and antioxidant defense during aging. *Int J Vitam Nutr Res* **71**: 332–338.
23. Hyland P, Duggan O, Turbitt J, Coulter J, Wikby A, Johansson B, Tompa A, Barnett C, Barnett Y (2002) Nonagenarians from the Swedish NONA immune study have increased plasma antioxidant capacity and similar levels of DNA damage in peripheral blood mononuclear cells compared to younger control subjects. *Exp Geront* **37**: 465–473.
24. Mecocci P, Polidori MC, Troiano L, Cherubini A, Cecchetti R, Pini G, Straatman M, Monti D, Stahl W, Sies H, Franceschi C, Senin U (2000) Plasma antioxidants and longevity: a study on healthy centenarians. *Free Rad Biol Med* **28**: 1243–1248.
25. Polidori M.C, Mecocci P, Cherubini A, Cecchetti R, Briviba K, Stahl W, Sies H, Senin U (1999) Plasma antioxidants peroxidation and vitamin C status in healthy centenarians. *J Am Geriatr Soc* **47**: 1038–1039.
26. Rabini RA, Moretti N, Staffolani R, Nanetti L, Franceschi C, Mazzanti L (2002) Reduced susceptibility to peroxidation of erythrocyte plasma membranes from centenarians. *Exp Geront* **37**: 657–663.
27. Polidori MC, Cherubini A, Senin U, Mecocci P (2001) Peripheral non-enzymatic antioxidant changes with human aging: a selective status report. *Biogeront* **2**: 99–104.
28. Rinaldi P, Polidori MC, Metastasio A, Mariani E, Mattioli P, Cherubini A, Catani M, Cecchetti R, Senin U, Mecocci P (2003) Plasma antioxidants are similarly depleted in mild cognitive impairment and Alzheimer's disease. *Neurobiol Aging* **24**: 915–919.

29. Courtière A, Cotte JM, Pignol F, Jadot G (1989) Etude de la lipidoperoxydation chez le malade âgé. *Thérapie* **44**: 13–17.
30. Lesgards JF, Durand P, Lassarre M, Stocker P, Lesgards G, Lanteaume A, Prost M, Lehucher-Michel MP (2002) Assessment of lifestyle effects on the overall antioxidant capacity of healthy subjects. *Environ Health Perspect* **110**: 479–486.
31. Freese R, Alfthan G, Jauhiainen M, Basu S, Erlund I, Salminen I, Aro A, Mutanen M (2002) High intake of vegetables, berries, and apples combined with a high intake of linoleic or oleic acid only slightly affect markers of lipid peroxidation and lipoprotein metabolism in healthy subjects. *Am J Clin Nutr* **76**: 950–960.
32. McCall MR, Frei B (1999) Can antioxidant vitamins materially reduce oxidative damage in humans? *Free Rad Biol Med* **26**: 1034–1053.
33. Enstrom JE, Kanim LE, Klein MA (1992) Vitamin C intake and mortality among a sample of the United States population. *Epidemiology* **3**: 194–202.
34. Sahyoun NR, Jacques PF, Russell RM (1996) Carotenoids, vitamins C and E, and mortality in an elderly population. *Am J Epidemiol* **144**: 501–511.
35. Fletcher AE, Breeze E, Shetty PS (2003) Antioxidant vitamins and mortality in older persons: findings from the nutrition add-on study to the Medical Research Council Trial of Assessment and Management of Older People in the Community. *Am J Clin Nutr* **78**: 999–1010.
36. Marniemi J, Järvisalo J, Toikka T, Räihä I, Ahotupa M, Sourander L (1998) Blood vitamins, mineral elements and inflammation markers as risk factors of vascular and non-vascular disease mortality in an elderly population. *Int J Epidemiol* **27**: 799–807.
37. De Waart FG, Schouten EG, Stalenhoef AFH, Kok FJ (2001) Serum carotenoids, α-tocopherol and mortality risk in a prospective study among Dutch elderly. *Int J Epidemiol* **30**: 136–143.
38. Losonczy KG, Harris TB, Havlik RJ (1996) Vitamin E and C supplement use and risk of all-cause and coronary heart disease mortality in older persons: the Established Populations for Epidemiologic Studies of the Elderly. *Am J Clin Nutr* **64**: 190–196.
39. Hercberg S, Galan P, Preziosi P, Bertrais S, Mennen L, Malvy D, Roussel AM, Favier A, Briançon S (2004) The SU.VI.MAX study: a randomized, placebo-controlled trial of the health effects of antioxidant vitamins and minerals. *Arch Intern Med* **164**: 2335–2342.
40. Ward JA (1998) Should antioxidant vitamins be routinely recommended for older people? *Drugs Aging* **12**: 169–175.
41. Trichopoulou A, Kouris–Blazos A, Wahlqvist ML, Gnardellis C, Lagiou P, Polychronopoulos E, Vassilakou T, Lipworth L, Trichopoulos D (1995) Diet and overall survival in elderly people. *Brit Med J* **311**: 1457–1460.
42. Osler M, Schroll M (1997) Diet and mortality in a cohort of elderly people in a North European community. *Int J Epidemiol* **26**: 155–159.
43. Strandhagen E, Hansson PO, Bosaeus I, Isaksson B, Eriksson H (2000) High fruit intake may reduce mortality among middle-aged and elderly men. The study of men born in 1913. *Eur J Clin Nutr* **54**: 337–341.
44. Gaziano JM, Manson JE, Branch LG, Colditz GA, Willett WC, Buring JE (1995) A prospective study of consumption of carotenoids in fruits and vegetables and decreased cardiovascular mortality in the elderly. *Ann Epidemiol* **5**: 255–260.
45. Fortes C, Forastiere F, Farchi S, Rapiti E, Pastori G, Perucci CA (2000) Diet and overall survival in a cohort of very elderly people. *Epidemiology* **11**: 440–445.

46. Singh PN, Sabaté J, Fraser GE (2003) Does low meat consumption increase life expectancy in humans? *AM J Clin Nutr* **78**: 526S–532S.
47. Renaud SC, Guéguen R, Siest G, Salamon R (1999) Wine, beer, and mortality in middle-aged men from eastern France. *Arch Intern Med* **159**: 1865–1870.
48. Orgogozo JM, Dartigues JF, Lafont S, Letenneur L, Commenges D, Salamon R, Renaud S, Breteler MB (1997) Wine consumption and dementia in the elderly: a prospective community study in the Bordeaux area. *Rev Neurol (Paris)* **153**: 185–192.
49. Lee IM, Paffenbarger RS (1998) Life is sweet: candy consumption and longevity. *Brit Med J* **317**: 1683–1684.
50. Warsama Jama J, Launer LJ, Witteman JCM, den Breeijen JH, Breteler MMB, Grobbee DE, Hofman A (1996) Dietary antioxidants and cognitive function in a population-based sample of older persons. *Am J Epidemiol* **144**: 275–280.
51. La Rue A, Koehler KM, Wayne SJ, Chiulli SJ, Haaland KY, Garry PJ (1997) Nutritional status and cognitive functioning in a normally aging sample: a 6-y reassessment. *Am J Clin Nutr* **65**: 20–29.
52. Perrig WJ, Perrig P, Stähelin HB (1997) The relation between antioxidants and memory performance in the old and very old. *J Am Geriatr Soc* **45**: 718–724.
53. Berr C, Richard MJ, Roussel AM, Bonithon–Kopp C (1998) Systemic oxidative stress and cognitive performance in the population-based EVA study. *Free Rad Biol Med* **4**: 1202–1208.
54. Berr C, Balansard B, Arnaud J, Roussel AM, Aplérovitch A (2000) Cognitive decline is associated with systemic oxidative stress: the EVA study. *J Am Geriatr Soc* **48**: 1285–1291.
55. Schmidt R, Hayn M, Reinhart B, Roob G, Schmidt H, Schumacher M, Watzinger N, Launer LJ (1998) Plasma antioxidants and cognitive performance in middle-aged and older adults: results of the Austrian prevention study. *J Am Geriatr Soc* **46**: 1407–1410.
56. Ortega RM, Requejo AM, Lopez–Sobaler AM, Andrés P, Navia B, Perea JM, Robles F (2002) Cognitive function in elderly people is influenced by vitamin E status. *J Nutr* **132**: 2065–2068.
57. Baker H, De Angelis B, Baker ER, Frank O, Jaslow SP (1999) Lack of effect of 1 year intake of a high-dose vitamin and mineral supplement on cognitive function of elderly women. *Gerontology* **45**: 195–198.
58. Chandra RK (2001) Effect of vitamin and trace-element supplementation on cognitive function in elderly subjects. *Nutrition* **17**: 709–712.
59. Grodstein F, Chen J, Willett WC. (2003) High-dose antioxidant supplements and cognitive function in community-dwelling elderly women. *Am J Clin Nutr* 77: 975–984.
60. Masaki KH, Losonczy KG, Izmirlian G, Foley DJ, Ross GW, Petrovitch H, Havlik R, White LR (2000) Association of vitamin E and C supplement uses with cognitive function and dementia in elderly men. *Neurology* **54**: 1265–1272.
61. Engelhart MJ, Geerlings MI, Ruitenberg A, van Swieten JC, Hofman A, Witteman JC, Breteler MMB (2002) Dietary intake of antioxidants and risk of Alzheimer's disease. *J Am Med Assoc* **287**: 3223–3229.
62. Morris MC, Evans DA, Bienias JL, Tangney CC, Bennett DA, Aggarwal N, Wilson RS, Scherr PA (2002) Dietary intake of antioxidant nutrients and the risk of developing Alzheimer disease in a biracial community study. *J Am Med Assoc* **287**: 3230–3237.
63. Helmer C, Peuchant E, Letenneur L, Bourdel–Marchasson I, Larrieu S, Dartigues JF, Dubourg L, Thomas MJ, Berberger–Gateau, P (2003) Association between antioxidant nutri-

tional indicators and the incidence of dementia: results from the PAQUID prospective cohort study. *Eur J Clin Nutr* **57**: 1555–1561.
64. Solfrizzi V, Panza F, Capurso A (2003) The role of diet in cognitive decline. *J Neural Transm* **110**: 95–110.
65. Ortega RM, Requejo AM, Andrés P, Lopez-Sobaler AM, Quintas ME, Redondo MR, Navia B, Rivas T (1997) Dietary intake and cognitive function in a group of elderly people. *Am J Clin Nutr* **66**: 803–809.
66. Ruitenberg A, van Swieten JC, Witteman JC, Mehta KM, van Duijn C.M, Hofman A, Breteler MMB (2002) Alcohol consumption and risk of dementia: the Rotterdam study. *Lancet* **359**: 281–286.

6

Hormone Therapy for Aging

Mahendra K. Thakur
Biochemistry & Molecular Biology Laboratory
Department of Zoology, Banaras Hindu University
Varanasi 221 005, India
E-mail: mkt_bhu@yahoo.com

Hormones are products of the endocrine system. They communicate between and coordinate the functions of hundred-trillion cells of the human body. The form of communication is chemical and hormones are carried in the blood throughout the body. The exceptions to the vascular medium include local intercellular hormonal messages (e.g. trophic factors), neurotransmitters (between neurons or between neurons and muscles), and intrapituitary hormonal transfer (e.g. hormonal releasing factors from the posterior to the anterior pituitary).

The endocrine system undergoes major changes during the aging process. There are abrupt or gradual changes in circulating hormone concentration. Such changes also include production, secretion, transport, clearance, temporal regulation, and target tissue response. Besides decline in hormone levels with increasing age, there also occurs deterioration in endocrine communication leading to an impairment of normal endocrine coordination of cellular and tissue functions. In all species examined to date, endocrine manipulations can slow aging.[1] The hormonal manipulation seems logical in principle because the blood levels of most hormones — among them growth hormone, insulin, melatonin, dehydroepiandrosterone (DHEA), androgen and estrogen — commonly decrease with age. While considering these changes, one should not view only single hormone function in isolation because hormones cross-talk with one another resulting in a complex cascade of causes and effects.[2]

The action of hormones is mediated through a signal transduction pathway (Fig. 1).[3] Some hormones (growth hormone, insulin) bind to their receptors in the membrane and exert their effects through a variety of second messengers existing in the cell. In contrast, other hormones (melatonin, steroids) bind to their specific receptors inside the cell and then interact with a host of coregulators and hormone response elements (HRE) located in the promoter of target genes. Such interaction influences the transcriptional regulation of specific genes and accordingly affects the physiological activities of the cell. The level of hormone and its receptor, a variety of factors that are involved in the process of signal transduction, and tissue responsiveness alter with advancing age.[2,4,5] As a

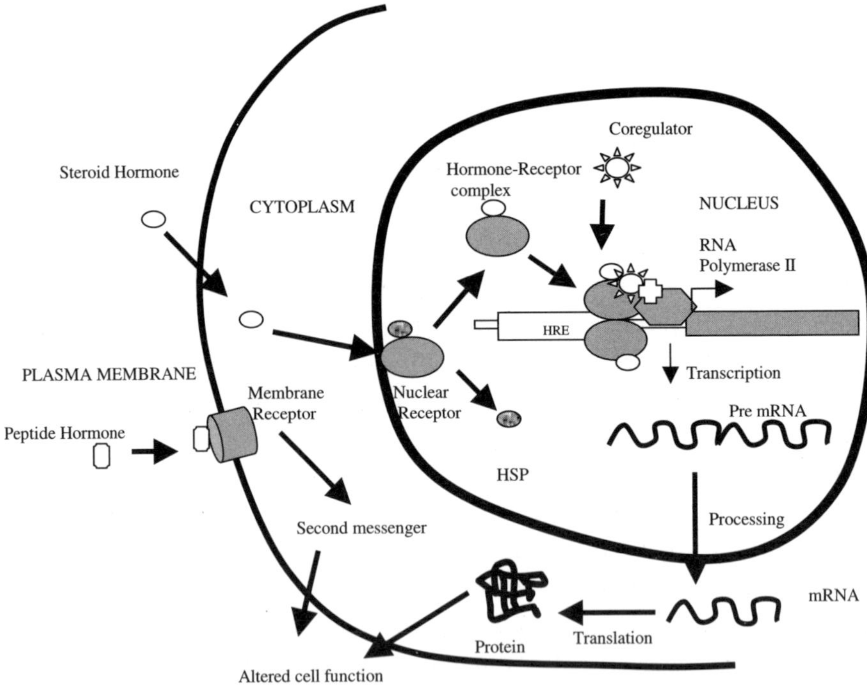

Fig. 1 Diagrammatic representation of mechanism of steroid hormone action through nuclear receptor and peptide hormone through membrane receptor. (HRE — Hormone responsive element, HSP — Heat shock protein.)

consequence, there are changes in the pattern of gene expression leading to alteration in the cellular function. Ultimately this may result in the development of hormone-related disorders and diseases in old age (Table 1). More details about changes in hormones involved during aging are mentioned below:

Table 1 Hormone-Related Diseases of Old Age.

Somatopause (decline in GH/IGF-1)
Menopause (decline in estrogen)
Andropause (decline in testosterone)
Adrenopause (decline in DHEA and DHEAS)
Decreased sexual activity
Deficits in memory
Cognitive impairment
Depression
Anxiety
Insomnia
Bone loss
Osteoporosis
Diabetes
Atherosclerosis
Cardiovascular disease
Cancer of different kinds (breast, prostate)

Growth Hormone

In mammals, growth hormone, insulin, insulin-like growth factor (IGF)-1 and thyroid hormone are interdependent. One of the effects of growth hormone released by pituitary gland is to stimulate liver cells, which release IGF-1 into the blood stream. IGF-1 then binds to receptors at the surface of target cells and stimulates the tyrosine kinase pathway, which principally regulates cell growth. Experiments of Holzenberger *et al.*[6] show that when tissue expression of IGF-1 receptor is halved in female mice, resistance to both oxidative stress and life expectancy increases, but energy metabolism, carbohydrate metabolism and fertility are unaltered.

Growth hormone generally declines with age,[7] and also with inactivity, changes in sleep cycles, malnutrition or disease.[8] The decline, on the average, occurs by 14% with each decade in normal adults after 20 years of age. This may be due to a decrease in growth hormone releasing factor,[9] or in other hormonal systems such as the thyroid axis.[10] Such decrease in turn leads to reduced circulating IGF-1 and changes in IGF-binding proteins[11] and a concomitant decline in immune function.[12] The decline in the function of GH-releasing hormone, growth hormone and insulin-like growth factor (GHRH-GH-IGF) axis has been termed the "somatopause".[13] Inhibition of GH-dependent pathway is thought to have an important role in lifespan extension because several premature aging phenomena are observed in GH-overexpressing mice.[14] For instance, ames dwarf mice, which are genetically deficient in growth hormone, live longer while those that overexpress the hormone live shorter than their normal counterparts.[15] The phenotypically similar snell dwarf mice with a mutation in *pit-1* gene also have a longer lifespan.[16] The mutant mice with a combined deficiency of growth hormone, prolactin and thyrotropin and knockout mice with growth hormone resistance, live upto 65% longer than their normal siblings.[17] Carter *et al.*[18] reported that GH/IGF-1-deficient dwarf mice live significantly longer than their wild-type counterparts. A cytokinin plant growth hormone kinetin retards senescence in plants and delays the onset and decreases the extent of many aging characteristics of human diploid fibroblast cells in culture.[19] It also slows down aging and prolongs the lifespan of the fruit fly *Zaprionus paravittiger.*[20]

As a consequence of reduced level of growth hormone, the physical and physiological attributes like muscle mass and skin elasticity show decline in old men. This might also play a role in age-related diseases, including atherosclerosis.[21] Interestingly, these effects are reversed after growth hormone replacement. Whereas some workers have suggested numerous beneficial effects of growth hormone on body composition, strength and quality of life; others have observed only marginal functional improvements. Moreover, it is clear that hormones can cause significant morbidity.[13] A number of studies have demonstrated that administration of growth hormone, made in pure form by genetic engineering technology, can reverse many age-associated changes in body composition. In aged men, growth hormone treatment for six months results in improvement of lean body mass and decrease in body fat.[22] Particularly, in the older adult, growth hormone may be necessary to maintain muscle mass and function.[23] In aging rats, growth hormone supplementation is shown to have beneficial effects upon myocardial blood flow by increasing the capillary density.[24] Thus these and other studies suggest that growth hormone plays an important role throughout the lifespan.

Although short-term treatment with growth hormone offers health benefits, excess hormone levels are known to have numerous detrimental effects. For example, Bouillanne *et al.*[25] have reported that short-term side-effects of growth hormone therapy include edema, carpal tunnel syndrome and arthralgia. The high level of serum IGF-1 is also associated with a greater risk of neoplastic disorders. Studies by Hauck and Bartke[26] suggest that the excess of growth hormone is associated with early decline in superoxide dismutase and glutathione peroxidase in the mice kidney and liver, which in turn increases the risk of free radical induced damage to these tissues and greater incidence of liver tumors and renal failure. Blackman *et al.*[27] also demonstrated that the replacement of growth hormone and sex steroids has adverse effects, including the development of diabetes mellitus. Hence there is less enthusiasm for the use of recombinant growth factors[23] to reverse the changes associated with the deficiency of growth hormone.

Insulin

Like other peptide hormones, insulin action is mediated through a receptor that is expressed in many tissues, including liver, muscle and adipose tissue. Insulin regulates the level of blood glucose by promoting the absorption of glucose by the cells of these organs and by activating metabolic pathways of carbohydrate storage. However, with increasing age, the resistance of insulin receptor arises progressively, contributing to an increase in circulating glucose levels and promoting glycation of long-life macromolecules like collagen. Glycation is accelerated in diabetes and is largely responsible for loss of elasticity of the large arterial trunks and for lesions in small vessels, which lead to microangiopathy.

Mutagenesis experiments have shown that the deletion of certain genes near the insulin receptor can alter the longevity of invertebrates. The reduced signaling of insulin-like peptides has been reported to increase the lifespan of nematodes, flies and rodents. For instance, the inactivation of daf-2 gene corresponding to an insulin receptor or of age-1 linked to the phosphoinositide 3-kinase pathway extends the longevity of *C. elegans* by 2–3-fold.[28] Similarly, mutation of *chico* gene, which encodes an insulin receptor that functions in an insulin/IGF signaling pathway, extends *Drosophila* lifespan by 48% in homozygotes and 36% in heterozygotes.[29] Mutation of another gene insulin-like receptor (*InR*), which is homologous to mammalian insulin receptor, also extends the longevity of females' upto 85% and reduces late age-specific mortality of male Drosophila.[30] Interestingly, treatment of such long-lived flies with a juvenile hormone analog restores life expectancy to that of wild-type controls. Bluher *et al.*[31] inactivated the expression of insulin receptor in adipose tissue in mice and found that they remain lean and live 15 to 20% longer than controls, even though they were eating the same amount of food everyday. Studies in aged populations (55–80 years) show that free IGF-1 levels do not decline with age and are even higher in individuals above 70 years old. This high level of IGF-1 is associated with a decreased risk of cardiovascular disease. In centenarians, the molar ratio of plasma IGF-1/IGF-binding protein 3 is higher than aged people,[32] suggesting that increased bioavailable IGF-1 might improve insulin action and plasma lipid profiles in centenarians.

Melatonin

Since detailed information about the biological effects of melatonin is discussed in a separate chapter in this book, only a brief discussion is given here. Melatonin is synthesized from tryptophan in the pineal gland and secreted at night. In normal people, endogenous melatonin levels are highest during the normal hours of sleep, increasing rapidly in the late evening, peaking after midnight and decreasing toward morning. The level of melatonin attains its peak during adolescence, and decreases thereafter in normal, healthy individuals.[33] Old people, who have lower serum concentration of melatonin, often complain irregular sleep-wake cycles. The secretion of melatonin is also severely reduced in Alzheimer's disease (AD).

Although melatonin levels decrease with age, there is no evidence that replacing these levels affects the aging process. It is generally accepted that the antioxidant properties of melatonin significantly contribute to its role as an anti-aging hormone. A melatonin precursor and metabolite, N-acetylserotonin (NAS), also shows anti-aging effects in female retired breeders and male mice.[34] Both NAS and melatonin administered with drinking water prolongs lifespan in male animals by about 20%. When C57BL mice are fed orally with this hormone from early age, their aging is delayed and lifespan is prolonged.[35] In the same study, it was observed that grafting of pineals from young mice into the thymus of syngenic old mice increased their survival and inhibited age-related structural changes in the thymus. Melatonin is remarkably effective in preventing death of cultured neuroblastoma cells as well as oxidative damage and intracellular Ca^{2+} increases induced by a cytotoxic fragment of amyloid peptide. Thus melatonin may be of therapeutic use in AD.[36] Replacement therapy with melatonin or melatonin-related compounds also proves useful in treating, preventing or delaying disturbances of circadian rhythmicity and/or sleep in older people.[37] Although some evidence suggests that melatonin supplementation has physiological benefits (at least in mice), including a marginal increase in lifespan, other evidence supports increased tumor incidence.[38] Therefore, melatonin is not recommended as a geroprotector for long-term use.

Dehydroepiandrosterone (DHEA)

The adrenals of humans and a few higher primates synthesize and secrete large amounts of DHEA and its sulfate (DHEAS) that are transformed in peripheral tissues into biologically active sex steroid hormones, androgens and estrogens. It is estimated that more than 30% of total androgen in men and over 90% of estrogen in post-menopausal women are derived from peripheral conversion of DHEA/DHEAS.[39] There exists a strong correlation between the DHEAS level and functional activities of 90–106 year-old healthy people.[40] Those individuals with very active functional status have the highest DHEAS level, whereas those with poor functioning capacity have the lowest DHEAS level.

Circulating levels of DHEA and DHEAS attain peak during early adulthood and then decline to less than 30% of peak level by the 6th decade of life. Falling levels of these hormones have been implicated as a contributory factor to the vascular disease, diabetes mellitus, malignancy, musculoskeletal disorder, dementia and cognitive impairment that occurs in old age.[41] DHEA influences dendritic structure and synaptic density in the hippocampus, a region of the brain vital for the processing of information. Hajszan *et al.*[42] showed that the subcutaneous injection of DHEA (1mg/day for two days) increased CA1 spine synapse density by more than 50% compared to vehicle-injected animals. The effect of DHEA on CA1 synapse density was abolished by pre-treatment with the non-steroidal aromatase inhibitor, letrozole. DHEA treatment, with or without letrozole, had no detectable uterotrophic effect. These observations are consistent with the hypothesis that DHEA treatment may be capable of reversing the decline in hippocampal spine synapse density resulting from the loss of ovarian steroid hormone secretion. The blockade of the synaptic response to DHEA by letrozole, despite the lack of uterotrophic response to this steroid, suggests that the hippocampal response to DHEA may be mediated via aromatization in the brain.

Researchers claim that DHEA might be protective against cardiovascular disease, cancer, immune-modulated diseases, dementia and aging.[10] Some suggest that DHEA might protect against glucocorticoid side effects, including diabetes, obesity, hypertension, immune suppression, myopathy, osteopenia, osteoporosis and avascular necrosis.[43] Therefore, elderly people may take DHEA as a food supplement in order to remain biologically young.[44,45] In a double-blind, placebo-controlled study conducted by Baulieu and colleagues,[46] 280 healthy individuals aged 60–79 years were orally given 50 mg DHEA or placebo daily for a year. The result showed improvement in bone turnover, decrease of osteoclastic activity and skin status in women above 70 years. Many libido parameters were also found to increase significantly in these old women. Interestingly, a number of biological indices confirmed the lack of harmful consequences of this dose of 50 mg/day DHEA administration over one year, indicating that this kind of replacement therapy normalized some effects of aging. Thus when taken as a dietary supplement, DHEA claims to improve mood and memory, counteract stress hormones, preserve muscle tone, and restore sexual vigor. However, high doses of DHEA confer a great risk for developing prostate and breast cancer.

ANDROGENS

The level of testosterone, which forms the major proportion of androgens, falls slowly and linearly with advancing age.[47] The mean total testosterone levels decrease by 30% between 25 and 75 years of age and mean free testosterone decreases by as much as 50%. Serum estradiol also fall due to a decrease in available testosterone to be aromatized to estrogen. Serum levels of adrenal androgens, DHEA/DHEAS decline most dramatically as men age.

Serum levels of total and bioavailable testosterone gradually decrease with age in men, and this condition is called "andropause". Also the level, structure, expression and other characteristics of androgen receptor (AR), through which testosterone acts, change with age.[48] Such changes have significant impact on several physiological processes and the affected individuals show the symptoms of diminished libido, sexual dysfunction, impotency, reduced muscle and bone mass, weakness, fatigue, depression, anemia, anxiety, irritability, insomnia, poor memory, cognitive impairment, decreased feeling of well-being and ultimately the appearance of a variety of age-associated disorders.[49,50] In aged people, the reduction of adrenal androgen secretion is accompanied by a host of neuroendocrine-metabolic dysfunctions that include decline in the GH-IGF-1 system, thyroid function, immune competence, fragmentation of sleep and neuronal loss.[39] Goodman–Gruen and Barrett–Connor[51] have suggested an inverse correlation between androgens and type 2 diabetes in men and a positive correlation between androgens and type 2 diabetes in women.

Some of the conditions resulting from androgen deficiency are reversed after supplementation with androgens. Over the past several decades, there has been an increasing interest in evaluating whether testosterone therapy is beneficial for older men in preventing or reversing some aspects of aging.[52] The major androgen target organs of interest with regard to beneficial effects of male HRT include bone, muscle, adipose tissue, cardiovascular system and central nervous system (libido and mood). As early as 1889, Brown–Sequard[53] administered a suspension of the dog testis to himself and reported that it rejuvenated him at age 72. Morley *et al.*[54] also used testosterone (or testicular extracts) to "rejuvenate" aging males. Sih *et al.*[55] examined effects of testosterone administration in hypogonadal men who were above 50 years and found considerable increase in grip strength and hemoglobin levels and a significant decrease in the level of leptin, which is the product of the obese (*ob*) gene. Testosterone replacement improves depressed moods in old men.[56] Short-term testosterone administration[57] and transdermal testosterone treatment[58] enhance cognitive function in healthy old men. Testosterone also increases bone density. It is further known to inhibit vascular cell adhesion molecule-1 mRNA and protein expression in human umbilical vein endothelial cells by its conversion to estradiol via aromatase.[59] Thus testosterone may have beneficial effects in atherosclerosis, which is more prevalent in old age.

Though testosterone replacement therapy may be beneficial for hypogonadal men, it has potential adverse effects on target organs such as the prostate. Therefore, older men receiving testosterone must be carefully monitored. More long-term studies and a substantially large group of older men are required for considering the long-term effect of testosterone treatment on the development of detectable prostate cancer or symptomatic benign prostatic enlargement and to examine its efficacy and safety.

Estrogens

Several workers have studied the level of estrogen, its receptor and responses in relation to aging using different experimental systems.[60] Estrogen exerts a number of effects in various target tissues throughout the entire lifespan. Usually it is observed that by the time a woman attains 65 years, the ovary becomes virtually devoid of follicles and is no longer the primary site of estrogen or progesterone synthesis. Consequently, estrogen level drops dramatically after menopause in females[61] and there occurs changes in several physiological processes leading to pathological conditions. Also, inhibin, a glycoprotein that is synthesized in granulosa and luteal cells of the ovary and that selectively suppresses follicle stimulating hormone (FSH) secretion, becomes undetectable in the blood. In response, the anterior pituitary gland secretes copious amounts of both FSH and luteinizing hormone (LH). The level of FSH increases by the time women are 45 to 50 years old, when they are still menstruating, whereas LH level increases later, when women become post-menopausal.[62]

Estrogen replacement therapy (ERT) has been available to post-menopausal women for a long time. The major benefits include relief of vasomotor symptoms and vaginal atrophy and prevention and treatment of osteoporosis.[63] Estrogen is known to increase the bone mineral density in prevention of osteoporosis. But at conventional doses, these benefits are achieved at the cost of adverse effects. However, low-dose estradiol increases bone mineral density with minimal adverse effects.[64]

Several studies suggest that estrogen plays a very important role in brain functions during the aging of rodents, primates and humans.[65,66] In rats, estrogen reduces the risk of neurodegeneration and induces cell survival and synaptic plasticity.[67] In the developing nervous system of rodents and primates, sex steroids exert a trophic influence by stimulating axonal and dendritic growth, modulating synapse formation and neurotransmitter activity. At later stages, estradiol treatment is associated with enhancement of short memory, anti-depressive effect, improvement of cerebral blood flow, direct stimulation of neuron, and development of gliacyte.

There is fairly consistent evidence from epidemiological studies that ERT significantly reduces the risk of AD in women and protects against the adverse effects, though the mechanism is not clearly understood. On the other hand, findings from controlled treatment trials of women diagnosed with probable AD failed to show that physiological doses of ERT ameliorate existing deficits in cognitive functioning and/or prevent further deterioration in memory that inevitably occurs in these women over time. Finally, an accumulating body of evidence is beginning to suggest that the immediate post-menopausal period may constitute a critical window for treatment with ERT that maximizes its potential to protect against cognitive decline with aging and/or to reduce the risk of AD.[68] Finally, estrogen's numerous neurotrophic effects might explain how this hormone could protect against decline in cognition with aging. The estrogen receptor through which estrogen mediates its action has been mapped to brain regions known to be highly susceptible to AD-type changes in humans. It is co-localized with low-affinity nerve growth factor (NGF) receptors in cholinergic neurons of the rat basal forebrain, suggesting that estrogen-NGF interaction may be important for survival of cholinergic neurons. This is further substantiated by the results of Smith *et al.*[69] that the extensive atrophy in subcortical cholinergic neuronal populations of aged rhesus monkeys is reversed when they are given grafts of NGF-secreting autologous fibroblast cells. Estrogen may also exert neuroprotective effects via modulation of molecules involved in apoptosis and via its action as an antioxidant. It also upregulates mouse brain apoE which has been implicated in neuronal protection and repair.[70]

Post-menopausal women experience permanent hypoestrogenicity and suffer from increased risk of brain injury associated with neurodegenerative diseases such as stroke and AD. The risk and severity of such neurodegenerative conditions is decreased by ERT. Moreover, estrogen promotes the breakdown of the amyloid precursor protein (APP) to fragments less likely to accumulate as amyloid beta peptide (Aβ).[71] This suggests that modulation of Aβ metabolism may be one of the ways by which ERT prevents or delays the onset of AD in post-menopausal women.[72] There are ample evidences from epidemiological and clinical studies suggesting that estrogen acts as a potent growth and protective factor and has beneficial and protective role in AD.[73] ERT also enhances working memory in aged monkeys despite many years of estrogenic deprivation.[74] It increases the concentration of choline acetyltransferase, the synthetic enzyme for acetylcholine, a neurotransmitter critically implicated in memory functions and the level of which is markedly reduced in AD. It improves cognitive function and reduces the risk of AD-type senile dementia in women. It protects verbal memory and possibly also frontal lobe mediated functions in older women. Lifetime HRT exposure is found to be associated with improved global cognition in a population of older women.[75] In contrast to the positive findings in women, endogenous sex steroids do not appear to be closely linked to better cognition in older men.[76]

Harmful Effects of Hormonal Therapy

Though hormonal therapy is beneficial, in some cases it can cause worrisome side effects. In mice, for instance, delivery of melatonin increases the risk of tumor development, and the overproduction of growth hormone leads to kidney problems, premature heart and lung failure, and an increased probability of early death. Human adults given growth hormone have suffered from acromegaly (excess bone growth) and carpal tunnel syndrome. ERT has also offered health benefits to some post-menopausal women; however, it has recently been challenged and its major drawback includes increased risk of breast cancer, uterine bleeding and tumor formation.[77] The Women's Health Initiative (WHI) study showed that administration of estrogen and progesterone to post-menopausal women increases the risk of stroke, particularly ischemic stroke.[78,79] In this study, two parallel randomized, double-blind, placebo-controlled clinical trials were conducted for hormone therapy in healthy post-menopausal women during 1991–1992. However, the trial was halted in July 2002 after a mean 5.2 years of follow-up because health risks exceeded benefits. There was increase of coronary heart disease (CHD), stroke and venous thromboembolic disease in all women assigned to active treatment with estrogen plus progestin. Breast cancer was also increased while colorectal cancer, hip fracture and other fractures were reduced.[79] In February 2004, after reviewing data obtained till 30 November 2003, the National Institutes of Health (NIH) decided to end the intervention phase of the trial prior to the scheduled tenure of October 2004 to March 2005.

As the hormonal therapy shows side effects, a new class of compounds called selective estrogen receptor modulators (SERM) is being considered as possible alternative to ERT.[80] The SERM raloxifene is particularly interesting because, like estrogen, it improves lipid metabolism and reduces bone loss, without adverse effects on the breast or uterus. It is believed that good nutrition and regular exercise do reduce the risk of various diseases and extend the duration of life and thereby serve as the best current prescription for a long and healthy life.

Summary and Prospectives

As a consequence of advancing age, there occurs a general impairment of different functions and failure of maintenance mechanisms leading to the development of a variety of disabilities and diseases. These conditions are caused due to changes in several genetic and non-genetic factors. One such contributing factor is alteration in the level and action of hormones, their receptors and responses with advancing age. Accordingly strategies are designed at different levels to delay or prevent the diseases prevalent in old age so that aging can be slowed or longevity can be extended. Out of several strategies employed for this purpose, none has been found perfectly satisfactory. Nonetheless, hormonal treatment is quite popular and has been in use for decades. It shows many beneficial effects, though these benefits are associated with risks. Recent reports show several side effects associated with hormonal therapy. The search of compounds with no harmful effects is on and it is likely that in near future suitable therapies will be available for the improvement of quality of elderly life.

Acknowledgments

The author's research work is supported by grants from the Department of Science and Technology, and Department of Biotechnology, Government of India.

References

1. Thakur MK (2003a) Hormonal interventions of aging and longevity. In: Rattan SIS (ed.) *Biology of Aging and its Modulation*, Vol 5 Modulating Aging and Longevity, pp. 219–238, Kluwer Academic Publisher, Great Britain.
2. Timiras PS, Quay WD, Vernadakis A (eds.) (1995) *Hormones and Aging*. CRC Press, Boca Raton.
3. Thakur MK (2003b) Sex steroid signaling during aging. *Proc Ind Natl Sci Acad* (special issue on Aging) **B69**: 179–190.
4. Kanungo MS (1980) *Biochemistry of Aging*. Academic Press, London.
5. Thakur MK (1988) Molecular mechanism of steroid hormone action during aging: a review. *Mech Ageing Dev* **45**: 93–110.
6. Holzenberger M, Dupont J, Ducos B *et al.* (2003) IGF-1 receptor regulates lifespan and resistance to oxidative stress in mice. *Nature* **421**:182–186.
7. Blackman MR (2000) Age-related alterations in sleep quality and neuroendocrine function: interrelationships and implications. *JAMA* **284**: 879–881.
8. Von Werder K (1999) The somatopause is no indication for growth hormone therapy. *J Endocrinol Invest* **22**: 137–141.
9. Russell–Aulet M, Jaffe CA, Demott–Friberg R, Barkan AL (1999) *In vivo* semiquantification of hypothalamic growth hormone-releasing hormone (GHRH) output in humans: evidence for relative GHRH deficiency in aging. *J Clin Endocrinol Metab* **84**: 3490–3497.
10. Tagawa N, Tamanaka J, Fujinami A, Kobayashi Y, Takano T, Fukata S, Kuma K, Tada H, Amino N (2000) Serum dehydroepiandrosterone, dehydroepiandrosterone sulfate and pregnenolone sulfate concentrations in patients with hyperthyroidism and hypothyroidism. *Clin Chem* **46**: 523–528.

11. Corpas E, Harman SM, Blackman MR (1993) Human GH and human aging. *Endocr Rev* **14**: 20–39.
12. Miller RA (1991) Aging and immune function. *Int Rev Cytol* **124**: 187–215.
13. Hoffman AR, Lieberman SA, Butterfield G, Thompson J, Hintz RL, Ceda GP (1997) Functional consequences of the somatopause and its treatment. *Endocrine* 7: 73–76.
14. Steger RW, Bartke A, Cecim M (1993) Premature aging in transgenic mice expressing different growth hormone genes. *J Reprod Fertil* **46**: S61–75.
15. Brown–Borg HM, Borg KE, Meliska CJ, Bartke A (1996) Dwarf mice and the aging process. *Nature* **384**: 33.
16. Flurkey K, Papaconstantinou J, Miller RA, Harrison DE (2001) Lifespan extension and delayed immune and collagen aging in mutant mice with defects in GH production. *Proc Natl Acad Sci USA* **98**: 6736–6741.
17. Bartke A, Coschigano K, Kopchick J, Chandrashekar V, Mattison J, Kinney B, Hauck S (2001) Genes that prolong life: relationships of growth hormone and growth to aging and life span. *J Gerontol* **56**: B340–B349.
18. Carter C.S, Ramsey MM, Sonntag WE (2002) A critical analysis of the role of GH and IGF-1 in aging and lifespan. *Trends Genet* **18**: 295–301.
19. Rattan SIS, Clark BFC (1994) Kinetin delays the onset of aging characteristics in human fibroblasts. *Biochem Biophys Res Commun* **201**: 665–672.
20. Sharma SP, Kaur P, Rattan SIS (1995) Plant growth hormone kinetin delays aging, prolongs the lifespan and slows down development of the fruitfly *Zaprionus paravittiger*. *Biochem Biophys Res Commun* **216**: 1067–1071.
21. O'Connor KG, Harman SM, Stevens TE, Jayme JJ, Bellantoni MF, Busby–Whitehead MJ, Christmas C, Munzer T, Tobin JD, Roy TA, Cottrell E, St Clair C, Pabst KM, Blackman MR (1999) Interrelationships of spontaneous growth hormone axis activity, body fat and serum lipids in healthy elderly women and men. *Metabolism* **48**: 1424–1431.
22. Rudman D, Feller AG, Nagraj HS (1999) Effects of human GH in men over 60 years old. *N Eng J Med* **323**: 1–6
23. Marcus R, Hoffman AR (1998) Growth hormone as therapy for older men and women. *Ann Rev Pharmacol Toxicol* **38**:45–61.
24. Khan AS, Lynch CD, Sane DC, Willingham MC, Sonntag WE (2001) Growth hormone increases regional coronary blood flow and capillary density in aged rats. *J Gerontol A Biol Sci Med Sci* **56A**: B364–371.
25. Bouillanne O, Rainfray M, Tissandier O, Nasr A, Lahlou A, Cnockaert X, Piette F (1996) Growth hormone therapy in elderly people: an age-delaying drug? *Fundam Clin Pharmacol* **10**: 416–430.
26. Hauck SJ, Bartke A (2001) Free radical defenses in the liver and kidney of human growth hormone transgenic mice. *J Gerontol* **56**: B153–B162.
27. Blackman MR, Sorkin JD, Munzer T, Bellantoni MF, Busby–Whitehead J, Stevens TE, Jayme J, O'Connor KG, Christmas C, Tobin JD, Stewart KJ, Cottrell E, Clair C, Pabst KM, Harman SM (2002). Growth hormone and sex steroid administration in healthy aged women and men: a randomized controlled trial. *JAMA* **288**: 2282–2292.
28. Kimura KD, Tissenbaum HA, Liu Y, Ruvkun G (1997) Daf-2, an insulin receptor-like gene that regulates longevity and diapause in *Caenorhabditis elegans*. *Science* **277**: 942–946.

29. Clancy DJ, Gems D, Harshman LG, Oldham S, Stocker H, Hafen E, Leevers SJ, Partridge L (2001) Extension of lifespan by loss of chico, a Drosophila insulin receptor substrate protein. *Science* **292**: 104–106.
30. Tatar M, Kopelman A, Epstein D, Tu MP, Yin CM, Garofalo RS (2001) A mutant Drosophila insulin receptor homolog that extends lifespan and impairs neuroendocrine function. *Science* **292**: 107–110.
31. Bluher M, Kahn BB, Kahn CK (2003) Extended longevity in mice lacking the insulin receptor in adiose tissue. *Science* **299**: 572–574.
32. Paolisso G, Ammendola S, Del Buono A *et al.* (1997) Serum levels of insulin-like growth factor-1 (IGF-1) and IGF-binding protein-3 in healthy centenarians: relationship with plasma leptin and lipid concentrations, insulin action and cognitive function. *J Clin Endocrinol Metab* **82**: 2204–2209.
33. Anisimov VN (2003) Effect of melatonin on longevity. In: Rattan SIS (ed.) *Modulating Aging and Longevity*, pp. 239–260. Kluwer Academic Publisher, Great Britain.
34. Oxenkrug G, Requintina P, Bachurin S (2001) Antioxidant and antiaging activity of N-acetylserotonin and melatonin in the *in vivo* models. *Ann NY Acad Sci* **939**: 190–199.
35. Pierpaoli W, Regelson W (1994) Pineal control of aging: effect of melatonin and pineal grafting on aging mice. *Proc Natl Acad Sci USA* **91**: 787–791.
36. Pappolla MA, Sos M, Omar RA, Bick RA, Hickson–Bick DLM, Reiter RJ, Efthimiopoulos, Robakis NK (1997) Melatonin prevents death of neuroblastoma cells exposed to the Alzheimer amyloid peptide. *J Neurosci* **17**: 1683–1690.
37. Van Reeth O, Weibel L, Olivares E, Maccari S, Mocaer EL, Turek FW (2001) Melatonin or a melatonin agonist corrects age-related changes in circadian response to environmental stimulus. *Am J Physiol* **280**: R1582–1591.
38. Anisimov VN, Zavarzina NY, Zabezhinski MA, Popovich IG, Zimina OA, Shtylick AV, Arutjunyan AV, Oparina TI, Prokopenko VM, Mikhalski AI, Yashin AI (2001) Melatonin increases both lifespan and tumor incidence in female CBA Mice. *J Gerontol* **56**: B311–B323.
39. Yen SSC (2001) Dehydroepiandrosterone sulfate and longevity: new clues for an old friend. *Proc Natl Acad Sci USA* **98**: 8167–8169.
40. Ravaglia G, Forti P, Maioli F, Boschi F, Bernardi M, Pratelli L, Pizzoferrato A, Gasbarrini G (1996) Relationship of dehydroepiandrosterone sulfate (DHEAS) to endocrine-metabolic parameters and functional studies in the oldest old: result from an Italian study on healthy free-living over-ninety years old. *J Clin Endocrinol Metab* **81**: 1173–1178.
41. Tilvis RS, Kahonen M, Harkonen M (1999) Dehydroepiandrosterone sulfate, diseases and mortality in a general aged population. *Aging* **11**: 30–34.
42. Hajszan T, MacLusky NJ, Leranth C (2004) Dehydroepiandrosterone increases hippocampal spine synapse density in ovariectomized female rats. *Endocrinology* **145**: 1042–1045.
43. Robinzon B, Cutolo M (1999) Should dehydroepiandrosterone replacement therapy be provided with glucocorticoids? *Rheumatology* **38**: 488–495.
44. Oelkers W (1999) Dehydroepiandrosterone for adrenal insufficiency. *New Eng J Med* **341**: 1073–1074.
45. Johnson MD, Bebb RA, Sirrs SM (2002) Uses of DHEA in aging and other disease states. *Ageing Res Rev* **1**: 29–41.

46. Baulieu EE, Thomas G, Legrain S *et al.* (2000) DHEA, DHEAS and aging: contribution of the DHEAge study to a sociobiomedical issue. *Proc Natl Acad Sci USA* **97**: 4279–4284.
47. Morley JE, Kaiser FE, Perry HM, Patrick P, Morley PM, Stauber PM, Vellas B, Baumgartner RN, Garry PJ (1997) Longitudinal changes in testosterone, luteinizing hormone, and follicle-stimulating hormone in healthy older men. *Metabolism* **46**: 410–413.
48. Kumar RC, Thakur MK (2004) Androgen receptor mRNA expression is inversely regulated by testosterone and estradiol in adult mouse brain. *Neurobiol Aging* **25**: 925–933.
49. Thakur MK (1995) Androgen receptor and the mechanism of androgen action. *Curr Sci* **68**: 806–812.
50. Lamberts SWJ, Van den BAW, Van der LAJ (1997) The endocrinology of aging. *Science* **278**: 419–424.
51. Goodman–Gruen D, Barrett–Connor E (2000) Sex differences in the association of endogenous sex hormone levels and glucose tolerance status in older men and women. *Diabetes Care* **23**: 912–918.
52. Tenover JL (1999) Testosterone replacement therapy in older adult men. *Int J Androl* **22**: 300–306.
53. Brown–Sequard CE (1889) Des effects produits chez Thomme par des injections sous-cutanees d'un liquide retire des testicles de cobaye et de chien. *CR Soc Biol* **41**: 415–418.
54. Morley JE, Perry IHM, Kaiser FE *et al.* (1993) Effects of testosterone replacement therapy in old hypogonadal males, a preliminary study. *J Amer Geriatr Soc* **41**: 149–152.
55. Sih R, Morley JE, Kaiser FE, Perr HM, Patrick P, Ross C (1997) Testosterone replacement in older hypogonadal men: a 12-month randomized controlled trial. *J Clin Endocrinol Metab* **82**: 1661–1667.
56. Barrett–Connor E, von Muhlen DG, Kritz–Silverstein D (1999) Bioavailable testosterone and depressed mood in older men: the Rancho Bernardo Study. *J Clin Endocrinol Metab* **84**: 573–577.
57. Cherrier MM, Asthana S, Plymate S, Baker L, Matsumoto AM, Peskind E, Raskind MA, Brodkin K, Bremner W, Petrova A, LaTendresse S, Craft S (2001) Testosterone supplementation improves spatial and verbal memory in healthy older men. *Neurology* **57**: 80–88.
58. Kenny AM, Bellantonio S, Gruman CA, Acosta RD, Prestwood KM (2002) Effects of transdermal testosterone on cognitive function and health perception in older men with low bioavailable testosterone levels. *J Gerontol* **57**: M321–M325.
59. Mukherjee TK, Dinh H, Chaudhuri G, Nathan L (2002) Testosterone attenuates expression of vascular cell adhesion molecule-1 by conversion to estradiol by aromatase in endothelial cells: implications in atheroscleosis. *Proc Natl Acad Sci USA* **99**: 4055–4060.
60. Asaithambi A, Mukherjee S, Thakur MK (1997) Expression of 112 kDa estrogen receptor in mouse brain cortex and its autoregulation with age. *Biochem Biophys Res Commun* **231**: 683–685.
61. Shifren JL, Schiff I (2000) The aging ovary. *J Womens Health Gend Based Med* **9**: S3–7.
62. Wise PM, Krajnak KM, Kashon ML (1996) Menopause: the aging of multiple pacemakers. *Science* **273**: 67–70.
63. Thakur MK, Ghosh S (2003) Hormone replacement therapy in aging women. *Aging and Society* **13**: 78–102.

64. Preswood KM, Kenny AM, Kleppinger A *et al.* (2003) Ultralow dose micronized 17 beta-estradiol and bone density and bone metabolism in older women. *JAMA* **290**: 1042–1048.
65. Thakur MK (1999) Estrogen and brain aging. *J Anti-Aging Med* **2**:127–132.
66. Thakur MK, Sharma PK, Ghosh S (2004) Estrogen effects in the brain: implications for aging and Alzheimer's disease. In: Thakur MK and Prasad S (eds.) *Neurobiology in the Post-genomic Era*. Narosa Publishing House, New Delhi (in press).
67. Wise PM, Dubal DB (2000) Estradiol protects against ischemic brain injury in middle-aged rats. *Biol Rep* **63**: 982–985.
68. Sherwin BB (2003) Estrogen and Cognitive Functioning in Women. *Endocrine Reviews* **24**: 133–151.
69. Smith JE, Roberts J, Gage FH, Tuszynski MH (1999) Age-associated neuronal atrophy occurs in the primate brain and is reversible by growth factor gene therapy. *Proc Natl Acad Sci USA* **96**: 10893–10898.
70. Thakur MK, Sharma PK, Ghosh S, Mani ST (2004) Estrogen intervention in aging and longevity: problems and prospectives. *Geriat Gerontol Int* **4**: S259–S261.
71. Zheng H, Xu H, Uljon SN, Gross R, Hardy K, Gaynor J *et al.* (2002) Modulation of Aβ peptides by estrogen in mouse models. *J Neurochem* **80**: 191–196.
72. Peranceaka SS, Nagy V, Brail D, Gandy S (2000) Ovariectomy and 17β-estradiol modulate the levels of Alzheimer's amyloid peptides in brain. *Neurology* **54**: 2212–2217.
73. Thakur MK (2000) Alzheimer's disease — a challenge in the new millennium. *Curr Sci* **79**: 101–108.
74. Lacreuse A, Wilson ME, Herndon JG (2002) Estradiol but not raloxifene improves aspects of spatial working memory in aged ovariectomised rhesus monkeys. *Neurobiol Aging* **23**: 589–600.
75. Carlson C, Zandi PP, Plassman BL *et al.* (2001) Hormone replacement therapy and reduced cognitive decline in older women. *Neurology* **57**: 2210–2216.
76. Wolf OT, Kirschbaum C (2002) Endogenous estradiol and testosterone levels are associated with cognitive performance in older women and men. *Horm Behav* **41**: 259–266.
77. Schairer C, Lubin J, Troisi R, Sturgeon S, Brinton L, Hoover R (2000) Menopausal estrogen and estrogen progestin replacement therapy and breast cancer risk. *J Amer Med Assoc* **283**: 485–491.
78. Wassertheil–Smoller S, Hendrix SL, Limacher M *et al.* (2003) Effect of estrogen plus progestin on stroke in postmenopausal women. *JAMA* **289**: 2673–2684.
79. Anderson GL, Limacher M, Assaf AR *et al.* (2004) Effects of conjugated equine estrogen in postmenopausal women with hysterectomy. *JAMA* **291**: 1701–1712.
80. Yaffe K (2001) Estrogens, selective estrogen receptor modulators, and dementia: what is the evidence? *Ann NY Acad Sci* **949**: 215–222.

7

Pineal Peptides as Modulators of Aging

Vladimir N. Anisimov and Vladimir Kh. Khavinson*

Department of Carcinogenesis and Oncogerontology
N.N. Petrov Research Institute of Oncology
Pesochny-2, St. Petersburg 197758, Russia

**St. Petersburg Institute of Bioregulation*
and Gerontology, 3, prospekt Dynamo
St. Petersburg 197110, Russia

INTRODUCTION

The role of the pineal gland in aging and cancer have been discussed intensively in the last decades.[1–6] Pinealectomized rats showed a reduced life-span [7,8] and stimulated tumor development[4,9] whereas an administration of pineal hormone melatonin to old mice or grafting of pineal glands from young donors into the thymus of old mice prolonged the life-span of the latter.[10,11] There are data on oncostatic effect of melatonin.[4] However, some of the effects of the pineal gland might have obviously resulted from pineal peptide secretion.[4,12] Some crude peptide extracts or purified peptides isolated from pineal glands were shown to have antigonadotropic, metabolic and antitumor activity.[13–19] Thirty years ago, we had published the first data describing the results of administration of low molecular weight pineal peptide extracts, the commercial drug form of which was later named Epithalamin.[20] The recovery of estrus cycle in old female rats with persistent estrous syndrome as well as

Corresponding author: Vladimir N. Anisimov, Tel.: 7 (812) 596-8607; Fax: 7 (812) 596-8947; E-mails: aging@mail.ru.

the lowering of the threshold sensitivity of the hypothalamo-pituitary complex to the feedback inhibition by estrogens in old animals were observed.[12,20] Since that time, the effect of Epithalamin on the function of the neuroendocrine and the immune system has been continuously studied by us and others demonstrating its high biological activity. Long-term treatment with the preparation prolonged the life-span of animals, slowed down aging of the reproductive system, improved parameters of immunity and inhibited the development of spontaneous, chemically and radiation-induced or transplanted tumors.[12,21–28]

In this chapter the results of pharmacological life-span extension with peptides regulating pineal function in experimental animals and in humans are summarized.

Pineal Polypeptide Preparation Epithalamin

Epithalamin is a complex peptide bioregulator isolated from the pineal gland, and is approved for medical application as a neuroendocrine system regulator by Order No. 250 of the USSR Ministry of Health of 19.06.1990 (Registration No. 90.250.1).[12,28]

The effect of Epithalamin on the life-span of C3H/Sn and SHR female mice was studied in two sets of experiments. Chronic treatment of C3H/Sn female mice with Epithalamin in a single dose of 0.5 mg per mouse, started at the age of 3.5 months, significantly prolonged their mean life-span (by 40%) and increased their maximum life-span by 3.5 months.[21] The survival curves of the mice exposed to Epithalamin were significantly shifted to the right. Long-term Epithalamin administration to SHR mice in a single daily dose of 0.1 mg per mouse slightly increased their mean life-span but failed to change their maximum life-span.[23] The exposure to Epithalamin was followed by a significant increase of life-span of tumor-free mice of both strains (by 41 and 26%, respectively). A comparative effect of Epithalamin long-term treatment on the survival of SHR female mice, starting at their age of 3.5 and 12 months, was estimated.[23] Epithalamin was shown to increase the mean life-span by 14 and 17% in young and middle-aged groups, respectively, as compared to controls treated with saline. Epithalamin treatment shifted survival curves of both age groups to the right distinctly, as compared to corresponding controls. The calculation of the aging rate in young and middle-aged groups treated with Epithalamin revealed a slowing down of aging in the young group in comparison with the middle-aged group. Administration of Epithalamin was followed by an increased life-span of tumor-free young and middle-aged mice by 20 and 36%, respectively.

Epithalamin was administered also to female outbred rats, beginning from the age of 3.5 months, for 20 months in daily doses of 0.1 or 0.5 mg per animal.[29] The mean life-span of rats exposed to Epithalamin increased proportionally to the doses of the drug, however, the maximum life-span increased by 2 months only in rats exposed to the highest dose of the preparation. The analysis of survival curves of rats treated with Epithalamin suggested a slowing down of the aging rate under the influence of pineal peptides. It is worth noting that the age-related dynamics of body weight in controls and treated rats was similar.

In another experiment, female rats were injected chronically with Epithalamin (a single dose of 0.5 mg per rat) starting at the age of 15 months.[25,30] It was shown that this treatment increased their life-span insignificantly by 6.2% when calculated from birth and by 18% when calculated from the onset of the experiment. At the same time, 23% of the Epithalamin-treated rats lived longer than the longest living control rat. The maximum life-span in the Epithalamin-treated group was 3 months longer than in controls. The calculation of the parameters of the Gompertz equation showed that Epithalamin treatment slowed down the aging rate. Thus, we observed the capacity of Epithalamin to increase the mean, and sometimes maximum, life-span of mice and rats and to slow down the aging rate of the treated population. These effects were less pronounced when Epithalamin treatment was started at the age of over 1 year in mice and 15 months in rats.

The antioxidative activity of Epithalamin was studied in series of experiments.[31,32] Administration of Epithalamin subcutaneously in the morning between 10:00 and 11:00 a.m. (0.5 mg per rat) to rats for five days produced a considerable inhibiting effect on the intensity of peroxide chemiluminescence. The impact of Epithalamin on lipid peroxide oxidation was estimated with respect to the content of diene conjugates and Schiff' bases in serum, while protein peroxide oxidation was estimated by the contents of carbonyl derivatives of amino acids in serum. Epithalamin considerably increased (by 35%, $p < 0.01$) the general antioxidative activity (AOA) of serum. The activity of Cu,Zn-superoxide dismutase (SOD) and the level of coeruloplasmin (CP) in serum increased by 20 and 6%, respectively, after Epithalamin treatment. SOD and CP levels were significantly higher (by 75 and 27%, correspondingly) in rats exposed to Epithalamin. In *in vitro* experiments, we studied the capacities of Epithalamin and melatonin to suppress the luminol-dependent chemiluminescence stimulated by different forms of oxygen (hypochlorite, superoxide radical and hydrogen peroxide) as well as by mixtures of various forms of oxygen generated by neutrophils under *in vitro* conditions. In case of hypochlorite, the greatest suppression was observed due to melatonin (up to 30%). Epithalamin was less efficient than melatonin (up to 17%), but more active than glutathione (up to 12%). In the presence of hydrogen peroxide, melatonin suppressed chemiluminescence up to 45%, Epithalamin up to 33%, and glutathione up to 20%. The range of capacities to suppress the formation of superoxide radical within the xanthinoxidase reaction was as follows: melatonin (up to 30%), Epithalamin (up to 22%) and glutathione (up to 16.5%). Chemiluminescence induced by activated neutrophils was also decreased (up to 20%) by melatonin, Epithalamin or glutathione. However, the effect of these substances was less pronounced in this case than in the experiments with separate distinct forms of oxygen.

The results of these experiments manifested a higher activity of melatonin than of Epithalamin and glutathione. These differences could probably be explained by the structural peculiarities of these compounds, the antioxidative properties of melatonin being related to the indole ring. Melatonin's antioxidative properties were stronger, as compared to Epithalamin under *in vitro* conditions, while *in vivo* the effect of the latter was much more pronounced. This could be explained by the ability of Epithalamin to stimulate synthesis and secretion of melatonin in animals,[12] and to affect enzyme systems of antioxidative defence.

There is every reason to suppose that the properties of melatonin and Epithalamin to increase animal life-span, to stimulate immune response and to inhibit the development of neoplasia may be, to some extent, are mediated by their inhibitory influence on free radical processes in the organism, which play a key role in the molecular mechanisms of aging and age-related pathologies, such as arteriosclerosis, immunodepression and cancer.

The effect of Epithalamin on the life-span of *Drosophila melanogaster* was also studied.[33,34] Epithalamin was given during the larval stage of fly life. The exposure to Epithalamin failed to significantly change any of the life-span parameters in male flies. However, in females, it significantly increased the mean life-span (by 17%, $p < 0.02$), medium (by 26%) and maximum longevity (by 14%), and decreased both population mortality rate (2.12 times, $p < 0.01$) and mortality rate doubling time (by 32%), as compared to controls. The survival curve was shifted to the right and the slope of the Gompertz plot was decreased only in female flies exposed to Epithalamin. We also studied the effect of Epithalamin on lipid peroxidation and antioxidative enzyme activity in *D. melanogaster*. The tissue levels of conjugated hydroperoxides (CHP) and ketodienes (KD) were significantly higher in control males than in females (by 40 and 49%, respectively), which inversely correlated with the mean life-span in flies. Most significantly the contents of CHP and KD were decreased in female flies exposed to Epithalamin (2.3 and 3.4 times, respectively, $p < 0.001$), as compared to controls. The activity of catalase was twice as high in control males than in control females, whereas the activity of SOD was the same in both sexes. The exposure to Epithalamin was followed by a significant elevation of catalase activity in males (by 20%) and in females (by 7%), and by an increase of the SOD activity in males (by 41%).

Thus, Epithalamin treatment significantly increased the mean and maximum life-span of female *D. melanogaster* and more than twice slowed down the aging rate of the population. These effects are in good agreement with the strong inhibitory effect of Epithalamin on lipid peroxidation and its stimulating effect on the activity of antioxidative enzymes.

It was shown that the threshold of sensitivity of the hypothalamo-pituitary complex to feedback suppression by estrogens increased with age in female rats.[35] This mechanism was suggested to be a crucial one for the age-related switching-off of reproductive function in rats.[35,36] Subcutaneous injections of Epithalamin to old rats increased the sensitivity of the hypothalamo-pituitary complex to estrogen suppression and restored cyclic estrous activity in rats with persistent estrous.[20,37] Injection of Epithalamin into the 3rd brain ventricle was followed by a similar effect. Pinealectomy or exposure to constant light were followed by an elevated threshold of the hypothalamic feedback sensitivity to inhibition by estrogen.[35] The study of the estrous function of female rats showed that 38% of the control animals aged 16–18 months had persistent estrous, anestrous or repeated pseudopregnancies, whereas only 7% of rats treated with Epithalamin from the age of 3.5 months onwards showed estrous cycle disturbances. Out of 16 month-old rats that remained sterile after mating with adult males, four females became pregnant and gave birth to 5–9 fetuses per litter after a 2-week course of Epithalamin administration.[29]

Daily cytological studies of vaginal smears in rats treated with Epithalamin from the age of 15 months onwards revealed a slowing down of the age-related switching-off of estrous function as compared to controls. While at the age of 18 months the number of rats with disturbances of estrous function (persistent estrous and anestrous) was the same in both groups, at the age of 27 months these disturbances were observed in 50% of controls and in 30% of females treated with the pineal preparation ($p < 0.05$).[30] Thus, we observed a slowing down of the age-related switching-off of estrous function in female rats exposed to Epithalamin. Epithalamin was also capable of restoring estrous function, ovulation and fertility in rats with persistent estrous. These effects were suggested to depend on the capacity of the preparation to prevent an age-related increase of the threshold sensitivity of the hypothalamo-pituitary complex to feedback regulation by estrogen.

Epithalamin also restored regular estrous cycles in young and old rats with persistent estrous syndrome induced by the exposure to constant illumination.[38] A significant decrease in the number of follicular cysts in rats treated with Epithalamin in comparison to untreated controls, as well as a partial luteinization of the follicular cyst wall and the appearance of corpora lutea in ovaries were observed under the influence of Epithalamin.

A single administration of Epithalamin in the morning was followed by decreased levels of serum luteinizing hormone (LH) and failed to influence serum testosterone levels in 4–5-month-old males, but increased the levels of both hormones in 16-month-old animals 30 mins after injection. Daily Epithalamin injections in the morning (between 10:00 a.m. and 11:00 a.m.) for 5 consecutive days failed to influence serum LH levels in young and old male rats and were followed by a significant decrease of serum testosterone levels in young rats at noon and 5:00 p.m. No influence on serum testosterone was registered in older males at any time of the day.[12] Epithalamin was tested for its effect on pineal androgen receptors. Epithalamin showed a trend to reduce cytosolic receptors but did not have any effect on the nuclear 5-alpha-dihydrotestosterone receptor in rat pineal gland.[39]

It was shown that the susceptibility of the hypothalamo-pituitary complex to the feedback inhibition by testosterone is elevated in aged male rats.[40] Our data support this observation. Thus, if Epithalamin decreases the threshold sensitivity of the hypothalamo-pituitary complex to the inhibitory effect of testosterone, it cannot be expected to decrease LH levels in response to elevated testosterone levels, but rather to lead to increased LH levels. This phenomenon was in fact observed in our experiment.

Administration of Epithalamin to adult rats during 5 consecutive days was followed by an increased activity of succinate-, alpha-ketoglutarate- and pyruvate-dehydrogenase in rat brain and by a 39% increase of learning capacity in labyrinth tests as well as of some other behavioral tests.[41] Arushanian *et al.*[42] observed an increase in the amplitude and shifts in the acrophase of circadian locomotor activity towards late hours in rats exposed to long-term administration of Epithalamin. Epithalamin treatment was also followed by changes in the time-course of forced swimming and by a decrease of the rhythmic index of depression. Intravenous administration of Epithalamin revealed a sedative effect in dogs.[43] Ten-day-long administration of Epithalamin to adult male rats was followed by a significant activation of neurosecretory elements in the hypothalamic nucleus paraventricularis, a lower activation in the nucleus supraopticus, and an increased content of neurosecretory substances in the neurohypophysis.[44] Ultra-microscopic studies showed that a single morning injection of Epithalamin into adult rats induced signs of an activated pinealocyte function.[12]

A single or 2-week-long administration of Epithalamin failed to change the levels of dopamine, norepinephrine, serotonin or 5-hydroxyindolyle acetic acid (HIAA) in the hypothalamus of adult male rats.[12] In other experiments, the effect of Epithalamin on serotonin metabolism in the pineal gland of young and old male rats was studied.[24,25,45,46] It appeared that a single injection of Epithalamin in the morning or a 5-day course of Epithalamin treatment of 4–5-month-old rats was followed by an increase of the night levels of serotonin, N-acetylserotonin and melatonin in the pineal gland. In 18–20-month old rats a similar treatment caused only a tendency to increase pineal melatonin level. At the same time, increased serum melatonin levels in young adult and in old rats treated with Epithalamin were observed. In young as well as in old rats, the exposure to Epithalamin failed to influence direct O-methylation reaction of serotonin into 5-methoxytryptamine and the oxidative deamination with subsequent O-methylation and 5-HIAA and 5-Methoxyindole acetic acid. These data suggest the existence of an ultra-short loop of regulation between pineal peptides and indoles. Pineal peptides may influence the metabolic reactions of tryptophan into serotonin and its subsequent transformation into melatonin, which declines with aging. It is worth noting that Epithalamin given at 6:00 p.m. entailed the decrease of night levels of serotonin, N-acetylserotonin and melatonin in the rat pineal gland.

We also observed a stimulating effect of a 5-day course of Epithalamin treatment on serum triiodothyronine (T3) levels and an inhibitory effect of this treatment on serum thyroxine (T4) in 4–5-month-old male rats. These data suggest that the pineal preparation influences the metabolism of T3 into T4. Direct measurement of the activity of 5′-deiodinase in the thyroid gland of rats treated with Epithalamin may help to explain these results. In 16-month-old male rats, the course of Epithalamin treatment was followed by a decrease of the levels of both T3 and T4 levels in serum.[12]

In male CBA/Ca mice we observed an age-related elevation of the serum levels of corticosterone and corticoliberin (CRF). In 2-month-old mice and 8-month-old mice the levels of CRF were 1.62 ± 0.45 pg/ml and 12.62 ± 4.4 pg/ml respectively, ($p < 0.05$) whereas the levels of corticosterone were 12.38 ± 3.44 μg/ml and 80.3 ± 18.9 μg/ml respectively, ($p < 0.05$). A 5-day-long course of Epithalamin given to younger mice was followed by a 3.2-fold decrease of serum corticosterone in comparison to controls.[12,47]

Treatment of rabbits with Epithalamin was followed by significantly increased glucose tolerance, but the tolerance to insulin loading remained unaffected.[48] The authors also observed that a 3-week-long exposure to Epithalamin led to decreased levels of both serum insulin and triglycerides. Slepushkin and Pashinski[43] showed a significant influence of Epithalamin on electrolyte and water metabolism: exposure of dogs or rats to the preparation was followed by an increase of urine excretion, accompanied by hyposodiumuria, hyperpotassium-, hypercalcium- and hypermagnesiumuria.

With respect to the action of Epithalamin on the immune system it was found that the pineal preparation increased the number of antibody-forming cells generated in the spleen as well as the level of serum hemagglutinins in response to immunization with sheep red blood cells.[49] Epithalamin treatment increased survival of CC57Br/Mv mice infected with 1.5×10^9 bacterium *S. typhimurium*, survival of skin allografts in mice and rats, stimulated the reaction of hypersensitivity of delayed type in guinea pigs, and stimulated the phagocytic activity of blood neutrophils.[50] Epithalamin was also shown to restore the level of proliferation of granulocytes and macrophages (CFU-GM) in pinealectomized rats up to the levels of intact animals.[39]

The treatment of both young (2-month-old) and 2-year-old mice with an intact thymus with Epithalamin lead to an increase of the level of thymic serum factor (TSF), of the compounds with thymosin-like activity (CTLA) and of thymic index and to an increase of spleen cellularity in old animals.[12,51] Epithalamin also induced an increase of TSF and CTLA titers in serum of old mice bearing spontaneous hepatomas discovered at autopsy, however, to a smaller extent than in old tumor-free animals. Melatonin administration to young mice was also followed by increased levels of TSF, CTLA, relative weight of thymus and its cellularity. A stimulating effect of melatonin on CTLA titer, thymic index and its cellularity, but not on TSF titer was observed in old mice. The CTLA titer of splenocyte supernatants taken from 2-month-old mice and cultivated *in vitro* with normal saline was 2.0 ± 0.30, but only 1.0 ± 0.32 in the culture of splenocytes from 2-year-old mice. Incubation of splenocytes from young mice with melatonin (5 pg/ml and 20 pg/ml) or Epithalamin (0.1 mg/ml and 1.0 mg/ml) was followed by an increase of the CTLA titer up to 6.7 ± 1.06; 6.3 ± 0.42; 6.5 ± 0.59 and 4.8 ± 0.71 respectively ($p < 0.05$). After *in vitro* incubation of splenocytes of old mice with melatonin (25 pg/ml), the levels of CTLA were significantly higher than in controls. The titers of CTLA in the supernatant of old splenocytes incubated with Epithalamin in doses of 0.1 mg/ml and 1.0 mg/ml were 1.7 ± 0.35 and 6.2 ± 1.47, respectively, demonstrating that Epithalamin induced an increase of serum titers of TSF and CTLA, which was accompanied by morphological features of a stimulation of thymus and spleen function.

Grinevich *et al.*[52] observed an activation of the proliferation and maturation of thymic cortical thymocytes, hyperplasia and medullar differentiation of epithelial cells in the thymus and epithelioid cells in the spleen of AKR mice under long-term Epithalamin treatment in comparison to controls. In female C3H/Sn mice, exposed to long-term treatment with Epithalamin starting at the age of 3.5 months, we observed a delay in the age-releated decrease of the phytohemagglutinin-induced blast transformation reaction of T-lymphocytes, in comparison to controls.[21]

Its application in senescent monkeys aged 20–26 years resulted in the complete normalization of initially disturbed melatonin and glucose blood levels and in the restoration of tissue sensitivity to insulin and cortisol circadian rhythm.[12] Thus, the comprehensive long-term investigation of Epithalamin in 6 animal species demonstrated its unique geroprotective activity. The application of Epithalamin was followed by reduction in spontaneous, radiation and chemical carcinogenesis.[12,21–23,28,30]

Epitalon (Ala–Glu–Asp–Gly)

Tetrapeptide Epitalon® (Ala–Glu–Asp–Gly) was designed on the basis of Epithalamin amino acid analysis and synthesized at the Laboratory of Peptide Chemistry of the St. Petersburg Institute of Bioregulation and Gerontology.[53] The geroprotective activity of Epitalon was studied in 3 strains of *D. melanogaster*.[33,53] Epitalon increased the life-span of imagoes significantly by 11–16% when applied at unprecedentedly low concentrations — from 0.001×10^{-6} to 5×10^{-6} wt% of the culture medium. The effective concentrations of Epitalon were 16,000–80,000,000 times lower than those of melatonin. In another experiment, the mean life-span of female VES and male NA^- flies increased under the influence of Epitalon by 26%. In NA^- females the mean life-span grew by 35% and in NA^- males, by 29%.

In our recent study the geroprotective effect of long-term Epitalon administration in female CBA mice was demonstrated.[54] The number of mice that survived the age of 24 months increased 4 times under Epitalon influence as compared to the controls treated with saline. The bioregulator slowed down aging of the reproductive function, decreased physical activity, inhibited free radical processes, and decreased total spontaneous tumor incidence 1.7 times in female CBA mice.[54] Epitalon inhibited mammary carcinogenesis and metastasis in *HER-2/neu* transgenic mice.[55] The expression of *HER-2/neu* mRNA in the mammary tumors of *HER-2/neu* transgenic mice treated with Epitalon reduced 3.7-fold in comparison with the control animals. The administration of Epitalon inhibited colon and small intestine carcinogenesis induced by 1,2-dimethylhydrazine in rats.[56]

Epitalon therapy of young (6–8-year-old) and senescent (20–26-year-old) female monkeys *Macaca mulatta* restored the evening level of melatonin and the circadian rhythm of cortisol in the blood serum of senescent monkeys.[57] However *in vitro* epitalon failed to induce the secretion of melatonin in the pineal gland of young (9 weeks) and old (27 months) rats.[58] Epitalon inhibited lipid and protein peroxidation in the brain and blood serum of young (3-month-old) and senescent (30-month-old) female rats.[12] It was shown that epitalon inhibited enkephalin-degrading enzymes from human serum and no interaction was observed between the peptide an mu- or delta-opioid receptors of the membrane fraction from the rat brain.[59] There is evidence of a slight but significant c-Fos increase in stress-exposed pinealocytes of rats after intranasal epitalon infusion.[60] Epitalon stress-protective effect at the level of interleukin-1beta dignal transduction via sphingomyelin pathway in the nerve tissues as well as at the level of target lymphocyte proliferation.[61] Epitalon also activated *in vitro* interleukin-2 mRNA synthesis in spleconcytes from CBA mice in the absence of specific inductors.[62] It is important that long-term treatment with epitalon inhibits age-related increase in the incidence of chromosome aberrations in senescence-accelerated (SAM) or outbred SHR mice.[63]

Addition of epitalon in telomerase-negative human fetal fibroblast culture induced expression of the catalytical subunit, enzymatic activity of telomerase, and telomerase elongation, which can be due to reactivation of telomerase gene in somatic cells and indicates the possibility of prolongation of life-span of a cell population and of the whole organism.[64]

Effects of Pineal Peptides on Aging in Humans

Thus, results of the comprehensive long-term experimental studies on Epithalamin allowed regarding them as potential geroprotectors. However, special investigations aimed at the assessment of their influence upon the main systems and functions and upon human longevity were necessary to confirm their geroprotective effectiveness. Certainly, this purpose could hardly be attained completely, yet a preliminary evaluation was possible. Epithalamin was clinically studied for its geroprotective activity at two gerontological research centers — in Russia and in the Ukraine. The study showed that the total antioxidation and antiradical activity of human blood decreased with age, which brought forth an imbalance of different indices of the pro- and antioxidation systems. The application of Epithalamin in the elderly and older persons resulted in the normalization of antioxidation indices (increased total antioxidation and antiradical activity of the blood serum, reduced content of peroxide lipid oxidation products, and intensified superoxide dismutase and glutathione peroxidase activity).[65]

The results of studying Epithalamin effectiveness in non-insulin-dependent diabetes mellitus patients (NIDDM) appeared extremely important. Epithalamin therapy applied in NIDDM patients exerted a prolonged normalizing effect upon their carbohydrate metabolism. The hypoglycemic effect of the bioregulator was conditioned by an increase in stimulated insulin secretion combined with intensified tissues sensitivity to insulin.[66,67] The administration of Epithalamin decreased the levels of low and very low density lipoproteins and raised the content of high density lipoproteins in the blood serum of NIDDM patients. This caused a decrease in the atherogenicity of lipid fractions maintained for 3 more months after the completion of Epithalamin therapy. NIDDM patients with hypertension disease revealed a decreased arterial pressure and restoration of the myocardium diastolic function after Epithalamin application. The hemodynamic basis of its hypotensive effect consisted in diminished peripheral vascular resistance, which was to a great extend explained by reduced basal hyperinsulinemia.[68] The administration of the drug in elderly NIDDM patients with purulent-inflammatory diseases also compensated for the disturbed carbohydrate metabolism and restoration of immune functions, thus, activating tissues regeneration and significantly cutting down the period of post-surgical wound healing.[69]

Epithalamin appeared effective in patients with vegetative dishormonal myocardiodystrophy. Its effect was manifested in significant clinical improvement, normalized ECG, increased tolerance to physical loads, normalized the content of FSH in the blood.[70,71]

The St. Petersburg Institute of Bioregulation and Gerontology of the Russian Academy of Medical Sciences started the clinical investigations of Epithalamin in 1996. The main group comprised 94 women aged 66–94 years. The most typical age-related pathologies in this group included ischemic heart disease (IHD) (angina of functional classes I–III), hypertension (Stage II), and deforming osteoarthrosis.[72] To evaluate the state of the main organism systems, the following parameters were studied:

1) 17 immune indices (leukocytes, lymphocytes, $CD3^+$, $CD4^+$, $CD8^+$, $CD4^+/CD8^+$, $CD20^+$, $CD56^+$, reaction of leukocyte migration inhibition with concanavalin A (RLMI with Con A), IgM, IgG, IgA, circulating immune complexes (CIC), phagocytic numbers and indices, and the indices of neutrophils and monocytes);
2) 8 endocrine indices (adenocortictropic (ACTH), thyroid-stimulating (TSH), luteinizing (LH), follicle-stimulating (FSH), and somatotropic (STH) hormones, insulin, cortisol, and estradiol in the blood);
3) 19 metabolic indices (cholesterol, triglycerides, glucose, urea, creatinine, total bilirubin, total protein, albumids, globulins, albumins/globulins ration, uric acid, alanine aminotransferase (ALT), aspartate aminotransferase (AST), alkaline phosphatase (AP), lactate dehydrogenase (LDH), γ-glutamyl aminotransferase (GGT), potassium, phosphor, iron).

   Homeostasis stability coefficient (HSC) was used as the integral parameter for the assessment of the investigated systems.[12,72] HSC was calculated by the formula:

$$\text{HSC} = \frac{\textit{number of indices corresponding to the norm}}{\textit{number of the studied indices}} \times 100\%$$

HSC equaling 100% corresponded to the norm.

After a complex examination of the patients they were divided into 4 groups by stratification randomization. The patients of Group I (control) received placebo, i.e. 0.9% sodium chloride solution, the patients of Group II were treated with Epithalamin. The drug was injected intramuscularly daily at 10 mg for 10 days (100 mg per course). The research was double blind. The bioregulators were applied in combination with a standard treatment for the corresponding indication. The patients were repeatedly examined in 10 days after the end of the therapy course and later in 4 months to assess the efficacy of the performed correction in different periods after the administration of Epithalamin. The bioregulation treatment was repeated in one year. Consequently, the patients underwent 2 courses of bioregulators over 2 years.

In Group II, Epithalamin normalized the same immune indices, as well as ACTH, TSH, cortisol, and insulin. The levels of cholesterol, uric acid, AP, GGT, and LDH came back to norm in 10 days and remained stable for a long time, except cholesterol. The application of Epithalamin twice reduced the clinical manifestations of IHD. Epithalamin relieved the manifestations of hypertension significantly.

However, the data on the patients' mortality over 6 years (1996–2001) appeared to be most interesting. The application of Epithalamin decreased mortality 1.95 and 1.79 times, correspondingly. The combined use of Epithalamin and thimic bioregulator, Thymalim decreased mortality 2 times and a half as compared to the control. The results of applying peptide bioregulators in patients for 6 years deserve a special attention. The rate of their mortality went down 4.1-fold as compared to the control.[72]

Consequently, the clinical results demonstrated the apparent geroprotective effectiveness of Epithalamin in elderly and older patients. Certainly, these results are preliminary, since the number of patients in the groups was not high, and the observation shall be continued.

The efficacy of Epithalamin has been studied at the Institute of Gerontology of the Ukrainian Academy of Medical Sciences (Kiev).[73] The investigation included 152 persons (71 men and 81 women) who were under continuous observation at the Institute outpatient department. The investigated group consisted chiefly of patients with ischemic heart disease (IHD) resulting from the accelerated development of age-related changes in the cardiovascular system and entire organism. The patients had no concomitant nervous, endocrine, and respiratory pathologies. Along with IHD patients the group of persons showing accelerated aging comprised 68 persons without any organic cardiovascular, endocrine or respiratory pathologies. Afterward, all the patients were divided into 3 groups by stratification randomization method. The patients of Group I (control) received placebo, i.e. 0.9% sodium chloride solution, patients of Group II were administered with Epithalamin. All the agents were injected at 10 mg intramuscularly daily at 2–3 days' intervals, with 5 consecutive doses (50 mg) per course. The interval between courses constituted 5–6 months. The patients were under treatment for 3 years (6 courses of the treatment in total). During the period of examination the patients normally did not take any strong medications. The efficacy of Epithalamin was assessed in dynamics according to the indices of the patients' subjective state of health, functional age of their physiological systems, physical and mental workability, immunity, state of their bone tissue, liver detoxication ability, blood lipid spectrum, tolerance to carbohydrates, oxygen tissue exchange, vegetative regulation, and the functional state of endocrine glands.[73]

The results of applying the bioregulators demonstrated a significant improvement in the state of the patients' health. The administration of the peptide bioregulator distinctly inhibited the development of age-related immune disorders in the patients with accelerated aging.[73] Prolonged application of the medications decreased the functional age of the central nervous system, thus, improving its functioning. Epithalamin normalized the initially raised level of cortisol in the patients. This effect remained stable for 3 years after applying the bioregulator. A stable decrease in cortisol concentration was considered a favorable factor for IHD patients, since it normalized lipid metabolism and improved certain immune functions. Epithalamin revealed its modulating effect on the functional state of sexual glands in elderly men and women. The ability of Epithalamin to normalize carbohydrate metabolism as well as to intensify bone tissue density appeared extremely important.

Consequently, the obtained data evidenced the normalizing effect of Epithalamin in case of various age-related homeostatic disorders in elderly persons with accelerated aging. The bioregulators were applied for 3 years with the subsequent observation period of 5 years. Homeostasis normalization in the patients was accompanied by a mortality decrease during 8 years — 1.6-fold, in case of Epithalamin.[73]

Thus, the clinical results on Epithalamin geroprotective effects were obtained at two gerontological research centers. The results provided by each of the two institutes revealed a strong similarity. Especially important was almost the same decrease in the patients' mortality over the entire observations period at both gerontological centers, which pointed at the high significant of the findings. It shall be emphasized that 30 years' experimental and clinical investigations of Epithalamin enabled a logically grounded conclusion on their undoubtedly high geroprotective effectiveness and on the expedience of their use in medical and social care as the means of health maintenance and age-related pathology prevention facilitating an increase in the working life-span of persons over 60.

Conclusions

It is increasingly becoming clear that the means normalizing age-related changes in hormonal status, metabolism, and immunity and, thus, slowing down the genetic program of aging (not postponing aging, but decelerating its rate) must be most effective in the protection from aging and in the prevention of cancer development.[74] Among these means there are pineal bioregulators Epithalamin and Epitalon. Protectors from the initiating action of damaging agents (antioxidants and antimutagens) may be important as the additional means against accelerated aging, especially under the increased risk of environmental hazardous factors.[74]

References

1. Armstrong SM, Redman JR (1991) Melatonin: a chronobiotic with anti-aging properties? *Medical Hypotheses* **34**: 300–309.
2. Pierpaoli W (1991) The pineal gland: a circadian or seasonal aging clock? *Aging* **3**: 99–101.
3. Trentini GP, De Gaetani C, Criscuolo M (1991) Pineal gland and aging. *Aging* **3**: 103–116.
4. Bartsch C, Bartsch H, Blask DE, Cardinali DP, Hrushesky WJM, Mecke D (eds.) (2001) *The Pineal Gland and Cancer. Neuroimmunoendocrine Mechanisms in Malignancy.* Springer Verlag, Berlin.
5. Reiter RJ, Tan DX, Allegra M (2002). Melatonin: reducing molecular pathology and dysfunction due to free radicals and associated reactant. *Neuroendocr Lett* **23**: 3–8.
6. Touitou Y (2001) Human aging and melatonin. Clinical relevance. *Exp Gerontol* **36**: 1083–1100.
7. Malm OJ, Skaug OE, Lingjaerde P (1959) The effect of pinealectomy on bobily growth, survival rate and 32P uptake in the rat. *Acta Endocrinol* **30**: 22–28.
8. Reiter RJ, Tan D-X, Kim SJ (1999) Augmentation of indices of oxidative damage in life-long melatonin-deficient rats. *Mech Ageing Dev* **110**: 157–173.
9. Anisimov VN, Reiter RJ (1990) The function of pineal gland in cancer and aging. *Vopr Onkol* **36**: 259–268.

10. Anisimov VN, Zavarzina NYu, Zabezhinsky MA *et al.* (2001) Melatonin increases both life-span and tumor incidence in female CBA mice. *J Gerontol Biol Sci* **56A**: B311–B323.
11. Pierpaoli W, Dall'Ara A, Pedrini E, Regelson W (1991) The pineal control of aging: the effects of melatonin and pineal grafting on the survival of older mice. *Ann NY Acad Sci* **621**: 291–313.
12. Khavinson VKh, Anisimov VN (1994) *Peptide Bioregulators and Ageing*. Nauka, St. Petersburg, Russia.
13. Lapin V, Ebels I (1976) Effect of some low molecular weight sheep pineal fractions and melatonin on different tumours in rats and mice. *Oncology* **33**: 11–13.
14. Benson B (1977) Current status of pineal peptides. *Neuroendocrinology* **24**: 241–248.
15. Blask DE, Vaughan MK, Reiter RE (1983) Pineal peptides and reproduction. In: Relkin R (ed.) *The Pineal Gland*, pp. 201–224, Elsevier, Amsterdam.
16. Bartsch H, Bartsch C (1988) Unidentified pineal substances with anti-tumor activity. In: Gupta D, Attanasio A, Reiter RJ (eds.) *The Pineal Gland and Cancer*, pp. 369–376. Brain Research Promotion, London & Tubingen.
17. Bartsch H, Bartsch C, Simon WE, Flehmig B, Ebels I, Lippert TH (1992) Antitumor activity of the pineal gland: effect of unidentified substances versus the effect of melatonin. *Oncology* **49**: 27–30.
18. Noteborn HPJM, Bartsch H, Bartsch C, Mans DRA, Weusten JJAM, Flehmig B, Ebels I, Salemnik CA (1988) Partial purification of a low molecular weight ovine pineal compound(s) with an inhibiting effect on the growth of human melanoma cells *in vitro*. *J Neural Transm* **73**: 135–155.
19. Parhon CI (1955) *Biologia Virstelor — Cercetari Clinici si Experimentali*. Acad R.P.R, Bucharest.
20. Anisimov VN, Khavinson VKh, Morozov VG, Dilman VM (1973) The lowering of the threshold of sensitivity of hypothalamo-pituitary system to estrogen action under influence of pineal extract in old female rats. *Dokl Acad Nauk USSR* **213**: 483–485.
21. Anisimov VN, Khavinson VKh, Morozov VG (1982) Carcinogenesis and aging IV. Effect of low-molecular factors of thymus, pineal gland and anterior hypothalamus on immunity, tumor incidence and life-span of C3H/Sn mice. *Mech Ageing Dev* **19**: 245–258.
22. Anisimov VN, Miretski GI, Morozov VG, Khavinson VKh (1982) Effect of polypeptide factors of thymus and epiphysis on radiation carcinogenesis. *Bull Exp Biol Med* **94**: 26–29.
23. Anisimov VN, Loktionov AS, Khavinson VKh, Morozov VG (1989) Effect of low-molecular-weight factors of thymus and pineal gland on life-span and spontaneous tumour development in female of different age. *Mech Ageing Dev* **49**: 245–257.
24. Anisimov VN, Bondarenko LA, Khavinson VKh (1990) The pineal peptides: interaction with indoles and the role in aging and cancer. In: Gupta D, Wollmann HA, Ranke MB (eds.) *Neuroendocrinology: New Frontiers*, pp. 317–325. Brain Research Promotion. London & Tubingen.
25. Anisimov VN, Bondarenko LA, Khavinson VKh (1992) Effect of pineal peptide preparation (epithalamin) on life-span and pineal and serum melatonin level in old rats. *Ann NY Acad Sci* **673**: 53–57.
26. Anisimov VN (1987) *Carcinogenesis and Aging*, Vol. 2. CRC Press, Inc., Boca Raton.

27. Anisimov VN, Mylnikov SV, Khavinson VKh (1998) Pineal peptide preparation Epithalamin increases the lifespan of fruit flies, mice and rats. *Mech Ageing Dev* **103**: 123–132.
28. Khavinson VKh, Morozov VG, Anisimov VN (2001) Experimental studies of the pineal gland preparation Epithalamin. In: Bartsch C, Bartsch H, Blask DE *et al.* (eds.) (2001). *The Pineal Gland and Cancer*, pp. 294–306. Springer Verlag, Berlin, Heidelberg.
29. Dilman VM, Anisimov VN, Ostroumova MN, Khavinson VKh, Morozov VG (1979) Increase in lifespan of rats following polypeptide pineal extract treatment. *Exp Pathol* **17**: 539–545.
30. Anisimov VN, Khavinson VKh (1991) Effect of polypeptide pineal preparation on life-span and spontaneous tumor incidence in old female rats. *Dokl Akad Nauk USSR* **319**: 250–254.
31. Anisimov VN, Prokopenko VM, Khavinson VKh (1995) Melatonin and epithalamin inhibit process of free radical oxidation in rats. *Dokl Russian Acad Sci* **343**: 557–559.
32. Anisimov VN, Arutyunian AV, Khavinson VKh (1996) Melatonin and epithalamin inhibit process of lipid peroxidation in rats. *Dokl Russian Acad Sci* **348**: 265–267.
33. Mylnikov SV, Lyubimova NE (2000) Effect of pineal peptides on mortality rate and antioxidant capacity in *Drosophila melanogaster*. *Adv Gerontol* **4**: 84–87.
34. Anisimov VN, Mylnikov SV, Oparina TI, Khavinson VKh (1997) Effect of melatonin and pineal peptide preparation epithalamin on life-span and free radical oxidation in *Drosophila melanogaster*. *Mech Ageing Dev* **91**: 81–91.
35. Dilman VM, Anisimov VN (1979) Hypothalamic mechanisms of ageing and of specific age pathology — I. Sensitivity threshold of hypothalamo-pituitary complex to homeostatic stimuli in the reproductive system. *Exp Gerontol* **14**: 161–174.
36. Dilman VM (1994) *Development, Aging and Disease. A New Rationale for an Intervention Strategy.* Harwood Academy Publ., Chur.
37. Anisimov VN, Dilman VM (1975) Increase of hypothalamic sensitivity to inhibiting effect of estrogens induced by administration of L-DOPA, diphenylhydantoin, epithalamin and phenformin in old rats. *Bull Exp Biol Med* **80**(11): 96–98.
38. Gadzhieva TS, Blinova TS (1980) Effect of epithalamin on gonadotropic function of pituitary and on ovaries. *Akush Gynekol* **9**: 15–18.
39. Gupta D, Haldar C, Coleveld M, Roth J (1993) Ontogeny, circadian rhythm pattern, and hormonal modulation of 5-alpha-dihydrotestosterone receptors in the rat pineal. *Neuroendocrinology* **57**: 45–53.
40. Meites J (1988) Neuroendocrine basis of aging in the rat. In: Everitt AV, Walton JR (eds.) *Regulation of Neuroendocrine Aging (Interdisciplinary Topics in Gerontology, Vol. 24)*, pp. 37–50. Karger, Bazel.
41. Belozertsev YA, Tertyshnik ID, Spinko AA, Grudinin AN (1987) Restoration by pineal cytomedins of learning and homeostatic disturbanses. In: Yakovlev GM (ed.) Abstracts of Scientific Conference *The Role of Peptide Bioregulators (Cytomedins) in Regulation of Homeostasis*, pp. 13–14. Military Medical Academy Publ., Leningrad.
42. Arushanian EB, Baturin VA, Ovanesov KB (1990) Chronic administration of pineal peptides change the circadian locomotor activity and time-course of forced swimming in rats. *J Pineal Res* **9**: 271–277.
43. Slepushkin VD, Pashinski VG (1982) *The Epiphysis and Adaptation of Organism.* Tomsk University Publ., Tomsk.

44. Khavinson VKh, Morozov VG, Grintzevich II, Knorre ZD (1977) Influence of polypeptide extracted from the anterior hypothalamus and the epiphysis on the cells of the hypothalamo-hypophyseal neurosecretory system. *Arch Anat Histol Embryol* **73**: 100–104.
45. Bondarenko LA, Anisimov VN (1990) Effect of polypeptide pineal preparation epithalamin on serotonin metabolism in the pineal gland of rats. *Bull Exp Biol Med* **110**: 150–151.
46. Bondarenko LA, Anisimov VN (1992) Age-related peculiarities of effect of epithalamin on metabolism of serotonin in pineal gland of rats. *Bull Exp Biol Med* **113**: 194–195.
47. Labunets IF, Butenko GM, Khavinson VKh, Magdich LV, Dragunova VA, Pishel IN, Azarskova MV (2003) The regulating influence of pineal gland's peptides on the development of T-lymphocytes in CBA mice at aging: role of the microenvironemnt imune system's organs and the neuroendocrine factors. *Adv Gerontol* **12**: 111–120.
48. Ostroumova MN, Vasilyeva IA (1976) Effect of pineal extract on the regulation of fat-carbohydrate metabolism. *Probl Endocrinol* **22**: 66–69.
49. Belokrylov GA, Morozov VG, Khavinson VKh, Sofronov BN (1976) Effect of low-molecular extracts from heterologous thymus, epiphysis and hypothalamus on the immune response of mice. *Bull Exp Biol Med* **81**: 202–204.
50. Slepushkin VD, Anisimov VN, Khavinson VKh, Morozov VG, Vasiliev NV, Kosykh VA (1990) *The Pineal Gland, Immunity and Cancer (Theoretical and Clinical Aspects)*. Tomsk University Publ., Tomsk.
51. Labunets IF, Butenko GM, Dragunova VA, Magdich LV, Kopulova GV, Rodnichenko AE, Michalskiy SA, Khavinson VKh, Azarskova MV, Maksyuk TV (2004) The pineal peptide factors and the rhythms of functions of the thymus and bone marrow in animals during aging. *Adv Gerontol* **13**: 81–89.
52. Grinevich YuA, Bendyug GD, Bobrov LI, Labunets IF, Kireeva ES (1992) Effect of polypeptide factors of thymus and pineal gland on spontaneous development of lymphomas in AKR mice. In: Yakovlev GM (ed.) Abstracts of the Symposium *Peptide Bioregulators Cytomedines*, pp. 47–48. Medical Military Academy Publ., St. Petersburg.
53. Khavinson VKh, Izmailov DM, Obukhova LK (2000) Effect of Epithalon on the lifespan increase in *Drosophila melanogaster*. *Mech Ageing Dev* **120**: 141–149.
54. Anisimov VN, Khavinson VKh, Mikhalski AI, Yashin AI (2001) Effect of synthetic thymic and pineal peptides on biomarkers of ageing, survival and spontaneous tumour incidence in female CBA mice. *Mech Ageing Dev* **122**: 41–68.
55. Anisimov VN, Khavinson VKh, Provinciali M, Alimova IN, Baturin DA, Popovich IG, Zabezhinski MA, Imynaitov EN, Mancini R, Franceschi C (2002) Inhibitory effect of the peptide epitalon on the development of spontaneous mammary tumors in HER-2/neu transgenic mice. *Int J Cancer* **101**: 7–10.
56. Anisimov VN, Khavinson VKh, Popovich IG, Zabezhinski MA (2002) Inhibitory effect of epitalon on colon cacinogenesis induced by 1,2-dimethylhydrazine in rats. *Cancer Lett* **183**: 1–8.
57. Khavinson VKh, Goncharova ND, Lapin BA (2001) Synthetic tetrapeptide epitalon restores disturbed neuroendocrine regulation in senescent monkeys. *Neuroendocrinol Lett* **22**: 251–254.
58. Djeridane Y, Khavinson VKh, Anisimov VN, Touitou Y (2003) Effect of a synthetic pineal tetrapeptide (Ala–Glu–Asp–Gly) on melatonin secretion by the pineal gland of young and old rats. *J Endocrinol Invest* **26**: 211–215.

59. Kost NV, Sokolov OI, Gabaeva MV, Zolotarev IA, Malinin VV, Khavinson VKh (2003) Effect of new peptide bioregulators livagen and epitalon on enkephalin-degrading enzymes in human serum. *Izv Akad Nauk Ser Biol* **4**: 427–429.
60. Sibarov DA, Kovalenko RI, Malinin VV, Khavinson VKh (2002) Epitalon influence pineal secreation in stress-exposed rats in the daytime. *Neuroendocrinol Lett* **23**: 452–454.
61. Khavinson VKh, Korneva EA, Malinin VV, Rybakyna EG, Pivanov IY, Shanin SN (2002) Effect of epitalon on interleukin-1beta signal transduction and the reaction of thymocyte blast transformation under stress. *Neuroendocrinol Lett* **23**: 411–416.
62. Kazakova TB, Barabanova SV, Khavinson VKh, Glushikhina MS, Parkhomenko EP, Malinin VV, Korneva EA (2002) *In vitro* effect of short peptides on expression of interleukin-2 gene in splenocytes. *Bull Exp Biol Med* **133**: 614–616.
63. Rosenfeld SV, Togo EF, Mikheev VS, Popovich IG, Khavinson VKh, Anisimov VN (2002) Effect of epitalon on the incidence of chromosome aberrations in senescence-accelerated mice. *Bull Exp Biol Med* **133**: 274–276.
64. Khavinson VKh, Bondarev IE, Butyugov AA (2003) Epitalon peptide induces telomerase activity and telomere elongation in human somatic cells. *Bull Exp Biol Med* **135**: 590–592.
65. Anisimov VN, Arutjunyan AV, Khavinson VKh (2001) Effect of pineal peptide preparation Epithalamin on free radical processes in animals and humans. *Neuroendocr Lett* **22**: 9–18.
66. Khavinson VKh, Shutak TS (2000) *Application of Epithalamin in Non-Insulin-Dependent Diabetes Mellitus*. Foliant, St. Petersburg.
67. Balin VN, Khavinson VKh, Madai DYu (2000) Application of Epithalamin in elderly patients with non-insulin-dependent diabetes mellitus and purulent-inflammatory diseases of the maxillofacial region. *Klin Gerontologia* **5–6**: 8–12.
68. Shustov SB, Khavinson VKh, Shutak TS (1998) Effect of Epithalamin on carbohydrate metabolism and the state of the cardiovascular system in the patients with non-insulin-dependent diabetes mellitus. *Klin Med* **9**: 45–48.
69. Khavinson VKh, Morozov VG (2001) *Thymic and Pineal Peptides in the Regulation of Aging*. Foliant, St. Petersburg.
70. Karpov RS, Slepushkin VD, Mordovin VF, Khavinson VKh, Morozov VG, Grischenko VI (1985) *The Use of Pineal Preparations in Clinical Practice*. Tomsk University Publ., Tomsk.
71. Komarov FI, Khavinson VKh, Simonenkova VA (1995) Application of Epithalamin in climacteric myocardiopathy. *Klin Meditsina* **73**: 40–42.
72. Khavinson VKh, Morozov VG, Solovieva DV (1998) Epithalamin in the prevention and treatment of genetically dependent age-related pathology. *Adv Gerontol* **2**: 103–106.
73. Korkushko OV, Khavinson VKh, Butenko GM, Shatilo VB (2002) *Thymic and Pineal Peptide Bioregulators in Accelerated Aging Prevention*. Nauka, St. Petersburg.
74. Anisimov VN (2001) life-span extension and cancer risk: myths and reality. *Exp Gerontol* **36**: 1101–1136.

8

Melatonin: Anti-Aging Perception and Current Perspectives

Arvind L. Bhatia
Department of Zoology
University of Rajasthan
Jaipur — 302004, India

INTRODUCTION

Several researchers are making genuine attempts to test and develop various means of intervention for the prevention and treatment of age-related diseases, for regaining the functional abilities and for prolonging the life-span of experimental organisms. At present, the main anti-aging approaches include supplementation with hormones including growth hormone, dehydro-epiandrosterone (DHEA), estrogen and melatonin, and nutritional supplementation with synthetic and natural antioxidants in purified form or in extracts prepared from plant and animal sources. Although some of these approaches have been shown to have some clinical benefits in the treatment of some diseases in the elderly, none of these really modulate the aging process itself. To date, no intervention, including drugs and hormones, has been proven to slow or reverse aging in human beings. On the other hand, remarkable progress has been made in the prevention and treatment of some age-related diseases, and especially the use of hormones constitutes a potentially new option. Of these, significant information is available regarding the so-called sleep hormone, melatonin with respect to its wide ranging biological effects, including gerontomodulatory effects. The aim of this article is to provide a brief introduction to the chemical nature and biological effects of melatonin, and to critically review its potential usefulness as an anti-aging compound.

Melatonin (N-acetyl 5-methoxytryptamine) is synthesized endogenously by the pinealocytes of the pineal gland. The essential amino acid L-tryptophan is a precursor in the synthesis of melatonin, which first gets metabolized to 5-hydroxytryptophan from which 5-hydroxytryptamine, also known as serotonin, is made. 5-hydroxytryptamine is then converted to melatonin in a two-step process, occurring mainly in the pineal gland. Melatonin is also known as N-acetyl-5-methoxytryptamine and N-[2-(5-methoxy-1H-indol-3-yl)ethyl]-acetamide. Its biosynthesis from tryptophan involves four well-defined intracellular steps catalyzed by tryptophan hydroxylase, aromatic amino acid decarboxylase, serotonin-N-acetyltransferase, and hydroxyndole-O-methyltransferase. Billions of years of evolution should have allowed organisms to use melatonin for different physiological purposes. However, contrary to the potential diversity in signaling, no such variation exists with regard to the physical and chemical properties of the molecule. Therefore, its amphiphilicity and its radical scavenging properties can be of importance in any aerobic organism from most distant taxa and may be assumed to represent a phylogenetically early function.[1]

Melatonin and Circadian Rhythm

Production of melatonin varies with periods of light and darkness in the environment. Even though in human beings the pineal gland lies near the centre of the brain, it obtains information about light in the environment through nerve pathways originating in the eyes. In general, light checks and darkness stimulates the pineal gland's production of melatonin. Therefore, the gland tends to secrete small amounts of melatonin during the day and larger amounts at night. Display of melatonin synthesis as a circadian rhythm is reflected in serum melatonin levels. The rhythm is generated by a circadian clock located in the suprachiasmatic nucleus (SCN) of the hypothalamus. The SCN clock is set to the 24-hour-day by the natural light-dark cycle. Light signals through a direct retinal pathway to the SCN. The SCN clock sends circadian signals over a neural pathway to the pineal gland. This drives rhythmic melatonin synthesis. The neural input to the gland is norepinephrine, and the output is melatonin. Specifically, the rhythm of the enzyme, arylalkylamine N-acetyl transferase (AA-NAT) is under SCN control, with the resulting melatonin rhythm characterized by high levels at night. Thus, the synthesis and release of melatonin are stimulated by darkness and inhibited by light.

Melatonin is involved in the regulation of seasonal and circadian fluctuations of other hormones and in the synchronization of many different aspects of circadian rhythmicity to the light-dark cycle. Recently, it has been suggested that a previously undetected, rhodopsin-based, visual pigment, located in some retinal ganglion cells and having peak sensitivity around 460 nm, may be responsible for light-induced melatonin suppression and, perhaps, maintenance of the circadian rhythm. As absorption in the crystalline lens for shorter visible wavelengths increases substantially with age, while the pupil diameter tends to decrease, the effective retinal exposure received under the same ambient lighting conditions by the pigment is almost 10 times lower in an old, as compared with a young eye. Interestingly, replacement of the old crystalline lens by an intraocular implant restores the exposure to youthful levels.[2] Though the pineal gland helps to regulate certain daily and seasonal body cycles even without visual cues, the level of melatonin in the blood rises and falls on a daily (circadian) cycle with peak levels occurring in the early hours of the morning. Evening naps and advanced illumination may play a role in the advance of the circadian system in aging. The relationships between the sleep-wake cycle and endogenous circadian rhythms in young and older adults have been established and the correlates between evening naps and circadian rhythms in older adults have been examined.[3] In addition to pineal gland, human lymphoid cells are an important physiological source of melatonin and that this melatonin could be involved in the regulation of the human immune system, possibly by acting as an intracrine, autocrine, and/or paracrine substance.[4]

In addition to above features, in human beings melatonin has been linked to the onset of puberty; the pineal gland's nightly secretion of melatonin decreases when a boy or a girl reaches puberty. Melatonin may also help regulate menstrual cycles in women and sperm production in men. In addition, researchers have also suggested a connection between melatonin levels and certain mental illnesses. Some of these features will be discussed later in this chapter.

Melatonin and Aging

Melatonin administration extends the life-span of mice. The prolongation of life-span by melatonin has been interpreted in favor of an upregulation of the immune system as well as due to anti-stress properties of melatonin acting via the brain opioid system. The circadian melatonin rhythm appears at the end of the neonatal period and persists thereafter.[5] However, it decreases with aging, and therefore melatonin has been attributed with its anti-aging role in the light of several perspectives. Anti-aging properties of melatonin as conclded by Pierpaoli and Regelson[6] are as follows:

1) The pineal gland shrinks with age, so melatonin production declines as we age.
2) In addition to the pineal gland, digestive tract is another source of melatonin production. Melatonin production is claimed to increase in the digestive tract when calories are restricted, which may explain why dietary restrictions have shown to increase life-span.
3) Being a powerful antioxidant, melatonin may act against age-related oxidative damage.
4) Being an immuno-stimulant, it may have protective effects.
5) Some melatonin-associated benefits *viz.*, prevention of cancer, protection of heart, and restful sleep may indirectly prolong life.
6) Experiments on mice resulted in 20% increase in average life-span.

An alternative explanation of melatonin's anti-aging effects has also been offered with the age dependent diminished amplitude as indexed by a decrease in circulating melatonin levels.[7] Exogenous melatonin increases the amplitude of the circadian pacemaker system by feedback onto that system. The hypothalamic SCN are thought to be the mammalian biological clock in the brain and have high concentrations of melatonin receptors. Therefore, melatonin administration in pharmacological doses may prevent aging symptoms by acting at the level of the circadian pacemaker's amplitude.

Although previous reports indicate that nocturnal plasma melatonin secretion declines with age, some recent findings do not support this point. Generally, similar circadian melatonin patterns that peaked at night and with very low levels during the daytime are noted. Nocturnal melatonin levels were significantly lower in the men in their 60's. However, nocturnal melatonin levels were higher among post-menopausal women than men. In the older age-groups, nocturnal melatonin levels did not differ between healthy controls and subjects with high blood pressure or ischemic heart disease.[8] Salivary melatonin measurement has been claimed to be a reliable, sensitive and easy method to monitor changes in the circadian rhythms of melatonin during the course of aging. The alterations in the circadian rhythms of salivary melatonin begin during middle-age. The decline in nocturnal peak levels in salivary melatonin was found in old and the oldest subjects. The off-set melatonin levels were more than two times higher in the oldest group than that in the other groups, indicating a delayed phase of salivary melatonin. A step-wise decrease in the circadian rhythms of saliva melatonin has been shown to occur early in life, around 40 years of age, and the middle-aged subjects had only 60% of the amplitude of the young subjects. In addition, the middle-aged subjects showed the longest peak levels duration and the lowest daytime melatonin levels.[9]

Melatonin's anti-aging action might be expected by the antioxidative characteristic of N-acetylserotonin (NAS), which is a melatonin precursor and metabolite. Both NAS and melatonin administered with drinking water prolonged life-span in male animals by about 20% but did not affect the life-span of female mice. NAS increased the antioxidant capacity of kidney. Both NAS and melatonin increased the antioxidant capacity of brain. NAS- and melatonin-treated mice had healthy and luxuriant fur coats with some gray fur in the melatonin group whereas mice in the saline group had large areas of baldness.[10] Some scientists advocate multihormonal replacement therapy and the use of antioxidant drugs that may favorably influence some of the pathological conditions in aging men. During the aging process, a number of morphological and neurochemical alterations have been found in the SCN, which are in part responsible for the age-dependent decrease in plasma testosterone [andropause or partial endocrine deficiencies of aging male (PEDAM)], DHEA (adrenopause), GH/IGF-I (somatopause) and melatonin that develops in most men at about fifty years of age.[11]

Since antioxidant levels are decreased in the aged brain, the greater predisposition to neuronal death in stroke leading to subsequent neurodegeneration in aged individuals may be related to changes in oxidant balance. Study on the effect of the endogenous antioxidant melatonin on excitotoxic injury resulting from N-methyl-D-aspartate (NMDA)-induced damage by developing an organotypic mouse brain slice model showed that melatonin significantly decreased redox active iron, heme-oxygenase (HO-1) induction and 8-hydroxyguanosine (8-OHG) following NMDA insult. The results support the hypothesis that melatonin is a neuroprotective antioxidant.[12] Levels of cerebellar nitric oxide synthase (NOS) and rates of generation of cortical reactive oxygen species (ROS) determined in mice of various ages showed a significant reduction in 9-month-old mice relative to 3-month-old mice. Dietary supplementation may aid in delaying onset of metabolic changes characteristic of the older brain.[13] The restoration of a more youthful gene profile in brains of aged animals by melatonin, to a large extent, involves reversal of age-induced elevation of basal inflammatory parameters.[14] Melatonin deprival in young rats induces alterations in cerebral arteriolar wall similar to those observed during aging — atrophy and a decrease in distensibility.[15]

Since aging is often associated with decreased ability of sleep maintenance, it has been hypothesized that the elderly experience a delayed timing of sleep period relative to the circadian phase of various sleep-promoting physiological functions. This possibly causes decreased sleep predisposition in the end part of their nocturnal sleep. The relationship between the sleep timing and circadian phase of melatonin secretion, which is known as a possible human sleep modulator as well as a stable marker of biological clock phase (BCP), was evaluated in healthy elderly volunteers with actigraph sleep recordings followed by the evaluation of melatonin phase under dim light. Remarkable clock time has been found to advance in both the midpoint of biological clock phase (BCP) and sleep timing, with a significant decrease in sleep maintenance in elderly subjects. Early morning awakening in the elderly appeared in a BCP for which sleep tendency remained sufficient to sustain sleep.[16]

Studies on the potential role of melatonin in the processes of aging, the prolongation of life-span and health in the aged suggest that exogenously administered melatonin may serve to extend life-span in some experimental animals.[17] Melatonin secretion is an endogenous synchronizer, and it is affected not only by age but also by the degree of cognitive impairment.

An age-related decline of the circadian amplitude of the melatonin rhythm occurred in old subjects, especially in demented individuals. The melatonin nocturnal peak was significantly correlated with the severity of the cognitive impairment. Urinary 6-hydroxymelatonin sulfate (aMT6s) excretion also declined with age.[18] Furthermore, the damage to the pituitary could also impair responses to stress and infection, and the release of NO during infection could be responsible for the degenerative changes in the pineal and diminished release of melatonin.[19]

Antioxidative Nature of Melatonin

Melatonin is an important antioxidant as an efficient neutraliser of hydroxyl radical (•OH). When melatonin interacts with the •OH, it becomes the indolyl (melatonyl) radical. This reactant has very low toxicity so there is a net gain when melatonin scavenges the •OH, and thus a highly toxic reactant is replaced by a radical with low toxicity. After some molecular rearrangement the indolyl radical scavenges a second •OH to form cyclic 3-hydroxymelatonin. This product has been identified by electron ionization mass spectrometry and protein nuclear magnetic resonance.[20] Melatonin also restricts lipid peroxidation by preventing the initiating events as well as interrupting the chain reaction.[21] Considering melatonin's protective effect against lipid peroxidation in the brain, this hormone could play a significant role in the antioxidative defense system of the brain.[22,23] The effects of pinealectomy on lipid peroxidation, antioxidant status and NMDA receptor subunits 2A and 2B concentrations in hippocampus showed that NR2A and NR2B concentrations in the rats maintained in dark were significantly increased as compared to the control and functional pinealectomy groups. But there was no significant increase in lipid peroxidation levels.[24]

Melatonin can also directly scavenge hydroxy radical, peroxyl radical, peroxynitrite anion and singlet oxygen protecting cell membrane, proteins in the cytosol and DNA in the nucleus. Cell membrane, proteins in the cytosol, and DNA in the nucleus. This tryptophan derivative further stimulates a number of antioxidative enzymes and stabilizes cell membranes. Melatonin's antioxidative actions include the protection of lipids in the morphophysiological barriers and enters all cells in the organism equally well.[25] Not only is melatonin an effective antioxidant but the products it generates also appear to reduce oxidative damage. For example, the product, N^1-acetyl-N^2-formyl-5-methoxykynuramine (AFMK), may also be an effective scavenger. This cascade of reactions contributes to the overall antioxidative potential of melatonin. Dioxetane is also one proposed intermediate product although there may be others as well.[20] Melatonin also stimulates brain glutathione peroxidase (GSH-Px) activity,[26] and may be the only known antioxidant to have such dual effects. The inhibitory effects of melatonin on free intracellular calcium in mouse brain cells have also been demonstrated.[27] Long-term administration with melatonin inhibited the overload of $[Ca^{2+}]$ in old mouse cerebral cortex and is an example of its anti-aging effect.

In vitro studies on the efficacy of 5 antioxidants compared with the effectiveness of melatonin (N-acetyl-5-methoxytryptamine) and the possible synergistic effects of melatonin with the other 5 molecules have been performed. The efficacies of three polyphenolic antioxidants, xanthurenic acid, resveratrol (3,4′,5-trihydroxy-trans-stilbene) and (−)-epigallocatechin-3-gallate (EGCG), and two classical non-polyphenolic antioxidants, vitamin C (ascorbic acid) and alpha-lipoic acid (LA, 1,2-dithiolane-3-pentanoic acid) in inhibiting •OH-induced oxidative DNA damage were examined.[28] Melatonin reversed the pro-oxidant effect of resveratrol and vitamin C, had an antagonistic effect when used in combination with EGCG and exhibited synergism in combination with vitamin C and with LA.[28]

Damage to mitochondria as a result of the intrinsic generation of free radicals is also involved in the processes of cellular aging. Whether acutely administered melatonin, due to its free radical scavenging activity, would influence mitochondrial metabolism has also been looked into. An age-related decrease in hepatic mitochondrial function in senescence-accelerated-prone mice (SAM) can be modified by an acute pharmacological injection of melatonin; and the indole stimulated mitochondrial respiratory chain activity would likely reduce deteriorative oxidative changes in mitochondria that normally occur in advanced age.[29]

Melatonin Against Radiation Damage

Melatonin administration prior to irradiation may prevent rat liver damage induced by irradiation, reflecting the antioxidant roles of different doses of melatonin against gamma-irradiation-induced oxidative damage after total body irradiation with a single dose of 6 Gy.[30] The liver tissue MDA levels in irradiated rats that were pre-treated with melatonin (5 or 10 mg/kg) were significantly decreased, while the SOD and GSH-Px activities were significantly increased. The prophylactic influence of melatonin against radiomimetic drug cyclophosphamide-induced oxidative stress in mouse tissues showed modulatory effect on lipid peroxidation, GSH, glutathione disulphide (GSSG), GSH-Px and serum phosphatase levels in brain, spleen liver, lungs, kidney and testes.[31]

The brain is particularly susceptible to free radical attack.[32] One of the potential major causes of age-related destruction of neuronal tissue is toxic free radicals that are a natural result of aerobic metabolism. Melatonin affords protection against age-induced oxidative stress in brain;[33,34] its antioxidative property affords protection against damage due to radiation and radiomimetic drug (cyclophosphamide). Vijayalaxmi *et al.*[35] studied the radioprotective ability of melatonin in mice exposed to an acute whole-body gamma radiation dose of 815 cGy (estimated LD50/30 dose) over a period of 30 days following irradiation. Irradiated mice which were pre-treated with 125 mg/kg melatonin exhibited an increase in their survival (60%), and 85% of irradiated mice pre-treated with 250 mg/kg melatonin were alive by end of 30 days. The lymphocytes in the blood samples collected at 1 and 2 h after melatonin ingestion and exposed *in vitro* to 100 cGy gamma radiation exhibited a significant decrease in the extent of primary DNA damage, as compared with similarly irradiated lymphocytes from the blood sample collected before melatonin ingestion. The possible mechanisms involved in its "radioprotective" effect of melatonin have also been discussed.[36]

Dose Discrepancy

Although compelling evidence suggests that melatonin may be effective for a variety of disorders, the mode and optimal dose of melatonin is still not clear. Most studies have used doses of supra-physiological blood levels. However, doses as low as 0.1 mg/kg given orally to mice proved the effective and appropriate minimum dose level of melatonin[33,34] at which age-induced changes in the levels of MDA, GSH, GSSG, GSH-Px and phosphatase activity have been shown to improve. This level of melatonin ameliorates the depletion of GSH as well as the fall of GSH-Px and alkaline phosphatase activities and reduces the level of GSSG, lipid peroxidation and acid phosphatase in mice. Pre-treatment with the same dose level (0.1 mg/kg/day) for 15 consecutive days affords potential protective effect against radiation induced oxidative stress and mortality in mice. It has been reported[35] that melatonin (at a dose as high as 250 mg/kg) is non-toxic and that high doses of melatonin are effective in protecting mice from lethal effects of acute whole-body irradiation. The effects on low-level chronic administration of melatonin against radiation-induced injury have also been shown to be quite effective.[33,34,37]

The effect of melatonin as a geroprotector is dose-dependent; low-dose melatonin (2 mg/L) significantly decreased spontaneous tumor incidence (by 1.9-fold), mainly mammary carcinomas in mice, whereas higher doses (20 mg/L) failed to influence tumor incidence as compared to controls. Melatonin in drinking water (2 or 20 mg/L) for 5 consecutive days every month to 3-month-old female Swiss-derived SHR mice until their natural death also showed that the treatment of melatonin did not significantly influence food consumption, but its administration at lower doses did decrease the body weight of mice, and it slowed down the age-related switching-off of estrous function. However, it did not influence the frequency of chromosome aberrations in bone marrow cells and mean life-span though it increased life-span of the last 10% of the survivors in comparison to controls.[38] However, it has not been possible to establish an LD50 for melatonin even though doses of 1200 mg/kg body weight have been administered for such purposes.[39]

Melatonin as an Anti-Carcinogen

Inhibition of cancer growth by melatonin has been observed. There are data which indicate that melatonin antagonizes the mitogenic effects of estrogens.[40–42] Inhibition of antioxidants by reducing agents such as glutathione eliminates the oncostatic effects of melatonin in certain human breast cancer cell lines.[43] As stated above, melatonin has been found to be the most effective scavenger of highly toxic free radicals,[44] which induce DNA damage. Melatonin may augment the antitumor activity of IL-2 (interleukin 2) by inhibiting tumor growth factor production.

These results suggest that low-dose IL-2 and melatonin may be a well-tolerated therapy for advanced endocrine tumors.[45] Melatonin may also enhance IL-2 anti-tumor immune effect.[46,47]

Melatonin reduces tumor growth in experimental models *in vivo*[48–50] and proliferation and invasive properties of cancer cells in culture.[51–53] Oral supplementation of melatonin reduced the viability and volume of the tumor Ehrlich ascites carcinoma cells (EAC) implanted intraperitoneally into female mice, delayed the progression of the cell cycle and reduced the DNA content of the cells.[54] The depressed cell viability suggests that melatonin might be inducing apoptosis in EAC cells. The pro-apoptotic action of melatonin in cancer cells is not restricted to breast tissue. Melatonin also suppresses the proliferation of a colon cell line both *in vitro* and *in vivo*.[55] In addition, the proliferating/apoptotic cell ratio in this cell line, originally developed from a 1,2-dimethylhydrazine-induced colonic tumor in mouse, was significantly lower when the animals were treated with melatonin, indicating that it also induces apoptosis.[56] Electromagnetic fields (EMF) have also been linked to tumors like gliomas, and EMFs is suggested to inhibit melatonin production in the pineal gland.[57]

Melatonin Against Depressive Disorders

There is evidence of a bidirectional relationship between cancer and depression, offering new opportunities for therapeutic intervention. The prevalence of depression among cancer patients increases with disease severity and symptoms such as pain and fatigue. Psychophysiological mechanisms linking depression and cancer progression include dysregulation of the hypothalamic-pituitary-adrenal axis, especially diurnal variation in cortisol and melatonin. Depression also affects components of immune function that may affect cancer surveillance.[58] The serum melatonin levels in patients suffering from endogenous depression and the effect of pharmacological therapy showed higher serum melatonin values. The possible correlation between the height from the mean sea levels and the patients suffering from endogenous depression showed that patients living at higher altitudes had higher serum values for the melatonin than those in patients suffering from endogenous depression.

Melatonin production is phase-shifted in major depression. Serum melatonin and urinary 6-sulfatoxymelatonin (aMT6s) measured in major depressive inpatients indicated a clear melatonin rhythm without significant difference in the mean level of melatonin or aMT6s, in the area under the curve of melatonin or in the melatonin peak. However, the time of the nocturnal melatonin peak secretion was significantly delayed in depressive subjects as compared to healthy controls. Moreover, the depressed patients showed urinary aMT6s concentrations enhanced in the morning compared to night time levels, while these concentrations were lowered from the night to the morning in the control group.[59]

Morning serum melatonin levels decreased with pharmacological therapies.[60] Atenolol strongly reduces melatonin levels in depressed and control subjects; this decrease relates inversely to the severity of depressive symptoms. Furthermore, responsiveness to atenolol differs between low melatonin depressed patients and low excretory controls, alluding to beta-adrenoceptor up-regulation in a subtype of depression.[61] Reports on melatonin secretion in depression are numerous but conflicting. There are very few studies relating the duration of the nocturnal melatonin peak to depression, and the results of those studies have been unequivocal. There might be a familial vulnerability in the endogenous melatonin signal in subjects prone to depression and an abnormality in the duration of the melatonin signal in those with current major depression.[62]

The role of melatonin in seasonal affective disorder (SAD) and endogenous depression has also drawn some attention. A sense of sadness, loss and lethargy accompanies the shortening days of fall and winter; tiredness, a bit of weight gain, difficulty getting out of bed and bouts of "the blues" are generally noticed as fall turns to winter; and some people experience an exaggerated form of these symptoms. The use of bright light therapy in patients with this disorder has been recommended,[63] and a large number of well-designed studies have confirmed and refined these findings. Researchers are still investigating the mode by which bright light can lift depression or reset a sleep cycle. One theory is that the area of the brain, near the visual pathway, the SCN responds to light by sending out a signal to suppress the secretion of melatonin. Brain studies suggest that there is impairment in serotonin function in neurons leading to the SCN. Though, initial theories suggested a pathway from the retina to the SCN, some recent research indicated that bright light applied to the back of an individual's knee could shift human circadian rhythms. This suggests that the bloodstream, not just the neurons of the visual pathways, might mediate the biological clock.

Melatonin and Endocrine Disorders

The hormonal basis of mood disorders, however, has not been investigated systematically and extensively. The major categories of proposed hormonal etiologies, including gonadal steroids, thyroid hormones, cortisol, prolactin, and melatonin are being studied in human subjects. As discussed above, melatonin is the mediator of external light to physiologic adaptation to day and night rhythms. Melatonin seems to be the natural hormone to facilitate sleep in insomniac patients and causes no hangover.

In peri-menopausal women melatonin administration did produce a change in LH, FSH and thyroid hormones. In Nordic countries, indigenous people suffer less from breast and prostate cancer, and winter darkness seems to protect. The supposedly increased melatonin levels created the "melatonin hypothesis". Epidemiological studies did show that blind people indeed have half the rate of breast cancers, supporting the hypothesis. Controversial results concerning melatonin and insulin resistance and glucose tolerance have been published. In post-menopausal women, application of melatonin reduced glucose tolerance and insulin sensitivity. Pregnant women should avoid melatonin since its teratogenic effect is not known. Patients suffering from non-hormone dependent tumors, like leukemia, should avoid melatonin, since tumor growth was promoted in animal experiments. It can be expected that melatonin will receive wide consideration for treatment of sleeping disturbances, jet lag, and fibromyalgia.[64]

External magnetic fields have also been found to synchronize melatonin secretion in experimental animals and humans and may be beneficial in the treatment of post-menopausal osteoporosis. Pineal melatonin has been shown in animals to be involved in the regulation of calcium and phosphorus metabolism by stimulating the parathyroid glands and by inhibiting calcitonin release and prostaglandin synthesis Menopause is associated with a decline in melatonin secretion and increased pineal calcification.[65]

Melatonin as a potential treatment for nocturia has been associated with bladder outflow obstruction in older men. Nocturia is a common condition often attributed in aging men to benign prostatic enlargement. Older adults are prone to nocturnal sleep disturbance, of which disturbed circadian rhythm may be a component since it improves with night time administration of melatonin. Melatonin treatment is associated with a significant nocturia response rate, improvement in nocturia-related bother and a good adverse effect profile. However, it is uncertain whether the observed changes in this study are clinically significant.[66] Puchalski *et al.*[67] reported that daily melatonin administration to middle-aged rats, restored nocturnal plasma melatonin to young adult levels, decreased body weight and suppressed visceral fat and plasma leptin. In some species, metabolic and some neuronal responses to melatonin are mediated or dependent at least in part on gonadal steroid levels. Thus, melatonin-induced changes in gonadal steroid secretion may have mediated the aging-dependent melatonin-induced metabolic responses. Melatonin treatment also decreased both intra-abdominal fat and plasma leptin levels in both intact and castrate rats, with no significant differences of percentage suppression in the intact versus castrate rats. These results demonstrate that suppression of body weight, visceral adiposity, and plasma leptin levels by daily melatonin administration to middle-aged rats was independent of gonadal function.

The pineal endocrine system may influence the outcome of stroke. The mechanism of action and the pathophysiological role of this system in aging should be further characterized.[68] Pinealectomy increased both the extent of DNA damage and the infarct volume, and administration of melatonin to pinealectomized rats reduced both these markers of brain injury.

Melatonin in Immunomodulation

Immunodepression has been counteracted by melatonin administration. Age-associated changes in the immune system render organisms more sensitive to infections, autoimmune diseases, and even to cancer. Cellular and humoral immunity decline with age and this is primarily related to the loss of lymphoid cells rather than to a loss of cell activity. Sainz and collaborators[69] underline the important role of glucocorticoids as immuno-depressors and melatonin as an immuno-stimulatory agent. Exogenous melatonin also induces a significant enhancement of murine antibody-dependent cellular cytotoxicity (ADCC).[70] Ablation of the pineal gland during the first week of life significantly reduces ADCC levels in adult mice and the administration of melatonin to pinealectomized mice restores ADCC levels regardless of the hour and seasonal time of injection.

Thymus is one of the main targets of melatonin and its immunoenhancing effects may be mediated by opioids derived from T-helper cells, lymphokines and possibly pituitary hormones. Lymphokines such as gamma-interferon and IL2 as well as thymic hormones can modulate the synthesis of melatonin in the pineal gland.[71] A relation between the pineal and puberty has been speculated for many years. Normal pubertal development does not appear to be linked to alterations in melatonin profile. However there is some evidence that delayed puberty, precocious puberty and hypothalamic amenorrhoea may have altered melatonin profiles.[72]

Melatonin treatment had beneficial effect on the proliferative activity, the rate of DNA synthesis, and the histopathological changes of splenic and thymic lymphocytes in old rats and caused stimulation of the lymphocyte activity, especially in old rats.[73] In the same manner melatonin distinctly reversed the age-related thymic involution as revealed by the notable increase of thymus weight, total number of thymocytes and percentage of thymocytes at G2+S phases.[74] Exogenous melatonin administered through the drinking water to 22-month-old female C57BL mice for 60 consecutive days caused recovery in spleen weight, total number of splenocytes and some peripheral immune capacity such as mitogen responsiveness and NK cell activity. Even when the melatonin supplementation begins late in life, the age-related thymic involution and peripheral immune dysfunctions can be restored, at least partially, in old mice.

Melatonin as a Dietary Supplement

The potential utility of dietary supplementation may prevent some of the oxidative and inflammatory changes that ocurs in the brain with age. Dietary supplementation cannot merely arrest but indeed reverse some age-related increases in markers of oxidative and inflammatory events occurring with the cortex.[75] The cerebral cortex of 27-month-old male

B6C3F1 mice had elevated levels of nitric oxide synthase-1 (nNOS) and peptide nitrotyrosine relative to cortices of younger animals. When old mice received basal diet together with 300 mg/L acetyl L-carnitine in the drinking water for 8 weeks, these levels were fully restored to those found in younger animals. A partial restoration was found when old animals received basal diet supplemented with 200 ppm melatonin in the diet. Behavioral indices indicative of exploratory behavior were also depressed in aged animals. Dietary supplementation with melatonin or acetyl L-carnitine partially reversed these changes.

Dietary supplementation with melatonin has been shown to result in a significant rise in levels of endogenous melatonin in the serum and other tissue samples tested, together with a reduction in the levels of amyloid beta-peptides in the murine cerebral cortex.[76] Melatonin supplementation did not significantly change cerebral cortical levels of NOS or synaptic proteins such as synaptophysin and SNAP-25. Increased brain melatonin concentrations, however, led to a significant reduction in levels of both short and long forms of toxic cortical amyloid beta, which are involved in amyloid depositions and plaque formation in Alzheimer's diseases.[76] Thus, melatonin supplementation may retard neurodegenerative changes associated with brain aging.

Melatonin supplementation also affected the ontogeny of immunity in the Large White turkey poultry. Melatonin (50 microg/mL) was administered via the drinking water from hatch to 28 days of age. The cutaneous basophil hypersensitivity (CBH) reaction to phytohemagglutinin (PHA-P) and primary antibody responses to Chukar red blood cells (CRBC) were determined at 5 intervals: 0, 1, 7, 14, and 21 days post-hatching. The bursal weight, but not thymus or spleen, and the body weight were elevated in melatonin-treated poultry as compared to controls. These data suggest that post-hatching melatonin supplementation is beneficial to neonatal immune parameters and growth responses of Large White turkey poultry.[77]

Dietary melatonin has been reported to influence the age-related changes in murine CNS mRNA gene expression. Brain cellular functions decline with normal aging, accompanied by a changing profile of gene expression. Melatonin deprival in young rats induces alterations in cerebral arteriolar wall similar to those observed during aging — atrophy and a decrease in distensibility.[15] Melatonin (200 ppm) included in the diet of aged mice for 8 weeks elevated serum and cortical melatonin and reversed 13 of the 25 genes altered with age. However, in no case did melatonin potentiated age-related changes in gene expression.[13]

Moderate-dosage nocturnal melatonin supplementation suppressed nocturnal hypothalamic pro-opiomelanocortin (POMC) gene expression in both middle-aged males and females, which suggests that melatonin supplementation during aging decreases the opiomelanocortinergic activity of the forebrain. These POMC responses were apparently not dependent on gonadal steroid responses and did not become refractory to melatonin treatment maintained until old age.[78] Both plasma melatonin levels and hypothalamic arcuate nucleus pro-opiomelanocortin (POMC), a biosynthetic precursor to the endogenous opioid ss-endorphin and other opiomelanocortins mRNA content, decrease with aging. Daily nocturnal melatonin treatment (50 μg/kg bw/night, in the night-time drinking water) for 7 months, starting at 13 months of age, did not significantly alter female arcuate nucleus POMC mRNA content determined at the end of the light period (i.e. before nightly melatonin administration), but suppressed by 24% at the end of the dark period (i.e. following melatonin administration). Likewise, nocturnal administration of 50 or 500 μg/kg bw/night to male rats for 7 months suppressed by 31% or 28%, respectively at the middle of the dark period at 20 months of age. Finally, 10 weeks of melatonin administration of 30 μg/kg bw/day suppressed POMC mRNA content by 31% in middle-aged male rats killed at the end of the dark period.[78]

Summary and Commentary

Research breakthroughs over the past decade have revealed some startling properties of melatonin. It is a ubiquitously acting direct free radical scavenger and indirect antioxidant. Importantly, numerous studies have shown melatonin to be safe over a very wide range of doses. Furthermore, a large number of reports on both animals and humans indicate its utility in certain disease states. Since melatonin may act as a modulator of intracellular signal transduction to enhance or suppress the responses of many different cells to other incoming signals, it is a drug hormone with multidimensional benefits and applications.

Although the longevity extending effects of melatonin have been demonstrated for some animals, its use for human being should be more thoroughly investigated at the clinical level in terms of specific disease processes. Furthermore, the mice experiments have certain flaws, the major one being that the mice strains used could not synthesize melatonin in the first place.[79]

The production of melatonin declines with increasing age, and circulating melatonin levels are affected by certain pharmacological or physiological manipulations. Animal and cell culture experiments suggest that melatonin may have beneficial effects on certain aspects of aging and age-associated diseases. Of particular interest in this respect are reports on the influence of melatonin on the brain and the immune system. However, more research data is urgently needed in order to clearly define the possible sites and mechanisms of actions. Clinical studies need to be performed in order to identify possible side effects of long-term melatonin treatment, especially in elderly and diseased subjects. Serious concerns are raised about the use of uncontrolled, impure, or partially degraded melatonin preparations.

The studies done to date are either laboratory studies or have been done on small numbers of patients. Hormones are generally required in small doses and the doses of melatonin recommended in some of the studies are generally much greater than physiologically indicated. Not enough is known about the effects in "normal" individuals. Once melatonin is on prescription, it will be possible to document side-effects more accurately. Some studies have been performed using very large doses but adverse effects by such doses can not be safely ruled out. There is an urgent need to find the most efficacious physiological dose.

Melatonin as sold at the present time is not a pure pineal extract and is therefore regarded by some workers to be impure. One format contains melatonin, herbs and other organic and inorganic constituents. Anti-aging claims will require detailed investigation and are difficult to measure if the individual memory recall and physical fitness are taken into consideration. The same applies to age-dependent decline in sexual potency. Whether melatonin administration or suppression of melatonin secretion can in any way improve this response has still to be shown.

The exact role of the pineal in normal daily physiology must be more clearly defined. An alteration in melatonin rhythm leading to altered sleep patterns requires further research and recommendations are needed as to how best to manipulate melatonin secretion to affect these rhythms and benefit depressed patients. The studies done to date indicate the vast number of systems affected by melatonin or leading to changes in melatonin levels in the plasma. The significance of these changes is not known and must be studied to test the claims made for melatonin as the wonder drug.

Acknowledgments

I expresses my gratitude to Drs. Russel J. Reiter, Suresh Rattan, Ruediger Hardeland, Namik Delibas, Thomas Erren for providing several reprints, books and other relevant material, as well as constant help, during the preparation of this article.

References

1. Hardeland R, Poeggeler B (2003) Non-vertebrate melatonin. *J Pineal Res* **34**: 233–241.
2. Charman WN (2003) Age, lens transmittance, and the possible effects of light on melatonin suppression. *Ophthalmic Physiol Opt* **23**: 181–187.
3. Yoon IY, Kripke DF, Elliott JA, Youngstedt SD, Rex KM, Hauger RL (2003) Age-related changes of circadian rhythms and sleep-wake cycles. *J Am Geriatr Soc* **51**: 1085–1091.
4. Carrillo–Vico A, Calvo JR, Abreu P, Lardone PJ, Garcia–Maurino S, Reiter RJ, Guerrero JM (2004) Evidence of melatonin synthesis by human lymphocytes and its physiological significance: possible role as intracrine, autocrine, and/or paracrine substance. *FASEB* **18**: 537–539.
5. Ardura J, Gutierrez R, Andres J, Agapito T (2003) Emergence and evolution of the circadian rhythm of melatonin in children. *Horm Res* **59**: 66–72.
6. Pierpaoli W, Regelson W (1995) *The Melatonin Miracle: Nature's Age-Reversing, Disease-Fighting, Sex-Enhancing Hormone*, Simon & Schuster, New York.
7. Armstrong SM, Redman JR (1991) Melatonin: a chronobiotic with anti-aging properties? *Med Hypotheses* **34**: 300–309.
8. Zhao ZY, Xie Y, Fu YR, Bogdan A, Touitou Y (2002) Aging and the circadian rhythm of melatonin: a cross-sectional study of Chinese subjects 30–110 yr of age. *Chronobiol Int* **19**: 1171–1182.
9. Zhou JN, Liu RY, Van Heerikhuize J, Hofman MA, Swaab DF (2003) Alterations in the circadian rhythm of salivary melatonin begin during middle-age. *J Pineal Res* **34**: 11–16.
10. Oxenkrug G, Requintina P, Bachurin S (2001) Antioxidant and antiaging activity of N-acetylserotonin and melatonin in the *in vivo* models. *Ann N Y Acad Sci* **939**: 190–199.
11. Qian SZ, Cheng Xu Y, Zhang J (2000) Hormonal deficiency in elderly males. **23** (Suppl 2): 1–3.
12. Clapp–Lilly KL, Smith MA, Perry G, Harris PL, Zhu X, Drew KL, Duffy LK (2002) Melatonin exhibits antioxidant properties in a mouse brain slice model of excitotoxicity. *Int J Circumpolar Health* **61**: 32–40.
13. Bondy SC, Yang YE, Walsh TJ, Gie YW, Lahiri DK (2002) Dietary modulation of age-related changes in cerebral pro-oxidant status. *Neurochem Int* **40**: 123–130.

14. Sharman EH, Sharman KG, Ge YW, Lahiri DK, Bondy SC (2004) Age-related changes in murine CNS mRNA gene expression are modulated by dietary melatonin. *J Pineal Res* **36**: 165–170.
15. Dupuis F, Regrigny O, Atkinson J, Liminana P, Delagrange P, Scalbert E, Chillon JM (2004) Impact of treatment with melatonin on cerebral circulation in old rats. *Br J Pharmacol* **141**: 399–406.
16. Tozawa T, Mishima K, Satoh K, Echizenya M, Shimizu T, Hishikawa Y (2003) Stability of sleep timing against the melatonin secretion rhythm with advancing age: clinical implications. *J Clin Endocrinol Metab* **88**: 4689–4695.
17. Reiter RJ, Tan DX, Mayo JC, Sainz RM, Lopez–Burillo S (2002) Melatonin, longevity and health in the aged: an assessment. *Free Radic Res* **36**: 1323–1329.
18. Magri F, Sarra S, Cinchetti W, Guazzoni V, Fioravanti M, Cravello L, Ferrari E (2004) Qualitative and quantitative changes of melatonin levels in physiological and pathological aging in centenarians. *J Pineal Res* **36**: 256–261.
19. McCann SM (1997) The nitric oxide hypothesis of brain aging. *Exp Gerontol* **32**: 431–340.
20. Reiter RJ Tan DX, Acuna–Castroviejo D, Burkhardt S, Karbownik M (2000) Melatonin: mechanisms and actions as an antioxidant, *Curr Topics Biophy* **24**: 171–183.
21. Reiter RJ (1995) The pineal gland and melatonin in relation to aging: a summary of the theories and of the data. *Exp Gerontol* **30**: 199–212.
22. Poeggeler B, Reiter RJ, Tan DX, Chen LD, Manchester LC (1993) Melatonin, hydroxyl radical-mediated oxidative damage and aging — a hypothesis. *J Pineal Res* **14**: 151–168.
23. Reiter RJ (1995) The pineal gland and melatonin in relation to aging: a summary of the theories and of the data. *Exp Gerontol* **30**: 199–212.
24. Delibas N, Tuzmen N, Yonden Z, Altuntas I (2002) Effect of functional pinealectomy on hippocampal lipid peroxidation, antioxidant enzymes and N-methyl-D-aspartate receptor subunits 2A and 2B in young and old rats. *Neuroendocrinol Lett* **23**: 367–372.
25. Reiter RJ, Tan D, Cabrera J, D'Arpa D (1999) Melatonin and Tryptophan derivatives as free radical scavengers and antioxidants. In: Huether G, Kochen W, Simat TJ, Steinhart H (eds.) *Tryptophan, Serotonin and Melatonin: Basic Aspects and Applications*, pp. 379–387. Kluwer Academic/Plenum Publishers, New York.
26. Barlow–Walden LR, Reiter RJ, RBE M *et al.* (1995) Melatonin stimulates brain glutathione peroxidase activity. *Neurochem Int* **26**: 497–502.
27. Zhang QZ, Zhang JT (1999) Inhibitory effects of melatonin on free intracellular calcium in mouse brain cells. *Zhongguo Yao Li Xue Bao* **20**: 206–210.
28. Lopez–Burillo S, Tan DX, Mayo JC, Sainz RM, Manchester LC, Reiter RJ (2003) Melatonin, Xanthurenic acid, resveratrol, EGCG, vitamin C and alpha-lipoic acid differentially reduce oxidative DNA damage induced by Fenton reagents: a study of their individual and synergistic actions. *J Pineal Res* **34**: 269–277.
29. Okatani Y, Wakatsuki A, Reiter RJ, Miyahara Y (2003) Acutely administered melatonin restores hepatic mitochondrial physiology in old mice. *Int J Biochem Cell Biol* **35**: 367–375.
30. Koc M, Taysi S, Buyukokuroglu ME, Bakan N (2003) Melatonin protects rat liver against irradiation-induced oxidative injury. *J Radiat Res* **44**: 211–217.

31. Manda K, Bhatia AL (2003) Prophylactic action of melatonin against cyclophosphamide-induced oxidative stress in mice. *Cell Biol Toxicol* **19**: 367–372.
32. Reiter RJ (1995) Oxidative processes and antioxidative defence mechanisms in the aging brain. *FASEB* **9**: 526–533.
33. Manda K, Bhatia AL (2003a) Melatonin-induced reduction in age-related accumulation of oxidative damage in mice. *Biogerontology* **4**: 133–139.
34. Manda K, Bhatia AL (2003b) Melatonin's anti-aging role: A study on LPO in mice tissues. *Ind. J Gerontol* **6**: 211–217.
35. Vijayalaxmi, Meltz ML, Reiter RJ, Herman TS, Kumar KS (1999) Melatonin and protection from whole-body irradiation: survival studies in mice. *Mutat Res* **425**: 21–27.
36. Vijayalaxmi, Reiter RJ, Tan DX, Herman TS, Thomas CR Jr. (2004) Melatonin as a radioprotective agent: a review. *Int J Radiat Oncol Biol Phys* **59**: 639–653.
37. Bhatia AL, Manda, Kailash (2004) Study on pretreatment of melatonin against radiation-induced oxidative stress in mice. *Environ Toxicol Pharmacol* **17**(4), in press.
38. Anisimov VN, Alimova IN, Baturin DA, Popovich IG, Zabezhinski MA, Rosenfeld SV, Manton KG, Semenchenko AV, Yashin AI (2003) Dose-dependent effect of melatonin on lifespan and spontaneous tumor incidence in female SHR mice. *Exp Gerontol* **38**: 449–461.
39. Barchas J, DaCosta F, Spector S (1967) Acute pharmacology of melatonin. *Nature* **214**: 919–920.
40. Blask DE, Pelletier DB, Hill SM, Lemus–Wilson A, Grosso DS, Wilson ST, Wise ME (1991) Pineal melatonin inhibition of tumor promotion in the N-nitroso-N-methylurea model of mammary carcinogenesis: potential involvement of antiestrogenic mechanisms *in vivo*. *J Cancer Res Clin Oncol* **17**: 526–532.
41. Blask DE (1993) Melatonin in oncology. In: Yu H-S, Reiter RJ (eds.) *Melatonin, Biosynthesis, Physiological Effects and Clinical Applications*, pp. 448–475. CRC Press, Boca Raton.
42. Wilson ST, Blask DE, Lemus–Wilson AM (1992) Melatonin augments the sensitivity of MCF-7 human breast cancer cells to tamoxifen *in vitro*. *J Clinical Endocrin Metabol* **75**: 669–670.
43. Blask DE, Leonard A, Sauer RT, Dauchy EW, Holowachuk MS, Ruhoff, Kopff HS (1999) Melatonin inhibition of cancer growth *in vivo* involves suppression of tumor fatty acid metabolism via melatonin receptor-mediated signal transduction events1. *Cancer Research* **59**: 4693–4701.
44. Tan DX, Poeggeler B, Reiter RJ, Chen LD, Chen S, Manchester LC, Barlow-Walden LR (1993) The pineal hormone melatonin inhibits DNA-adduct formation induced by the chemical carcinogen safrole *in vivo*. *Cancer Lett* **70**: 65–71.
45. Lissoni P, Barni S, Tancini G, Mainini E, Piglia F, Maestroni GJ, Lewinski A (1995) Immunoendocrine therapy with low-dose subcutaneous interleukin-2 plus melatonin of locally advanced or metastatic endocrine tumours. *Oncology* **52**: 163–166.
46. Lissoni P, Barni S, Cazzaniga M, Ardizzoia A, Rovelli F, Brivio F, Tancini G (1994) Efficacy of the concomitant administration of the pineal hormone melatonin in cancer immunotherapy with low dose IL-2 in patients with advanced solid tumours who had progressed on IL-2 alone. *Oncology* **51**: 344–347.
47. Erren TC, Reiter RJ, Piekarski C (2003) Light, timing of biological rhythms, and chronodisruption in man. *Naturwissenschaften* **90**: 485–494.

48. Bartsch H, Bartsch C, Simon WE, Flehmig B, Ebels I, Lippert TH (1992) Antitumor activity of the pineal gland: effect of unidentified substances versus the effect of melatonin. *Oncology* **49**: 27–30.
49. Blask DE, Sauer LA, Dauchy RT, Holowachuk EW, Ruhoff MS, Kopff HS (1999) Melatonin inhibition of cancer growth *in vivo* involves suppression of tumor fatty acid metabolism via melatonin receptor-mediated signal transduction events. *Cancer Res* **59**: 4693–4701.
50. Blask DE, Sauer LA, Dauchy RT (2002) Melatonin as a chronobiotic/anticancer agent; cellular, biocheming and molecular mechanisms of action and their implications for circadian-based cancer therapy. *Curr Topics Med Chem* **2**: 113–132.
51. Hill S, Blask DE (1988) Effects of the pineal hormone melatonin on the proliferation and morphological characteristics of human breast cancer cells (MCF-7) in culture. *Cancer Res* **48**: 6121–6126.
52. Cos S, Fernandez R, Guezmes A, Sanchez-Barcelo EJ (1998) Influence of melatonin on invasive and metastatic properties of MCF-7 human breast cancer cells. *Cancer Res* **58**: 4383–4390.
53. Cos S, Mediavilla MD, Fernandez R, Gonzalez-Lamuno D, Sanchez-Barcelo EJ (2002) Does melatonin induce apoptosis in MCF-7 human breast cancer cells *in vitro*? *J Pineal Res* **32**: 90–94.
54. EI-Missiry MA, EI-Aziz Abd AF (2000) Influence of melatonin on proliferation and antioxidant system in Ehrlich ascites carcinoma cells. *Cancer Lett* **151**: 119–125.
55. Karasek M, Winczyk K, Kunert–Radk J, Wiesenberg I, Pawlikowski M (1998) Antiproliferative effects of melatonin and CGP52608 on the murine colon 38 adenocarcinoma *in vitro* and *in vivo*. *Neuroendocrinol Lett* **19**: 71–78.
56. Melen–Mucha G, Winczyk K, Pawlikowski M (1998) Somatostatin analogue oxtreotide and melatonin inhibit bromodeoxyuridine incorporation into cell nuclei and enhance apoptosis in the transplantable murine colon 38 cancer. *Anticancer Res* **18**: 3615–3620.
57. Hughes JT (1994) Teratogenesis. *Carcinogenesis Mutagenesis* **14**: 213–217.
58. Spiegel D, Giese–Davis J (2003) Depression and cancer: mechanisms and disease progression. *Biol Psychiatry* **54**: 269–282.
59. Crasson M, Kjiri S, Colin A, Kjiri K, L'Hermite–Baleriaux M, Ansseau M, Legros JJ (2004) Serum melatonin and urinary 6-sulfatoxymelatonin in major depression. *Psychoneuroendocrinology* **29**: 1–12.
60. Varma A, Kaul RK, Varma P, Kalra V, Malhotra V (2002) The effect of antidepressants on serum melatonin levels in endogenous depression. *J Assoc Physicians India* **50**: 1262–1265.
61. Paparrigopoulos T (2002) Melatonin response to atenolol administration in depression: indication of beta-adrenoceptor dysfunction in a subtype of depression. *Acta Psychiatr Scand* **106**: 440–445.
62. Tuunainen A, Kripke DF, Elliott JA, Assmus JD, Rex KM, Klauber MR, Langer RD (2002) Depression and endogenous melatonin in postmenopausal women. *J Affect Disord* **69**: 149–158.
63. Rosenthal NE (1998) *Winter Blues: Seasonal Affective Disorder — What It Is and How to Overcome It*. Guilford Press, New York.
64. Rohr UD, Herold J (2002) Melatonin deficiencies in women. *Maturitas* **41** (Suppl 1): S85–S104.

65. Sandyk R, Anastasiadis PG, Anninos PA, Tsagas N (1992) Is postmenopausal osteoporosis related to pineal gland function? *Internatl J Neuroscie* **62**: 215–225.
66. Drake MJ, Mills IW, Noble JG (2004) Melatonin pharmacotherapy for nocturia in men with benign prostatic enlargement. *J Urol* **171**: 1199–1202.
67. Puchalski SS, Green JN, Rasmussen DD (2003) Melatonin effects on metabolism independent of gonad function. *Endocrine* **21**: 169–173.
68. Joo JY, Uz T, Manev H (1998) Opposite effects of pinealectomy and melatonin administration on brain damage following cerebral focal ischemia in rat. *Restor Neurol Neurosci* **13**: 185–191.
69. Sainz RM, Mayo JC, Reiter RJ, Tan DX, Rodriguez C (2003) Apoptosis in primary lymphoid organs with aging. *Microsc Res Tech* **62**: 524–539.
70. Vermeulen M, Palermo M, Giordano M (1993) Neonatal pinealectomy impairs murine antibody-dependent cellular cytotoxicity. *J Neuroimmunol* **43**: 97–101.
71. Maestroni GJ (1993) The immunoendocrine role of melatonin. *J Pineal Res* **14**: 1–10.
72. Cavallo A (1993) Melatonin and human puberty: current perspectives. *J Pineal Res* **15**: 115–121.
73. El-Sokkary GH, Reiter RJ, Abdel-Ghaffar SKH (2003) Melatonin supplementation restores cellular proliferation and DNA synthesis in the splenic and thymic lymphocytes of old rats. *Neuroendocrinol Lett* **24**: 215–223.
74. Tian YM, Zhang GY, Dai YR (2003) Melatonin rejuvenates degenerated thymus and redresses peripheral immune functions in aged mice. *Immunol Lett* **88**: 101–104.
75. Sharman EH, Vaziri ND, Ni Z, Sharman KG, Bondy SC (2002) Reversal of biochemical and behavioral parameters of brain aging by melatonin and acetyl L-carnitine. *Brain Res* **13**(957): 223–230.
76. Lahiri DK, Chen D, Ge YW, Bondy SC, Sharman EH (2004) Dietary supplementation with melatonin reduces levels of amyloid beta-peptides in the murine cerebral cortex. *J Pineal Res* **36**: 224–231.
77. Moore CB, Siopes TD (2002) Effect of melatonin supplementation on the ontogeny of immunity in the Large White turkey poultry. *Poult Sci* **81**: 1898–1903.
78. Rasmussen DD, Marck BT, Boldt BM, Yellon SM, Matsumoto AM (2003) Suppression of hypothalamic pro-opiomelanocortin (POMC) gene expression by daily melatonin supplementation in aging rats. *J Pineal Res* **34**: 127–133.
79. Ebihara S, Marks T, Hudson DJ, Menaker M (1986) Genetic control of melatonin synthesis in the pineal gland of the mouse. *Science* **231**: 491–493.

9

Zinc and Other Micronutrients for Healthy Aging

Eugenio Mocchegiani and Marco Malavolta
Immunology Ctr. (Section Nutrition, Immunity and Aging)
Res. Dept., INRCA, Ancona, Italy

Efstathios S. Gonos
National Hellenic Research Foundation
Inst. Biol. Res. & Biotechnol.
Athens, Greece

INTRODUCTION

Aging is an inevitable biological process that is accompanied with gradual and spontaneous biochemical and physiological changes including increased susceptibility to diseases, adverse environmental conditions and loss of mobility and agility. Alterations in the neuroendocrine-immune interactions as well as in the antioxidant capacity also play a fundamental role in aging.[1] The inability of an organism in remodelling these changes may lead in the appearance of age-related pathologies, such as cancer, infections, neurodegenerative and cardiovascular diseases, diabetes, osteoporosis and osteoarthritis. As a result, the "remodelling theory of aging" has been proposed.[2] Various nutritional factors are directly linked with these phenomena as for instance they are involved in restoring neuroendocrine-immune network, the metabolic harmony and the capacity of an organism to respond to oxidative stress.[3]

Corresponding author: Dr. Eugenio Mocchegiani PhD, Immunology Ctr. (Section Nutrition, Immunity, Aging), Res. Dept. INRCA, Via Birarelli 8, 60121, Ancona, Italy. Tel.: +39-071-8004216; fax: +39-071-206791; e-mail: e.mocchegiani@inrca.it.

Approximately, 40 micronutrients (vitamins, essential minerals and other compounds required in small amount for normal metabolism) have been reported as essential components in the diet. The dietary intake of essential macro and micronutrients is usually inadequate in the elderly due mainly to loss of appetite, to a lack of teeth as well as to decreased requirement of energy.[4] Many authors have reported that the deficiency of macro and micronutrients in aging is strictly related to global impairments of immune functions, metabolic harmony and antioxidant defence by external noxae with subsequent appearance of age-related diseases.[5] In contrast, recent longitudinal studies in dietary daily intake in human nonagenarian/centenarians (i.e. individuals who represent the best examples of successful aging) have shown that an adequate consumption of micro and macronutrients as well as a satisfactory content of some trace elements in the cells lead to good performances in several immune functions, to a metabolic compensation and to a preservation of antioxidant activity.[6] Therefore, nutritional factors may play a pivotal role for healthy aging and longevity. We herein report on the role of zinc, selenium and niacin in aging and in "successful aging". We have focused on these specific nutritional factors for three specific reasons: The relevance of zinc in immune functions, metabolic harmony and antioxidant activity;[7] the importance of selenium in the release of zinc by zinc-bound metallothioneins, via reduction of glutathione peroxidase (GPx);[8] and the relevance of niacin in conferring NAD+ substrate for the activity of DNA repair enzyme, PARP-1,[9] which in turn it encodes two zinc finger motifs.[10] Additionally, zinc is also involved in cellular homeostasis, via the regulation of metallothionein (MT).[11]

ZINC

Zinc biology

Zinc is one of the most important trace elements in the body, although its presence in nature does not exceed 0.02%.[7] The major characteristics of zinc include a highly concentrated charge, a small radius (0.65 Å), no variable valence (low risk of free radical production), ready passage from one symmetry in its surroundings to another without exchange, rapid exchange of ligands (on and off reactions), and binding mostly to 5- and

N-donors in biological systems. These properties enable zinc to play a major biological role as a catalyst.[12] In particular, zinc is required for the biological function of more than 300 enzymes. Removal of the catalytic zinc results in an active apoenzyme that usually retains the native tertiary structure. The catalytic zinc is bound by three protein ligands and a water molecule signifying an open coordination site, which is considered essential for the nucleophile function of zinc. Because of its flexible coordinator geometry, zinc ions act as template to bring together the substrate and the nucleophile.[12]

Zinc also regulates the balance between gene expression of metalloproteinases (MMPs) and the tissue inhibitors of matrix metalloproteinases (TIMPs).[13] Expression of MMPs genes are under the control of the TIMPs gene products α-2M and 13-amyloid precursor protein, which encode zinc-finger motifs.[13] As a result, a balance in the expression of the metalloproteinases (activators and inhibitors) is necessary for an optimal function of many biological systems. Examples of altering the balance between MMPs and TIMPs or α-2M have been recorded in cancer, in infections and in aging.[13] Zinc also regulates G0/G1 phase of cell cycle by means of cyclins/CDK complexes in a dose dependent manner. Specifically, high doses of zinc (900 μM) result in cell cycle arrest,[14] whereas low doses of zinc (150 μM) inhibit apoptosis.[15,16] Zinc is present in "zinc finger domains" of many proteins, peptides, enzymes, hormones, transcriptional factors and cytokines, who act in maintaining body homeostasis.[17] Zinc also regulates mRNA stability[18] and extracellular matrix.[12] Moreover, zinc binds enzymes, proteins and peptides with different binding affinity (kd) ranging from 10^{-2} to 10^{-14} mol/L.[16] These compounds display low biological activity when zinc-binding doesn't occur.[7] Finally, zinc plays a critical role in structure, function, stabilization and fluidity of biomembrane due to its binding to sulphydryl groups forming mercaptides.[12]

Zinc is also required to maintain enzymatic activity of inducible nitric-oxide synthase (iNOS).[19] Zinc is bound to two cysteine residues, which are part of the structures of the heme domain of iNOS.[20] As 1) NO is involved in metallothionein (MT) mRNA expression in protecting against oxidative stress;[21] and 2) NO prevents cell death activating PARP, the structural task of zinc in NO production is of a great importance. In addition, NO is involved in zinc release from MT to antioxidant enzymes.[22]

Taking into account the previously reported roles, zinc is an essential trace element for many biological functions. In this context, zinc-bound MT plays a peculiar role because these proteins are deputed in zinc transport within the cell and in the release of zinc. The role of MT becomes pivotal in aging because low zinc ion bioavailability is a constant event during aging and it associates with high MT expression, impaired immune functions and decreased antioxidant activity,[11,16,23,24] phenomena that relate to various age-related pathologies.

Zinc, metallothioneins and aging

Metallothioneins (MT), are a group of low-molecular-weight metal-binding proteins who have high affinity for zinc (kd = 1.4×10^{-13} M).[23] MT contain 20 cysteines, all in reduced form, and bind seven zinc atoms through mercaptide bonds that have the spectroscopy characteristics of metal thiolate clusters.[24] The zinc/cysteine clusters are of two different types: In the beta-domain cluster, three bridging and six terminal cysteine thiolates provide a coordination environment that is identical for each of the three zinc atoms. In the alpha-domain clusters, there are two different zinc sites: two of them have one terminal ligand and three bridging ligands respectively, while the other two have two terminal and two bridging ligands.[24]

Following these biochemical characteristics, MT distribute intracellular zinc as zinc undergoes rapid inter- and intracluster exchange.[23] Moreover, MT act as antioxidants since zinc-sulfur cluster is sensitive to changes of cellular redox state and oxidizing sites in MT (reduced thiol groups) induce the transfer of zinc from its MT binding sites to those of lower affinity in other proteins.[23] This transfer confers biological activity to antioxidant metalloenzymes. Therefore, the redox properties of MT and their effect on zinc in the clusters are crucial for the protective role of MT in transient stress, as it may occur in young adult age.[25] In contrast, this role is harmful in aging because the stress-like condition is permanent. Such an assumption is supported by the constant oxidative damage during aging,[26] despite MT increase in these conditions.[11] Therefore, MT by sequestering zinc and it cannot be transferred to other proteins during aging. Indeed, in aging, MT bind zinc preferentially over copper, and low zinc bioavailability, but not copper[16] is present in these conditions.

Abnormal high MT and low zinc ion bioavailability are observed in syndromes of accelerated aging such as the Down's syndrome. In contrast, low MT and satisfactory zinc ion bioavailability occur in successful aging.[27] Therefore, the different role of zinc-bound MT in aging may be crucial for both the immune and the antioxidant efficiency. This different role of MT in aging may be also relevant for metabolic harmony because zinc also confers biological activity to some hormones involved in these phenomena.[1]

Immune efficiency

MT induce the secretion from macrophages of some cytokines (IL-1, IL-6, IFN-α, TNF-α), which, in turn, provoke new synthesis of MT in the liver but, at the same time, an alteration in the zinc status.[28] These findings suggest the existence of interplay between zinc, MT and the immune system. IL-1 affects MT mRNA in thymic epithelial cells (TECs) by means of PKC, which is, in turn, zinc-dependent[29] and participates in metal-induced MT mRNA.[30] Moreover, MT are donors of zinc for thymulin reactivation in TECs.[29] MT protect during zinc deficiency and zinc toxicity or act as a type of zinc reservoir or they sequester excess of zinc.[31] However, the existence of high MT and low zinc ion bioavailability in atrophic thymus[16] suggests a different role played by MT in immunosenescence: turning from protective to harmful. The presence of low MT in the liver and in the thymus from very old mice coupled with enhanced liver NK cell activity supports this assumption.[27] In support to these findings, liver NKT cells, which are the first sentinel for the host defense, preserve their cytotoxicity in very old mice as compared to old mice. This preservation occurs in TCRγδ than αβ cells and it is strictly related to good zinc ion bioavailability and low MT.[32]

The high MT in lymphocytes from old and Down's syndrome subjects and the low MT in lymphocytes from centenarians[27] further support to the different role of MT in aging by provoking an impaired immune response at the thymus.[33] In addition, atrophic thymus is present in transgenic mice over-expressing MT.[33] Consistent with these findings, MT are not donors of zinc in aging but rather they sequester zinc. On the other hand, high MT induce down-regulation of many other biological functions related to zinc, such as metabolism, gene expression and signal transduction.[23] However, the inability of MT to release zinc during aging is still unresolved. In this context, a correct balance between iNOS and cNOS provokes the zinc release by MT.[22] During aging iNOS increases as compared to cNOS inducing an unbalance of NO synthases with limited zinc release by MT.[11] "*In vitro*" zinc (10^{-6} M) in old lymphocytes corrects the unbalance of NO synthases (Mocchegiani E, unpublished results). This occurs because NO synthases are codified by zinc finger motifs.[10] Therefore, zinc-bound MT may be considered a biological and genetic marker of immunosenescence.

Metabolic harmony and antioxidant activity

Many data indicate that zinc is involved in maintaining metabolism harmony and antioxidant activity.[34] Additionally, some diseases like diabetes and cardiovascular diseases display low zinc ion bioavailability and increased MT. High MT in diabetes are protective against ROS, avoiding the development of congestive heart failure.[35] However, such a situation is also harmful since MT sequester zinc ions. Indeed, diabetes type II displays low zinc ion bioavailability and impaired immune efficiency correlated to an incorrect metabolic compensation.[36] Moreover, the activity of SOD decreases in diabetes with a risk of myocardial infarction.[35] Zinc deficiency is also observed in diabetes type I (juvenile diabetes insulin-dependent).[37] These findings suggest that zinc, via MT homeostasis, plays a key role in diabetes for immune efficiency and antioxidant activity. On the other hand, old atherosclerosis patients display high MT, low zinc bioavailability and impaired immune response.[38] Zinc also affects thyroid hormones turnover. In hypothyroidism (observed in Down's syndrome), the low thyroid hormones production (T3 and T4) correlate to high MT, low zinc ion bioavailability and impaired immune efficiency.[27] Conversely, hyperthyroid patients display abnormal increments of zinc and immune efficiency.[39] In support to these findings, human centenarians and very old mice display low MT, satisfactory thyroid metabolism and antioxidant activity as well as good zinc ion bioavailability.[27]

An intriguing role of zinc in metabolic harmony may come from mitochondria functions. Correct functions of mitochondria, in particular the oxidative phophorylation (OXPHOS) enzymes, are related to successful aging.[40] Although, no direct evidences exist on the role of zinc in

OXPHOS enzymes and mtDNA, indirect evidences in mice lacking mitochondrial SOD[41] suggest a possible role of zinc in mitochondria functions. On the other hand, zinc preserves nuclear DNA integrity.[17] Therefore, zinc might also preserve mtDNA integrity. Based on these studies it is tempting to assign a pivotal role to MT and zinc to reach healthy aging and longevity.

Zinc supplementation

Zinc supplementation in old mice restores cell-mediated immune deficiency and increases the rate of survival. Physiological zinc prevented thymocytes apoptosis in old mice by acting on endonuclease enzymes.[42] In human prostate carcinoma, zinc (at physiological dose of 100 μg/dL) inhibits tumor cell growth by inducing a G2/M arrest and by enhancing p21 expression.[43] Therefore, increased zinc ion bioavailability through zinc supplementation might regenerate the role of MT protection and, additionally, it might also cause apoptosis in cancer cells.

Zinc supplementation in the elderly yielded contradictory data regarding its beneficial effect. This may be due to different doses of zinc used and to variations in the period of treatment. Delivery of physiological dose of zinc for a long period or high doses of zinc for short periods may induce a zinc accumulation in various organs and tissues with subsequent toxic effects on the immune functions.[44] In our experience, zinc treatment at the dose of 15 mg Zn^{2+}/day (RDA) for one month in Down's syndrome subjects, in elderly and in old infected patients restores thymic functions, peripheral immune efficiency[45] and DNA-repair.[46] At the clinical level, significant reductions of infection relapses occur in Down's syndrome,[47] in elderly and in old infected patients.[45] Zinc supplementation is also useful in reducing the oxidative stress in patients with both diabetes type II[48] and type I.[49] In type I patients, zinc supplementation also inhibits NF-kB activation and decreases inducible NO synthase. As such, the generation of ROS decreases, thus zinc provides a protective effect on β cells against death.[49]

All these studies demonstrate the pivotal role of zinc in maintaining immune functions during aging and in the alleviation of age-related disease. However, one may ask the question, whether or not zinc supplementation may further increase MT, causing major harmful effects. Old mice treated with zinc exhibit no further increments of liver MT mRNA suggesting that the high MT in aging are completely saturated by pre-existing zinc ions.[27] In conclusion, the possible harmful effect of MT might be avoided by physiological zinc supplementation during aging.

Zinc interaction with other micronutrients and zinc toxicity

The beneficial effect of physiological zinc supplementation must be, however, related to the levels of other cations such as cadmium, lead, calcium, iron, manganese and copper. The beneficial effects of zinc on ameliorating toxicity of cadmium and lead, accentuation of zinc deficiency by administration of calcium and phytate, and production of hypocupremia by excessive zinc intake in humans and animals, are some examples of competition phenomena between these cations.[50] Such a competition occurs because these ions have similar valence shell electronic structure and, therefore could be antagonistic to each other. Copper, zinc and cadmium have similar orbitals, configurations and coordination numbers to interact to each other.[51] Such interactions occur between zinc and iron (Fe^{2+}) at the level of cysteine-histidine ligands for the formation of iron or zinc "fingers" proteins.[52] If iron is in excess, a preferential binding of iron than zinc to the metal free-protein occurs. Excess of zinc as well as zinc deficiency impairs DNA-protein interactions of zinc-fingers domains with their cognate DNA target sites. In these conditions the production of some transcriptional factors like SP1 or TFIIIA is impaired.[53] The same impairment of zinc fingers DNA domains occurs in excess or deficiency of copper.[53] This reinforces the notion of the relevance of interactions between zinc and copper as well as with other metals in immunity efficiency.[44] Thus a limited range of bioavailability exists for each metal. As such, immune responses are optimum. Indeed, the beneficial effect of zinc is strictly dependent by the dose and the length of treatment. Zinc accumulation or imbalanced zinc-to-copper ratio may occur despite low doses of zinc.[54] As such, harmful side effects in the cardiovascular system and in the brain may appear with increased low-density lipoprotein and cholesterol[55] and neural cell-death.[56] Therefore, caution in zinc supplementation is necessary for avoiding undesirable and harmful unexpected side effects. Zinc supplementation must not exceed 2–3 times the RDA/day, for short periods (1–2 months) and in periodical cycles. This treatment doesn't interfere in copper absorption.[54] Zinc picolinate form may be the best supplement.[57]

OTHER MICRONUTRIENTS

Niacin

Niacin is the generic description for nicotinic acid (pyridine-3-carboxilic acid, NA) and derivatives exhibiting the biological activity of nicotinamide (nicotinic acid amide, NAM). NAM and NA are the substrates for the synthesis of the nicotinamide nucleotide coenzymes nicotinamide adenine nucleotide (NAD^+) and the phosphorylated derivative $NADP^+$.[58] The ratio of NAD to NADP is a measure of niacin status in humans.[58] Niacin is therefore a precursor of NAD^+, which is the sole substrate for the nuclear enzyme poly(ADP-ribose) polymerase-1 (PARP-1). This enzyme is involved in base excision DNA-repair, cellular differentiation, gene expression, genomic stability and cell death. PARP-1 is activated by DNA strand breaks, and catalyses the formation of extensive linear and branched polymers of poly(ADP-ribose) attached to a protein acceptor using NAD^+ as a substrate.[9] As a consequence of PARP-1 activation, cellular NAD^+ is consumed. Overactivation of PARP-1 during a chronic stress-like condition, as it occurs during aging, leads to rapid NAD^+ and ATP depletion, and ultimately cell death.[59] In contrast, in transient stress-like conditions, as it happens in young-adult age, the consumption of NAD^+ is limited with subsequent base excision DNA-repair by PARP-1.[59] Following that, PARP-1 is also involved in maintaining genomic stability. Therefore, the intracellular NAD^+ pool and a sufficient content of niacin via the diet are important for PARP-1 function and for genomic stability.[60] Tryptophan is also a dietary precursor of NAD^+. Dietary tryptophan (especially in meats, cereals and legumes) is converted to quinolinate and via the kynurinic pathway it is further converted in nicotinic acid ribonucleotide.[9] In humans, the biosynthesis of niacin from tryptophan is an important route for meeting the body's niacin requirement.[61] The recommended dietary daily allowance for niacin is 6.6 mg per 1000 kcal. Studies in humans suggest that an average 60 mg of tryptophan is converted to 1 mg of niacin. Since most proteins contain 1% of tryptophan, a daily intake of 100 g protein is sufficient to maintain and adequate niacin status.[58]

In elderly people, for reasons reported earlier (poor appetite, lack of teeth, intestinal malabsorption), the daily intake of trypthophan or niacin is not sufficient.[58] As a consequence, the conversion to NAD^+ may be diminished or maintained, but is totally consumed by PARP-1 due to the presence of chronic stress-like condition by high levels of pro-inflammatory cytokines. Indeed, the production of NAD^+ decreases in old lymphocytes stimulated with mitogens[62] and the ratio of total PARP-1 to endogenous PARP-1 (that is an index of the status of DNA damage) is low in elderly as compared to higher values recorded in young subjects. This means that more PARP-1 is more available to contribute to DNA-repair in young than in old people.[45] Although no data is available for NAD^+ and the niacin status in centenarians, human nonagenarian/centenarians display an increased ratio of total PARP-1/endogenous PARP-1 in their lymphocytes, suggesting increased base excision DNA-repair by PARP-1 in comparison with normal elderly.[45] It has been reported that a niacin deficiency dramatically inhibits DNA repair in cell culture models after exposure to chemical agents[63] as well as in rats treated with propylnitrosourea (treatment that causes a spectrum of leukemias very similar to those induced in humans by alkylation-based chemotherapy).[64] Niacin supply in rats (0.03 g/kg for 1-month) induces a more capacity of poly(ADP-ribose) metabolism and a latency of carcinogenesis, mediated by p53, leading to an increased rate of survival in comparison to untreated rats.[65] p53 gene expression is also regulated by Sir2 protein (SIRTs in humans), which is in turn a NAD^+-dependent histone deacetylase that plays a critical role in transcriptional gene silencing, genome stability and longevity.[66] Recent finding reports that a SIRT3 gene (genotype G477T) is involved in human longevity.[67] Moreover, the role of p53 and niacin against cancer development is well-recognized.[9] All these findings clearly indicate the relevance of niacin, via NAD^+ production, against DNA damage by oxidative stress.

Taking into account the requirement of two zinc finger motifs for PARP-1 gene expression[10] and the zinc-dependence of p53,[68] the role of zinc and niacin in the effects of genotoxic agents or in chronic stress conditions like in aging, becomes relevant. In this context, the combination of niacin and zinc treatment is more efficient than niacin alone in inducing resistance against oxidative DNA damage in peripheral blood lymphocytes exposed to H_2O_2.[69] Therefore, although niacin is crucial for NAD^+ requirement, many molecular mechanisms related to niacin are zinc-dependent, involving PARP-1 and p53. Thus, the lack of free zinc ions due to high MT may also play a key role for the complete beneficial effects of niacin. In summary, the combination of zinc and niacin may have beneficial roles for healthy aging.

Selenium

Biological function

Selenium is an essential dietary element for the prevention of several diseases.[70] In food, selenium derives from vegetables and animal products and in particular from the consumption of seafood, liver, and cereals. However, in vegetables and cereals, the amount of selenium varies in soil in different countries and geographical regions.[71] Indeed, selenium deficiency and related diseases have been well-documented in geographic regions where the soil content is low, such as the Chinese province of Keshan.[72]

Mammals can use both inorganic and organic selenium as a nutrient. Most of the biological functions of selenium are attributed to selenoproteins, which contain selenocysteine residues responsible for their specific activity. The human selenoproteome consists of 25 selenoproteins, mostly involved in antioxidant defence systems.[73] Glutathione peroxidases (GPxs), a family of the selenoproteins, protect cells against oxidative damage by catalysing the reduction of hydrogen peroxide and other hydroperoxides by glutathione (GSH).[74] Five GPx isoforms exist in humans[74]: GPx1 is found in the cytosol of almost all cells and catalyses the reduction of free hydroperoxides; GPx2 is expressed in the gastrointestinal tract and has a substrate specificity similar to GPx1; GPx3 is an extracellular enzyme found in plasma and reduces membrane-bound phospholipid hydroperoxides, thus playing a key role in the protection of lipoprotein from oxidative damage; GPx4 is expressed in various tissues, especially in testis and it is believed to be mostly cytosolic and, in part, associated with mitochondria. It reduces phospholipid hydroperoxide and hydrogen peroxide using also thiols, such as 2-mercaptoethanol, cysteine and homocysteine, other than GSH as reductant agents;[75] and GPx5 is absent in human tissues,[76] whereas the isoform GPx6 could be expressed in embryonic tissues and olfactory epithelium.[73]

Selenium is also involved in the thioredoxin system, which is a major enzymatic system modulating ROS. This system is highly complementary to the GSH system in protecting against oxidative stress.[77] It comprises with thioredoxin, thioredoxin peroxidase and the selenoprotein, thioredoxin reductase (TR) and uses the reducing power of NADPH to act as a potent antioxidant system as well as a general disulfide redox system.[78] Three isoforms of TR (TR1, TR2 and TR3) exist in humans.[79] All these isoforms reduce thioredoxin, which is oxidized to reduce oxidative changes of proteins, including protein disulfides, methioninyl sulfoxides and cysteinyl sulfenic acids. Moreover, TR is critical for redox regulation of protein function and signal transduction[80] and is involved in restoring the reduced form of several antioxidant compounds, including ascorbic acid, selenium-containing substances, lipoic acid, and ubiquinone.[81]

Another selenoprotein, which reduces phospholipid hydroperoxides in the presence of thiols, is the Selenoprotein P (SeP).[82] Sep is expressed in many tissues and represents the major plasma selenoprotein, which contains 50% of the total plasma selenium in the form of selenocysteine. SeP protects endothelial cells against damage from peroxynitrite and transports selenium from the liver to peripheral tissues.[82] Selenium is also essential for thyroid hormone metabolism and immune function. A class of selenoproteins (iodothyronine deiodinase enzymes), which catalyze the peripheral deiodination of thyroxin (T4) to 3,3′5-triiodothyronine (T3), play crucial roles in determining the circulating and intracellular levels of T3 and, consequently, the control of growth, development, differentiation and metabolism.[83] The influence of selenium on the immune function can be, in part, attributed to the same selenoproteins involved in the protection against oxidative damage (cited above) and, in part, to still undefine biochemical pathways. The antioxidant GPxs have probably a role in protecting neutrophils from ROS that are produced during inflammation.[84] Selenium supplementation, in mice, increases the expression of subunits alpha (p55) and/or beta (p70/75) of IL-2 receptor (IL-2R) from activated lymphocytes and NK cells, thereby enhancing proliferation and clonal expansion of cytotoxic precursor cells. *In vitro*, selenium enhances the release of tumor necrosis factor (TNF), IL-1 and IL-6 from LPS stimulated macrophages (see review).[85] Taking into account all these mechanisms, selenium deficiency has been extensively studied in relation to aging and some age-related diseases, which their pathogenesis is related to preservation of membrane integrity and to oxidative damage of biomolecules such as lipids, lipoproteins and DNA.

Selenium, aging and age-related diseases

Selenium deficiency is a rare condition, mainly attributed to low selenium content in the soil or to long-term parenteral nutrition. Selenium is essential for several biochemical mechanisms and selenium blood decline concentrations relate to chronic age-related disease such as cancer, cardiovascular disease and immune dysfunctions.[86] During aging, selenium deficiency may occur in relation to intestinal malabsoprtion. However, there are not many reports on marked selenium deficiency in old subjects.[86] Other authors demonstrate selenium deficiency in elderly people in relation to hypothyroidism.[87] Interestingly, human healthy centenarians display selenium values quite similar to normal elderly.[88] As very few trials have been carried out, it is difficult to report a specific beneficial effect of selenium in human aging, even if beneficial effects of selenium supplementation on immune cell function have been reported in old animals.[89] The major evidence of the beneficial effects of selenium relate to age-associated diseases. A lot of studies have investigated the effects of selenium in carcinogen-exposed animals showing a reduction in tumor incidence and/or pre-neoplastic endpoints.[90] A supplementation with 200 μg/day of organic selenium in randomized subjects showed preventive effects in the incidence and the mortality from various types of cancer.[90]

The relevance of selenium in the etiology of cardiovascular diseases is more studied. Selenium metabolism is potentially involved in several protective biochemical pathways related to cardiovascular disease, such as reduction of LDL levels and lipoprotein oxidation, inhibition of foam-cell formation and shift in prostaglandin production from prostacyclin to tromboxane.[91] However, the existence of a clear link between selenium deficiency and cardiovascular disease remains to be defined.

Finally, an intriguing point is the association between selenium and immune function. Selenium deficiency seems to be associated with the incidence of infections in adults and elderly. Patients with systemic inflammatory response syndrome display a strong impairment in immune efficiency, a decrease of 40% in plasma selenium concentrations coupled with increased morbidity and mortality rates.[92] Selenium supplementation (200 mg/d of sodium selenite) enhanced immune functions in cancer patients with an emphasis in cytotoxic activity.[93] The interrelationships between selenium deficiency and impaired immune response have been clearly shown in animals. An inoculated avirulent virus in selenium deficient animals turn into a virulent one due to genomic changes within the virus.[94] An enhanced oxidative stress, caused by selenium deficiency, is the reason of these genetic changes. Additionally it is of interest that: 1) the emergence of new influenza virus strains (SARS) from China, where there are significant areas of overt selenium deficiency; 2) the decline of selenium levels during the progression of HIV infection;[95] and 3) the increased risk of enhanced virulence of influenza virus in elderly[96] associated with a possible selenium deficiency.[86]

Interrelationship between zinc and selenium: implications for healthy aging

Dietary zinc and selenium are important nutritional factors for the immune response and the protection against age-related diseases. The regulation of zinc ion bioavailability by selenium and selenoproteins has been recently investigated.[97] Zinc/thiolate coordination occurs in MT affecting the binding and release of zinc from MT. Zinc/thiolate cluster of MT can be oxidized by glutathion disulfide (GSSG) or other disulphides in order to release zinc. However, the efficiency of this chemical reaction seems very low even at high concentrations of GSSG in the absence of selenium. In contrast, this chemical process following the addition of selenium compounds releases zinc from MT very rapidly.[97] These compounds have the capacity to form selenol(ate)s, which interact with GSH/GSSG and zinc-MT/thionein redox pairs in order to release or bind zinc. In presence of t-butylhydroperoxide, GPx catalyzes the MT oxidation with subsequent zinc release. This suggests that this redox chemistry may be employed in biological system for the zinc release when peroxides are generated during oxidative stress.[8] Therefore, the assessment of zinc ion bioavailability, MT and selenium concentrations could represent useful tools for studying the physiology of successful aging. Indeed, a recent study shows that 84.4% of the "healthy" nonagenarian/centenarians display both zinc and selenium levels equal or greater than the lowest values in the elderly.[88] Moreover, healthy nonagenarians display low MT, good zinc ion bioavailability[45] and satisfactory GPx activity,[98] thus indicating that adequate zinc and selenium intake are useful to escape age-related diseases and, in turn, to reach successful aging.

Conclusions and Future Perspectives

Based on the reported data, the micronutrients zinc, niacin and selenium all play a pivotal role for healthy aging and longevity. However, zinc plays the major role because some biochemical mechanisms involved in the action of both niacin and selenium are under the control of zinc ion bioavailability, which in turn is affected by MT homeostasis. One of the most relevant biochemical pathways is the interrelationship between zinc-bound MT and PARP-1 activity. PARP-1 is codified by two zinc finger motifs and is involved in base excision DNA repair and in genomic stability and, consequently, in maintaining immune efficiency, metabolic harmony and antioxidant activity. Niacin is relevant in conferring NAD^+ substrate to PARP-1, whereas selenium is important in conferring activity to GPx, which is fundamental for the zinc release by MT via oxidation of glutathione. During aging, PARP-1-(ADP) ribosylation, NAD^+ and GPx decrease. The decrement in PARP-1 is largely due to low zinc ion bioavailability because the zinc release by MT, via gluathione peroxidation, is very limited. As a consequence, PARP-1 activity is altered and cell death occurs rather than base excision DNA-repair. Conversely, a good zinc ion bioavailability by low MT is correlated to satisfactory PARP-1 function in human centenarians. This leads to good performances in immune efficiency and antioxidant activity as well as to sufficient metabolic compensation in these exceptional individuals. Therefore, zinc, niacin and selenium in the diet during aging are relevant in order to improve immune, antioxidant and metabolic functions, which can lead to healthy aging and longevity. Alternatively, a supplementation of these micronutrients may be recommended taking into account the beneficial effects of zinc and selenium in improving the immune functions of old individuals. However, as the gap between the estimated average requirement of selenium and the upper limit of safe intake is relatively narrow[99] and excessive zinc levels are dangerous,[44] supplementation with both zinc and selenium should be recommended with caution in elderly and appropriate doses have to be carefully evaluated with specific studies. With regard to zinc and niacin supplementation in elderly, clinical trials are currently in progress in INRCA.

Acknowledgments

Experimental work presented in this review article was supported by INRCA, Italian Health Ministry (R.F. 216/02 to E.M) and European Commission (Project ZINCAGE n. FOOD-CT-2003-506850, Coordinator Dr. E. Mocchegiani).

References

1. Mocchegiani E, Giacconi R, Muti E, Muzzioli M, Cipriano C (2004) Plasticity of neuroendocrine-thymus interactions during ontogeny and ageing: role of zinc. *NeuroImmune Biol* **4**: 305–327.
2. Prolisso G, Barbieri M, Bonafe M, Franceschi C (2000) Metabolic age modelling: the lesson from centenarians. *Eur J Clin Invest* **10**: 888–894.
3. Chandra RK (2002) Nutrition and the immune system from birth to old age. *Eur J Clin Nutr* **56** (Suppl 3): S73–S76.
4. Tucker KL, Buranapin S (2001) Nutrition and aging in developing countries. *J Nutr* **9**: 2417S–2423S.
5. Failla ML (2003) Trace elements and host defense: recent advances and continuing challenges. *J Nutr* **5** (Suppl 1): 1443S–1447S.
6. Chernoff R (2001) Nutrition and health promotion in older adults. *J Gerontol A Biol Sci Med Sci* **56**: 47–53.
7. Mills CF (1989) *Zinc in Human Biology*. Springer, London.
8. Jacob C, Maret W, Vallee BL (1999) Selenium redox biochemistry of zinc-sulfur coordination sites in proteins and enzymes. *Proc Natl Acad Sci USA* **96**: 1910–1914.
9. Hageman GJ, Stierum RH (2001) Niacin, poly(ADP-ribose) polymerase-1 and genomic stability. *Mutat Res* **475**: 45–56.
10. Trucco C, Flatter E, Fribourg S, de Murcia G, Menissier–de Murcia J (1996) Mutations in the amino-terminal domain of the human poly(ADP-ribose) polymerase that affect its catalytic activity but not its DNA binding capacity. *FEBS Lett* **399**: 313–316.
11. Mocchegiani E, Muzzioli M, Giacconi R (2000) Zinc and immunoresistance to infection in aging: new biological tools. *Trends Pharmacol Sci* **21**: 205–208.
12. Vallee BL, Falchuk KH (1993) The biochemical basis of zinc physiology. *Physiol Rev* **73**: 79–118.
13. Nagase H, Woessner JF Jr (1999) Matrix metalloproteinases. *J Biol Chem* **274**: 21491–21494.

14. Paramanantham R, Bay BH, Sit KH (1996) Flow cytometric evaluation of the DNA profile and cell cycle of zinc supplemented human Chang liver cells. *Acta Paediatr Jpn* **38**: 334–338.
15. Provinciali M, Di Stefano G, Stronati S (1998) Flow cytometric analysis of CD3/TCR complex, zinc, and glucocorticoid-mediated regulation of apoptosis and cell cycle distribution in thymocytes from old mice. *Cytometry* **32**: 1–8.
16. Mocchegiani E, Muzzioli M, Cipriano C, Giacconi R (1998) Zinc, T-cell pathways, aging: role of metallothioneins. *Mech Ageing Dev* **106**: 183–204.
17. Berg JM, Shi Y (1996) The galvanization of biology: a growing appreciation for the roles of zinc. *Science* **271**: 1081–1085.
18. Taylor GA, Blackshear PJ (1995) Zinc inhibits turnover of labile mRNAs in intact cells. *J Cell Physiol* **162**: 378–387.
19. Bogdan C, Rollinghoff M, Diefenbach A (2000) The role of nitric oxide in innate immunity. *Immunol Rev* **173**: 17–26.
20. Li H, Raman CS, Glaser CB, Blasko E, Young TA, Parkinson JF, Whitlow M, Poulos TL (1999) Crystal structures of zinc-free and -bound heme domain of human inducible nitric-oxide synthase. Implications for dimer stability and comparison with endothelial nitric-oxide synthase. *J Biol Chem* **274**: 21276–21284.
21. Arizono K, Kagawa S, Hamada H, Ariyoshi T (1995) Nitric oxide mediated metallothionein induction by lipopolysaccharide. *Res Commun Mol Pathol Pharmacol* **90**: 49–58.
22. Zangger K, Oz G, Haslinger E, Kunert O, Armitage IM (2001) Nitric oxide selectively releases metals from the amino-terminal domain of metallothioneins: potential role at inflammatory sites. *FASEB J* **15**: 1303–1305.
23. Kagi JH, Schaffer A (1988) Biochemistry of metallothionein. *Biochemistry* **127**: 8509–8515.
24. Maret W, Vallee BL (1998) Thiolate ligands in metallothionein confer redox activity on zinc clusters. *Proc Natl Acad Sci USA* **95**: 3478–3482.
25. Cai L, Satoh M, Tohyama C, Cherian MG (1999) Metallothionein in radiation exposure: its induction and protective role. *Toxicology* **132**: 85–98.
26. Ashok BT, Ali R (1999) The aging paradox: free radical theory of aging. *Exp Gerontol* **34**: 293–303.
27. Mocchegiani E, Giacconi R, Cipriano C, Muzzioli M, Gasparini N, Moresi R, Stecconi R, Suzuki H, Cavalieri E, Mariani E (2002) MTmRNA gene expression, via IL-6 and glucocorticoids, as potential genetic marker of immunosenescence: lessons from very old mice and humans. *Exp Gerontol* **37**: 349–357.
28. Bui LM, Dressendorfer RH, Keen CL, Summary JJ, Dubick MA (1994) Zinc status and interleukin-1 beta-induced alterations in mineral metabolism in rats. *Proc Soc Exp Biol Med* **206**: 438–444.
29. Coto JA, Hadden EM, Sauro M, Zorn N, Hadden JW (1992) Interleukin 1 regulates secretion of zinc-thymulin by human thymic epithelial cells and its action on T-lymphocyte proliferation and nuclear protein kinase C. *Proc Natl Acad Sci USA* **89**: 7752–7756.
30. Yu CW, Chen JH, Lin LY (1997) Metal-induced metallothionein gene expression can be inactivated by protein kinase C inhibitor. *FEBS Lett* **420**: 69–73.
31. Kelly EJ, Quaife CJ, Froelick GJ, Palmiter RD (1996) Metallothionein I and II protect against zinc deficiency and zinc toxicity in mice. *J Nutr* **126**: 1782–1790.

32. Mocchegiani E, Giacconi R, Cipriano C, Gasparini N, Bernardini G, Malavolta M, Menegazzi M, Cavalieri E, Muzzioli M, Ciampa AR, Suzuki H (2004) The variation during the circadian cycle of liver CD1d-unrestricted NK1.1+TCRγδ+ cells lead to successful ageing. Role of metallothionein/IL-6/gp130/PARP-1 interplay in very old mice. *Exp Gerontol* **39**: 775–788.
33. Mocchegiani E, Giacconi R, Cipriano C, Gasparini N, Orlando F, Stecconi R, Muzzioli M, Isani G, Carpene E (2002) Metallothioneins (I+II) and thyroid-thymus axis efficiency in old mice: role of corticosterone and zinc supply. *Mech Ageing Dev* **123**: 675–694.
34. Cousins RJ, Blanchard RK, Moore JB, Cui L, Green CL, Liuzzi JP, Cao J, Bobo JA (2003) Regulation of zinc metabolism and genomic outcomes. *J Nutr* **133** (Suppl 1): 1521S–1526S.
35. Ye G, Metreveli NS, Ren J, Epstein PN (2003) Metallothionein prevents diabetes-induced deficits in cardiomyocytes by inhibiting reactive oxygen species production. *Diabetes* **252**: 777–783.
36. Niewoehner CB, Allen JI, Boosalis M, Levine AS, Morley JE (1986) Role of zinc supplementation in type II diabetes mellitus. *Am J Med* **81**: 63–68.
37. Mocchegiani E, Boemi M, Fumelli P, Fabris N (1989) Zinc-dependent low thymic hormone level in type I diabetes. *Diabetes* **38**: 932–937.
38. Giacconi R, Cipriano C, Albanese F, Boccoli G, Saba V, Olivieri F, Franceschi C, Mocchegiani E (2004) The -174G/C polymorphism of IL-6 is useful to screen old subjects at risk for atherosclerosis or to reach successful ageing. *Exp Gerontol* **39**: 621–628.
39. Fabris N, Mocchegiani E, Mariotti S, Pacini F, Pinchera A (1986) Thyroid function modulates thymic endocrine activity. *J Clin Endocrinol Metab* **62**: 474–478.
40. De Benedictis G, Carrieri G, Varcasia O, Bonafe M, Franceschi C (2000) Inherited variability of the mitochondrial genome and successful aging in humans. *Ann N Y Acad Sci* **908**: 208–218.
41. Hinerfeld D, Traini MD, Weinberger RP, Cochran B, Doctrow SR, Harry J, Melov S (2004) Endogenous mitochondrial oxidative stress: neurodegeneration, proteomic analysis, specific respiratory chain defects, and efficacious antioxidant therapy in superoxide dismutase 2 null mice. *J Neurochem* **88**: 657–667.
42. Barbieri D, Troiano L, Grasselli E, Agnesini C, Cristofalo EA, Monti D, Capri M, Cossarizza A, Franceschi C (1992) Inhibition of apoptosis by zinc: a reappraisal. *Biochem Biophys Res Commun* **187**: 1256–1261.
43. Liang JY, Liu YY, Zou J, Franklin RB, Costello LC, Feng P (1999) Inhibitory effect of zinc on human prostatic carcinoma cell growth. *Prostate* **40**: 200–207.
44. Chandra RK (1984) Excessive intake of zinc impairs immune responses. *JAMA* **252**: 1443–1446.
45. Mocchegiani E, Muzzioli M, Giacconi R, Cipriano C, Gasparini N, Franceschi C, Gaetti R, Cavalieri E, Suzuki H (2003) Metallothioneins/PARP-1/IL-6 interplay on natural killer cell activity in elderly: parallelism with nonagenarians and old infected humans. Effect of zinc supply. *Mech Ageing Dev* **124**: 459–468.
46. Chiricolo M, Musa AR, Monti D, Zannotti M, Franceschi C (1993) Enhanced DNA repair in lymphocytes of Down syndrome patients: the influence of zinc nutritional supplementation. *Mutat Res* **295**: 105–111.
47. Fabris N, Mocchegiani E, Albertini G (1993). Psychoendocrine-immune interactions in Down's Sindrome: role of Zinc. In: Castells S, Wisniewski KE (eds.) *Growth Hormone Treatment in Down's Syndrome*, pp. 203–218. John Wiley, New York, USA.

48. Roussel AM, Kerkeni A, Zouari N, Mahjoub S, Matheau JM, Anderson RA (2003) Antioxidant effects of zinc supplementation in Tunisians with type 2 diabetes mellitus. *J Am Coll Nutr* **22**: 316–321.
49. Ho E, Quan N, Tsai YH, Lai W, Bray TM (2001) Dietary zinc supplementation inhibits NF-kappaB activation and protects against chemically induced diabetes in CD1 mice. *Exp Biol Med* **226**: 103–111.
50. Hill CH (1976) Mineral interrelationship. In: Prasad AS (ed.) *Trace elements, human health and disease*, pp. 281–300. Academic Press, New York, USA.
51. Hurley LS, Keen CL, Lonnerdal B (1983) Aspects of trace element interactions during development. *Fed Proc* **42**: 1735–1739.
52. Prasad AS (1993) *Biochemistry of Zinc.* Plenum Press, New York, USA.
53. Thiesen HJ, Bach C (1991) Transition metals modulate DNA-protein interactions of SP1 zinc finger domains with its cognate target site. *Biochem Biophys Res Commun* **176**: 551–557.
54. Fabris N, Mocchegiani E (1995) Zinc, human diseases and aging. *Aging (Milano)* 7: 77–93.
55. Fosmire GJ (1990) Zinc toxicity. *Am J Clin Nutr* **51**: 225–227.
56. Kim YH, Kim EY, Gwag BJ, Sohn S, Koh JY (1999) Zinc-induced cortical neuronal death with features of apoptosis and necrosis: mediation by free radicals. *Neuroscience* **89**: 175–182.
57. Wapnir RA, Khani DE, Bayne MA, Lifshitz F (1983) Absorption of zinc by the rat ileum: effects of histidine and other low-molecular-weight ligands. *J Nutr* **113**: 1346–1354.
58. Jacob RA, Swendseid ME (1996) Niacin. In: Ziegler EE, Filer LJ (eds.) *Present Knowledge in Nutrition*, 7th edn., pp. 184–190. ILSI Press, Washington DC, USA.
59. Szabo C, Dawson VL (1998) Role of poly(ADP-ribose) synthetase in inflammation and ischaemia-reperfusion. *Trends Pharmacol Sci* **19**: 287–298.
60. Zhang JZ, Henning SM, Swendseid ME (1993) Poly(ADP-ribose) polymerase activity and DNA strand breaks are affected in tissues of niacin-deficient rats. *J Nutr* **123**: 1349–1355
61. DiPalma JR, Thayer WS (1991) Use of niacin as a drug. *Annu Rev Nutr* **11**: 169–187.
62. Chapman ML, Zaun MR, Gracy RW (1983) Changes in NAD levels in human lymphocytes and fibroblasts during aging and in premature aging syndromes. *Mech Ageing Dev* **21**: 157–167.
63. Durkacz BW, Omidiji O, Gray DA, Shall S (1980) (ADP-ribose)n participates in DNA excision repair. *Nature* **283**: 593–596.
64. Shisa H, Hiai H (1985) Genetically determined susceptibility of Fischer 344 rats to propylnitrosourea-induced thymic lymphomas. *Cancer Res* **45**: 1483–1487.
65. Boyonoski AC, Spronck JC, Gallacher LM, Jacobs RM, Shah GM, Poirier GG, Kirkland JB (2002) Niacin deficiency decreases bone marrow poly(ADP-ribose) and the latency of ethylnitrosourea-induced carcinogenesis in rats. *J Nutr* **132**: 108–114.
66. Bitterman KJ, Anderson RM, Cohen HY, Latorre–Esteves M, Sinclair DA (2002) Inhibition of silencing and accelerated aging by nicotinamide, a putative negative regulator of yeast sir2 and human SIRT1. *J Biol Chem* **277**: 45099–45107.
67. Rose G, Dato S, Altomare K, Bellizzi D, Garasto S, Greco V, Passarono G, Feraco E, Mari V, Barbi C, Bonafe M, Franceschi C, Tan Q, Boiko S, Yashin AI, De Benedictis G (2003) Variability of the SIRT3 gene, human silent information regulator Sir2 homologue, and survivorship in the elderly. *Exp Gerontol* **38**: 1065–1070.

68. Cho Y, Gorina S, Jeffrey PD, Pavletich NP (1994) Crystal structure of a p53 tumor suppressor-DNA complex: understanding tumorigenic mutations. *Science* **265**: 346–355.
69. Sheng Y, Pero RW, Olsson AR, Bryngelsson C, Hua J (1998) DNA repair enhancement by a combined supplement of carotenoids, nicotinamide, and zinc. *Cancer Detect Prev* **22**: 284–292.
70. Schwarz S (1976) Essentiality and metabolic functions of selenium. *Med Clin North Am* **60**: 745–758.
71. Wasowicz W, Gromadzinska J, Rydzynski K, Tomczak J (2003) Selenium status of low-selenium area residents: Polish experience. *Toxicol Lett* **137**: 95–101.
72. Li GS, Wang F, Kang D, Li C (1985) Keshan disease: an endemic cardiomyopathy in China. *Hum Pathol* **16**: 602–609.
73. Kryukov GV, Castellano S, Novoselov SV, Lobanov AV, Zehtab O, Guigo R, Gladyshev VN (2003) Characterization of mammalian selenoproteomes. *Science* **300**: 1439–1443.
74. Brigelius–Flohe R (1999) Tissue-specific functions of individual glutathione peroxidases. *Free Radic Biol Med* **27**: 951–965.
75. Roveri A, Maiorino M, Nisii C, Ursini F (1994) Purification and characterization of phospholipid hydroperoxide glutathione peroxidase from rat testis mitochondrial membranes. *Biochim Biophys Acta* **1208**: 211–221.
76. Hall L, Williams K, Perry ACF, Frayne J, Jury JA (1998) The majority of human glutathione peroxidase type 5 (GPX5) transcripts are incorrectly spliced: implications for the role of GPX5 in the male reproductive tract. *Biochem J* **333**: 5–9.
77. Watson WH, Yang X, Choi YE, Jones DP, Kehrer JP (2004) Thioredoxin and its role in toxicology. *Toxicol Sci* **78**: 3–14.
78. Rundlof AK, Arner ES (2004) Regulation of the mammalian selenoprotein thioredoxin reductase 1 in relation to cellular phenotype, growth, and signaling events. *Antioxid Redox Signal* **6**: 41–52.
79. Sun QA, Wu Y, Zappacosta F, Jeang KT, Lee BJ, Hatfield DL, Gladyshev VN (1999) Redox regulation of cell signaling by selenocysteine in mammalian thioredoxin reductases. *J Biol Chem* **274**: 24522–24530.
80. Arner ES, Holmgren A (2000) Physiological functions of thioredoxin and thioredoxin reductase. *Eur J Biochem* **267**: 6102–6109.
81. Nordberg J, Arner ES (2001) Reactive oxygen species, antioxidants, and the mammalian thioredoxin system. *Free Radic Biol Med* **31**: 1287–1312.
82. Burk RF, Hill KE, Motley AK (2003) Selenoprotein metabolism and function: evidence for more than one function for selenoprotein P. *J Nutr* **133** (Suppl 1): 1517S–1520S.
83. Kohrle J (2000) The deiodinase family: selenoenzymes regulating thyroid hormone availability and action. *Cell Mol Life Sci* **57**: 1853–1863.
84. Arthur JR, McKenzie RC, Beckett GJ (2003) Selenium in the immune system. *J Nutr* **133** (Suppl 1): 1457S–1459S.
85. Beckett GJ, Arthur JR, Miller SM, McKenzie RC (2003) Selenium immunity and disease. In: Hughes DA, Bendick A, Darlington G (eds.) *Dietary enhancement of human immune functions*, pp. 250–280. Humana Press, Totowa NJ, USA.
86. Seiler WO (2001) Clinical pictures of malnutrition in ill elderly subjects. *Nutrition* **17**: 496–498.

87. Olivieri O, Girelli D, Stanzial AM, Rossi L, Bassi A, Corrocher R (1996) Selenium, zinc, and thyroid hormones in healthy subjects: low T3/T4 ratio in the elderly is related to impaired selenium status. *Biol Trace Elem Res* **51**: 31–41.
88. Savarino L, Granchi D, Ciapetti G, Cenni E, Ravaglia G, Forti P, Maioli F, Mattioli R (2001) Serum concentrations of zinc and selenium in elderly people: results in healthy nonagenarians/centenarians. *Exp Gerontol* **36**: 327–339.
89. Roy M, Kiremidjian-Schumacher L, Wishe HI, Cohen MW, Stotzky G (1995) Supplementation with selenium restores age-related decline in immune cell function. *Proc Soc Exp Biol Med* **209**: 369–375.
90. Reid ME, Duffield-Lillico AJ, Garland L, Turnbull BW, Clark LC, Marshall JR (2002) Selenium supplementation and lung cancer incidence: an update of the nutritional prevention of cancer trial. *Cancer Epidemiol Biomarkers Prev* **11**: 1285–1291.
91. Alissa EM, Bahijri SM, Ferns GA (2003) The controversy surrounding selenium and cardiovascular disease: a review of the evidence. *Med Sci Monit* **9**: RA9–RA18.
92. Forceville X, Vitoux D, Gauzit R, Combes A, Lahilaire P, Chappuis P (1998) Selenium, systemic immune response syndrome, sepsis, and outcome in critically ill patients. *Crit Care Med* **26**: 1536–1544.
93. Kiremidjian-Schumacher L, Roy M, Glickman R, Schneider K, Rothstein S, Cooper J, Hochster H, Kim M, Newman R (2000) Selenium and immunocompetence in patients with head and neck cancer. *Biol Trace Elem Res* **73**: 97–111.
94. Beck MA (1999) Selenium and host defence towards viruses. *Proc Nutr Soc* **58**: 707–711.
95. Daniels LA (2004) Selenium: does selenium status have health outcomes beyond overt deficiency? *Med J Aust* **180**: 373–374.
96. Ellis SE, Coffey CS, Mitchel EF Jr, Dittus RS, Griffin MR (2003) Influenza and respiratory syncytial virus-associated morbidity and mortality in the nursing home population. *J Am Geriatr Soc* **51**: 761–767.
97. Maret W (2003) Cellular zinc and redox states converge in the metallothionein/thionein pair. *J Nutr* **133** (Suppl 1): 1460S–1462S.
98. Mecocci P, Polidori MC, Troiano L, Cherubini A, Cecchetti R, Pini G, Straatman M, Monti D, Stahl W, Sies H, Franceschi C, Senin U (2000) Plasma antioxidants and longevity: a study on healthy centenarians. *Free Radic Biol Med* **28**: 1243–1248.
99. Johnson LJ, Meacham SL, Kruskall LJ (2003) The antioxidants — vitamin C, vitamin E, selenium, and carotenoids. *J Agromedicine* **9**: 65–82.

10

Calorie Restriction as a Potent Anti-Aging Intervention: Modulation of Oxidative Stress

Byung Pal Yu
Department of Physiology, University of Texas Health Science Center
San Antonio, TX, USA
E-mail: bpyu@usa.net

INTRODUCTION

Aging is time-dependent, inevitable process that affects all living organisms. Despite the enormous consequences associated with age-related deterioration, relatively little systematic research has been expended on the basic mechanisms of aging. The ever-increasing aged population is challenging scientists in various disciples to discover solutions to nature's most complex problems of aging.[1,2]

Ways to intervene or retard the aging process, commonly called anti-aging interventions, are not new products of modern scientific advancement.[3] Throughout human history, people have tried a variety of interventions to extend life. Just as survival is an organism's strongest, most basic instinct, for humankind, extending one's life seems equally crucial. Modern research has produced convincing data that lifestyle patterns, such as nutrition and physical activity, are the most important factors in influencing life-span and aging processes.[4,5]

In the past, nutritional science focused mainly on the eradication of deficiencies and the prevention of diseases by establishing nutritional requirements for the proper maintenance of health. Numerous studies have evaluated specific nutritional ingredients in their ability to prevent particular diseases. However, very few systematic studies have explored the nutritional implications in modulating the aging process.[6,7] At present, there is no scientific basis to claims of specific dietary supplement or additive that leads to the prolongation of life by retarding aging processes, although some show beneficial effects on various age-related phenomena. Indeed, researchers have produced an overwhelming body of evidence firmly establishing caloric restriction (CR) as the most effective anti-aging intervention known to retard the aging process in mammals.[8,9] One practical reason for the success of the CR paradigm is its reproducibility, effectiveness, and simplicity.[8]

Because numerous papers on calorie restriction (CR) are cited else where,[8–10] in this chapter, coverage here is confined to CR's salient effect on oxidative stress during aging, focusing on the modulation of cellular membranes and major redox-sensitive transcription factors. General information on CR's ability to modulate various physiological systems and functions are cited in the reference section.[8] For those who want to gain wider access to the most up-to-date, documented data on aging in relation to oxidative stress, a newly published bioinformatic database should be useful (see Park *et al.*[11]).

A Brief Historical Note on the Beginnings of CR Studies

It might be of interest to know the origins of the dietary paradigm that eventually led to calorie restriction, the most popular model among gerontologists and nutritionists. Studies with caloric restriction began in early 1920's, based on an old notion that the life-span can be extended through growth retardation, i.e. slower growth leads to a longer life, which was later disputed. To test this notion, McCay fed rats a diet containing cellulose and was as 10 to 20% of total diet, which reduced the calorie density.[12] Rats fed this diet were found to grow slower and live longer than control rats. Thus, McCay's findings did confirm the growth retardation hypothesis experimentally, although the results were contradictory to the earlier results of Robertson and Ray,[13] who observed that the mice that grew more rapidly lived longer. Robertson and Ray's findings that those mice that grew at a slow rate, died prematurely, were apparently caused by unintended malnutrition. McCay therefore blamed unequivocal data on the animals' growth pattern and body weight gains.

In 1935, McCay *et al.*[12,14] published the first seminal paper that showed results of a systematic examination of this growth retardation hypothesis by controlling food intake, i.e. dietary restriction. In this work, McCay and his group sought to feed animals as little as possible to control heterogeneous growth patterns as much as possible while maintaining their health. To control the growth and body weight gains of the animals, McCay and colleagues used what is termed the "stair-step" method to achieve the desired level of growth retardation. In this experiment, McCay and co-workers had shown the ability of growth retardation to increase longevity, but later experiments discredited the validity of an inverse relationship between growth and longevity. Since then, researchers working with growth hormone-deficient transgenic rats have found interesting data that implies CR's life extension effect might be mediated partly through growth hormone-related activities.[15]

To characterize these calorie-restricted animals, McCay's group launched a study on the basal metabolism of restricted animals. Interestingly, these animals were found to have higher cal/kg/h metabolisms than their *ad libitum*-fed counterparts. Furthermore, the McCay team observed no age-related change in the basal metabolic rate of animals at ages between 850 to 1150 days. Although the question of reduced metabolic rate as the underlying mechanism for CR's anti-aging action is still unsatisfactorily settled among modern researchers, these early findings have laid the foundation of modern CR research.[16]

One of the frequently asked questions about CR's efficacy is the length of treatment, i.e. long-term vs. a short-term feeding, timing, and degree of restriction. Data have established that the longest restriction leads to the maximum extension of life-span with robust beneficial efficacies.[16] However, earlier studies[9,15] have shown that even short-term restriction can bring about some benefits. A recent report by Judge *et al.*[17] showed that short-term CR of two months suppressed the mitochondrial production of superoxide and glutathione peroxidase, while boosting mitochondrial and cytosolic catalase, slightly mimicking the long-term effects of CR.[17] A report from Spindler's laboratory,[18] when looking at the genomic levels, also observed a short-term CR effect on gene expression, revealing how quickly restricted animals can adapt to the new dietary deprivation. The ultimate question would be how can theses changes in gene expression be translated into the final outcome of life-span extension by CR, and the suppression of age-related disease processes, which have yet to be delineated.

A Model for Cross-Talk between Oxidatively Modified Aging and Disease Processes

One fundamentally important question on aging concerns the interrelation between biological processes and the aging-related diseases. Aging is often defined as a time-dependent biological process, manifested by the loss in ability to withstand stresses, which leads to an increased risk to disease.[19] To conceptualize this interaction, a cross-talk model is diagramed in Fig. 1. This model emphasizes the molecular process that underlies the altered biological functions that lead to the pathological status. Thus, this cross-talk model represents the connection between age-associated vulnerability and the increased risk for disease. A recent proposal on the oxidative stress-based molecular inflammation hypothesis of aging offered reasonable molecular data to support the nature of this cross-talk.[20] The

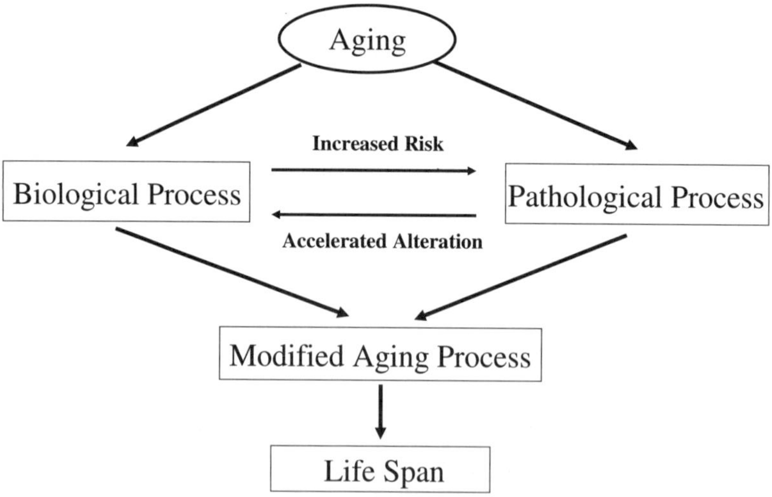

Fig. 1. Model for cross-talk between the aging and the pathological processes.

retardation of aging, and life extension, can be achieved by modulating either the biological process or the pathological process, but the suppression of the cross-talk can bring about significant beneficial anti-aging effects.[20,21]

To date, CR is the only known measure to suppress the biological, pathological, and cross-talk processes, resulting in the most powerful and diverse anti-aging intervention measure known to exist experimentally.[8–10] The extension of the maximum life-span by CR is accepted by many scientists as the strongest evidence for claiming the retardation of aging. This distinction between the extension of maximum and mean life-spans is therefore important in differentiating between CR and other life-extending treatments like therapeutic drugs and dietary supplements.

Oxidative Stress and CR

Oxidative stress refers to a condition whereby imbalances in the redox state cause cellular dysfunctions and oxidatively modified constituents. Data from various sources consistently indicate that overall oxidative stress is elevated during aging.[22,23] Among the most relevant and impressive mechanisms underlying CR's action in retarding the aging process, are its abilities to resist age-related oxidative stress and to maintain a proper redox balance. It has been said that if CR has no ability to modulate age-related oxidative stress, then free radicals and other reactive species, as caustic factors in the aging process, would lose its strong support.[24]

Harman proposed the original free radical theory of aging in 1956.[25] The basic tenet of this theory is that superoxide and hydrogen peroxide, generated from the aerobic metabolism of organisms, are intrinsic factors capable of destroying both cellular structure and function, which lead to the pathogenesis of many diseases that leads to aging. However, information generated during the last several decades necessitates the modification of the original free radical theory, because when the original theory was proposed, our knowledge of both aging and free radicals were in their infancy. Therefore, understandably, what we know now as most the essential aspects of free radical biology and aging were not addressed in the original form. Thus, it is prudent to know why the oxidative stress hypothesis was proposed.[26,27]

Major Distinctions between Free Radical Theory and Oxidative Stress Hypothesis of Aging

Current knowledge gained from both the fields of aging and free radical biochemistry have revealed the shortfalls in the original free radical theory.[27] Three major aspects of the free radical theory need amendments: 1) inclusion of various reactive species, such as nitrogen-derived NO and its derivatives, singlet oxygen, and reactive lipid products, such as lipid hydroperoxides and aldehydic species, which were not included in the original free radical theory; 2) essentiality of defense systems and the maintenance of the redox balance; and 3) consideration of biological aging as primary process and age-related disease as secondary process, not the reverse as was described in the original version.

To fulfill these deficiencies, an oxidative stress theory of aging was proposed in 1996.[26,27] This theory is particularly meaningful when one considers possible interventions and antioxidant strategies. For instance, because of the multiple types of reactive species, no single antioxidant is capable of fending off the large a variety of reactive species; the most effective antioxidative measures require multi-functional efficacies and the ability to neutralize a broad spectrum of reactive species.[28] The defense systems emphasized in the oxidative stress theory are essential components for the proper maintenance of redox balance, as important as intracellular pH in regulating signal transduction activity.[20,27]

As intervention strategies against aging and age-related disease, the possibility of boosting either the overall the defense system and/or any specific functional component of the defense system may be logical approaches.[28] The last point of oxidative stress theory making an important distinction is its stance that biological aging and pathological processes have separate underlying mechanisms. This distinction signifies that aging is not a disease; as we well know, the aging process occurs in the absence of disease. Thus, according to this view, therapeutic treatments on any specific disease would have only limited success as an anti-aging intervention. It is important to recognize that CR is the only experimental paradigm known to modulate both processes.[3]

CR as a Powerful Protector Against Oxidative Stress

Accumulated evidence has proven CR to be the most effective modulator of oxidative stress.[3,27] The oxidative stress hypothesis offers explanations on how CR can build up defenses and resistance against oxidative stress. Investigations looking into the oxidatively stressed animals, consistently found a disrupted redox status, which was conversely well preserved in CR animals. Studies show that free radical generation[25] and lipid peroxidation[21,28] are suppressed by CR, while boosting the defense systems.[24]

Weakened defenses and redox imbalance cause alterations in cellular structural constituents, such as DNA, membrane lipids, and proteins.[29–32] In caloric-restricted organisms, all the oxidative alternations are shown to be blunted.[33] The chemical modifications of other cellular constituents, like collagen by oxidative-induced glycation during aging, is shown to be suppressed by CR in a nonhuman primate study, offering another good illustration of CR's anti-aging effects.[34]

Studies[22,27] show that damage to both nuclear and mitochondrial DNA and the membrane instability of mitochondria and microsomes occurs with age are all attenuated by CR. Functionally, many uncontrollably activated redox-sensitive transcription factors are shown to be modulated by CR,[20] resulting in well-regulated cellular signal transduction pathways. Because many forms of age-related diseases, neoplasms in particular, have been linked to oxidatively modified cellular constituents and functions,[22] the suppression of oxidative stress by CR may be the likely mechanism for its beneficial action in the prevention of many age-related disease processes as produced by the molecular inflammation hypothesis of aging.[21,23,24]

Protection of Membrane Integrity by CR

The dynamic, yet tightly structured cell membrane, has been the subject of many continuing research endeavors.[35] The importance of membrane deterioration *in* aging was highlighted in the membrane hypothesis of aging proposed by Zs–Nagy.[36] In a recent publication, Hulbert[37] proposed the membrane pacemaker hypothesis of aging that is based mainly on CR's action in modifying the acyl composition of membrane bilayers, decreasing membrane lipid peroxidation, and extending life-span.

Although the membrane damage by free radicals has been the topic of many reports in the past, not much attention was given to the membrane lipid peroxidation, *per se*, and its consequences in the aging membrane. Because of the membrane's high lipid content, membrane stability is strongly dependent on the integrity of the redox status, as unstable polyunsaturated bonds are preferential targets of oxidative stress.[38] The most convincing data showing an age-related increase in oxidative stress comes from life-long, longitudinal measurements comparing levels of pentane, as a marker for *in vivo* lipid peroxidation, in exhaled air of CR rats and *ad libitum*-fed (AL) rats. CR rats showed much lower pentane levels than the AL rats.[39]

Protection by Reduction of Lipid Peroxidation

The gerontological literature shows substantial evidence that the membranes of aging organisms lose their resilience and become rigid with time due to the loss of membrane fluidity, probably from mainly lipid peroxidation and its reactive byproducts. This change in membrane fluidity seems to be widespread, occurring in most tissues, whether they are mitotic or postmitotic.[40–44]

Generally, changes in membrane fluidity can be explained in two ways: 1) changes caused by increased saturation or decreased unsaturation of fatty acid composition, and 2) changes caused by increased membrane cholesterol content causing the membrane to become condensed, i.e. rigid.[45] Several old data recorded in gerontological papers show such occurrences in aged membranes. One popular misconception points to increased cholesterol as the culprit in age-related membrane rigidity.[12] However, later findings from membrane lipid peroxidation studies[45,46] revealed that the age-related changes in membrane fluidity that cause membrane rigidity develop from lipid peroxidation and its peroxidized lipids, not from increased saturation or cholesterol. Evidence shows that lipid peroxidation and peroxidized lipids, when compared to cholesterol, are far better inducers of membrane rigidity.[45,46]

Further, because mitochondria are major sites of reactive species production, both mitochondrial structure and its membrane-associated functions are exquisitely sensitive to lipid peroxidation. It is interesting to note that deleterious changes in the mitochondrial membrane (e.g. rigidity) take place as young as 6 and 12 months in *ad libitum*-fed animals, as opposed to CR animals.[40–43] As expected from its antioxidative action, CR can effectively suppress age-related lipid peroxidation-induced membrane rigidity and maintain membrane integrity throughout life.

Protection by Changing Membrane Fatty Acid Profile

The membrane's resistance to oxidative stress is especially important to biological homeostasis because of the essential roles that the membrane-transport system, permeability, membrane potential, and the receptor and signal transduction systems play in membrane maintenance.[44] CR as a powerful membrane protector exhibits a clever manipulation by rearranging the membrane fatty acid composition profiles to resist oxidative attack. This is a surprising revelation because no such changes have been reported from simply reducing calorie intake, although many dietary lipids have been known to influence the membrane fatty acid composition.[38,42,43]

It was found that CR rearranges the membrane composition by decreasing polyunsaturated fatty acids (PUFA) such as 22:4 and 22:5, while replacing PUFA with the less peroxdizable, 18:2, fatty acid.[38,45] By this manipulation, proper membrane fluidity is maintained without risking for peroxidation. This interesting manipulation is deemed as part of the organism's innate defense strategy to protect the membrane against increased peroxidizability.[33] This survival strategy seems to be an adaptive trait gained through the evolutionary process, as shown in the work of Pamplona *et al.*,[42] who found an inverse relationship between the amount of PUFA (i.e. increased peroxidizability) and the life-spans of several animal species, including mice, rats, pigeons, horses, and long-living humans: the higher the PUFA content in the membranes, the shorter the life-span — similar to findings with *ad libitum*-fed and CR rats.[38]

Additional data support-membrane fatty acid damage with age, as evidence by the increased membrane phospholipase A_2 in *ad libitum*-fed rats (probably in response to increased peroxidized lipids) and the reduced levels of this enzyme in CR rats.[41,46] Additional corroborating data comes from a study in which delta-6 desaturase, a rate-limiting enzyme in the conversion of linoleic acid (18:2) to PUFA, is significantly lower in CR rats than in *ad libitum*-fed rats.[46] Thus, it seems that organisms with a restricted calorie intake has multiple ways to reduce age-related lipid peroxidation, i.e. by replacing PUFA with more stable, but less rigid 18:2 unsaturated fatty acids, and by regulating selective desaturases, as shown by delta-9 desaturase.

Protection of Mitochondria by CR

The first sign of oxidative modification occurring in the mitochondrial structure during aging is increased levels of peroxidized mitochondrial lipids. The amount of lipid hydroperoxide was observed to be much higher in mitochondria taken from old *ad libitum*-fed rats compared to the levels found in CR rats.[21] The significance of these high levels of peroxidized lipids is that not only can they cause structural disruption, but more dangerously, peroxidized lipids generate various reactive lipid species such as aldehydic 4-hydroxynanenol (HNE).[47] Evidence shows increased accumulations of degraded lipid products are more potent within the mitochondrial membrane during aging,[48] because they are long-acting, easy to diffuse, and have no specific scavengers, unlike most free radical derivatives. Again, the diverse actions of CR effectively prevented damage from HNE.[48]

Mitochondrial sensitivity to reactive aldehydes can be shown in mitochondrial transcription.[49] Proper transcription function requires well-maintained mitochondrial gene expressions of many respiratory enzyme proteins. Although the susceptibility of these functions to oxidative stress by oxygen-derived free radicals has been suspected, the effect of lipid peroxidation has not been well examined until recently. Published data indicate that HNE perturbs the inhibition of mitochondrial transcription activity as effectively as peroxyl radicals. CR's anti-peroxidative action certainly has been shown to work well in the maintenance of transcription function by reducing the amount of reactive aldehydic compounds.[50,51]

Exposure of mitochondria to various oxidants causes mitochondrial deterioration with concurrent functional loss. One functional change sensitive to oxidant treatment is mitochondrial permeability transition (MPT).[52] An age-related increase in MTP of liver mitochondria isolated from male Fischer 344 rats, 6–24 months of age, was shown to be due to deteriorated membranes, and was reversed by CR,[53] showing that CR regimens greatly delay the opening of the MPT Ca^{2+} mega channel upon exposure to various oxidants like tert-butyl hydroperoxide, thereby indicating resistance of mitochondria to oxidative stress. The increased resistance to MPT induction was maintained through 24 months of age in CR animals, showing the resilient nature of the membrane of CR animals.[53]

The significance of mitochondrial membrane protection and well-regulated MTP by CR could provide possible clues for the differences noted between *ad libitum*-fed and CR animals in mitochondrial membrane potential and ROS generation, as reported in recent interesting papers.[54,55]

A most recent paper raised interesting questions regarding the counteraction of insulin against the inhibitory effect of CR on the mitochondrial hydrogen peroxide generation.[54] The authors found that CR was able to decrease the mitochondrial proton motive force that resulted from an increased proton leak activity and decreased substrate oxidation. Interestingly, a 2-week treatment of insulin attenuated the CR effect. Although the exact mechanism for the insulin counteraction is unknown, the authors speculate that under CR conditions, a reduced substrate and increased proton leak (due to reduced insulin) could lead to a lower the electrochemical energy gradient and a lower production rate of reactive species.[55] This interesting finding on a possible link between oxidative stress and insulin-related mitochondrial bioenergetics in CR should vigorously be explorer.[56]

Apoptosis, a programmed cell death, is a normal function of an organism's tissue development and maintenance, and one important point of apoptotic induction is its responsiveness to reactive species and oxidative stress. This relatively new area of study has attracted the attention of aging researchers, who have documented some interesting data on age-related apoptosis,[57,58] although there is some debate on exactly how big a role apoptosis plays in the aging process *per se*.

Mitochondrial participation in apoptosis seems to be essential, and many consider apoptosis to plays an important role as part of a defense mechanism in the protection against cell or tissue damage. Accumulated evidence indicates that age-related apoptosis is tissue- and organ-specific as seen in increased apoptosis of mitotic tissues and cells.[56]

Several mitochondria-associated factors, including cytochrome c, proteolytic capases, and redox-sensitive genes, like Bcl-2, Bcl-x, and pro-apoptotic Bax, all play specific roles in regulating the apoptotic activity, all of which seems to be influenced by age and oxidative status.

The effect of CR's mitochondrial protection can be seen in age-related apoptotic activity. Rajini *et al.* reported that the incidence of apoptosis increased with age in rat brain and that CR attenuated this activity.[59] However, in the liver, animals on short-term CR (two months) showed significantly higher apoptotic activity than controls. Age-related mitochondrial membrane dysfunction, which has well been documented, can influence the regulation of the apoptotic process by releasing cytochrome c or anti-apoptotic Bcl-2 from mitochondria in coordination with caspases and pro-apoptotic Bax. In the kidney, it was found that CR effectively prevents the release of cytochrome, up-regulates Bcl-2, and reduces Bax, leading to the suppression of the apoptosis process.

Age-related DNA damage and modifications have been linked to many cellular dysfunctions and age-related diseases. Rao suggested recently that one of the life-extending mechanisms of CR may stem from its ability to channel limited energy resources to maintain essential processes, like DNA repair, rather than towards reproductive and anabolic activities.[60]

Mitochondrial DNA (mtDNA) is also subject to oxidative stress during aging. Several laboratories have already hypothesized that compared to nuclear DNA, mtDNA is probably more susceptible to oxidative stress. Similar conclusions were reached based on reports showing oxidatively modified DNA in rodent models. Fraga *et al.*[61] and Richter *et al.*[62] have shown that the levels of oxidatively damaged mtDNA found in rat liver are about 10-fold higher than those found in nuclear DNA. Additional data on CR's ability to protect DNA from both nuclear and mitochondria damage were reported by Chung *et al.*,[63] who found about a 15-fold higher level of 8-OHdG in mitochondrial DNA compared to nuclear DNA, for which CR attenuated damage in both.

Mitochondrial DNA deletions causing various genetic abnormalities are also well documented. One interesting finding by Hruskewycz[64] on the possible causality of the abnormalities cross-links mtDNA to lipid peroxidation. Data supporting their findings on mtDNA alterations are reported.[65] For example, in aging rats, mtDNA deletions clearly showed a 6-fold increase in liver mitochondrial between ages 6 and 24 months.

One salient point of this finding is that the CR's anti-oxidative action effectively suppressed the age-related mtDNA deletions.[65]

One interesting paper recently reported[66] negative data, showing that reactive oxygen spices derived from the mitochondrial respiratory chain are not responsible for the basal levels of oxidative modifications to nuclear DNA. Further investigation could be significant because, as mentioned above, endogenous long-acting reactive lipid species like 4-HNE, rather than short-lived free radicals and be less diffusible free radicals or charged oxidants, and are thereby expected to cause more serious damage with a long-lasting age-associated consequences.[46]

One word of caution is the sensitive measurement on any membrane-associated activities like proton leaks or cytochrome c releases, which are all dependent on how well the mitochondria were prepared and whether integrity was maintained during the isolation procedure. We already know that aged mitochondria from *ad libitum*-fed animals are more fragile than young mitochondria. Damage due to improper technical procedures could lead to erroneous and exaggerated results.

Modulation of Gene Expression by CR

With interest to gene modulation during aging, several lower organisms have been used to study age-related genomic changes during aging.[67] The consensus obtained from these studies is that aging, characterized by structural and functional alterations of physiological systems, exhibit the same degree of dynamic change in the molecular expression.[68] As a potent anti-oxidative modulator, CR exhibits remarkable effects in how genes are regulated, particularly the redox-responsive genes. In the following sections, some of major findings involving the CR paradigm studies are described.

With the advent of the high-density oligonucleotide array method for a screening more than 7000 genes, researchers were able to show that aging leads to the selective activation of transcripts known to be involved in oxidative stress and inflammation, and that the suppressed genes involved in mitochondrial electron transport and oxidative phoshphorylation were attenuated in CR mice.[69] In a more recent study, microarray analysis of 11,000 genes in the liver was reported.[18] It is interesting to note that many of the genomic responses to CR were adjusted quickly within a matter of a couple of weeks. Modulated genes were interestingly associated with inflammation, oxidative stress, and DNA replication, similar to those data from muscle of monkeys.[68]

Transcription factors are so named by their function in the transduction of cellular signals to the transcriptional process. They are bound to specific DNA sequences, and many of them are redox-responsive. Among many transcription factors, NF-kB is known to be exquisitely sensitive to oxidative stress, and the chronically activated NF-kB during aging is readily downregulated by CR.[70] The significance of the age-related activation of NF-kB is that it regulates unusually a large number of genes involved in many normal and pathological processes like immune and inflammatory processes. The inflammatory process is now recognized as part of many age-related, chronic disease processes, including arthritis, vascular aging, cardiovascular diseases, cancer, and dementia. Molecular probing of the age-related chronic activation of NF-kB supports this conclusion.

Kim *et al.* found the age-related activation of NF-kB is enhanced by the degradation of its subunits, IkB, allowing NF-kB to translocate into the nucleus during aging.[71] Moreover, in another recent study, it was found that the CR's effective suppression of the age-related NF-kB activation was carried out by the inhibiting the dissociation of IkB from NF-kB, thereby, leading to the suppression of the inflammatory process.[72]

The Anti-Inflammatory Action of CR

Recent research has produced convincing data implicating oxidative stress as a major causative factor underlying the inflammatory process.[20] Although the precise mechanism has yet to be worked out, the involvement of redox-responsive transcription factors play a major part in creating the pro-inflammatory state seen with aging.[73] During aging, the up-regulation of NF-kB, IL-6, TNFa, and iNO synthase occurs, and CR suppresses this age-related change.[71]

The suppression of inflammatory process by CR is further demonstrated by the attenuation of the age-related up-regulation of COX-2 gene expression and pro-inflammatory cytokine synthesis, which are shown to be suppressible under reduced oxidative stress, as in the case with CR.[73]

Based on common observations of the pro-inflammatory status in the aged, the broad involvement of inflammation in many major chronic diseases,[74–76] and CR's prevention against the activation of the pro-inflammatory transcription factors, Chung *et al.*[20,77] proposed the molecular inflammation hypothesis of aging. This hypothesis identifies the age-related inflammatory process as a possible molecular cross-talk mechanism that bridges biological and pathological processes. The anti-inflammatory action of CR can be attributed more likely to its ability to suppress age-related oxidative stress, than to a chronically elevated glucocorticoid level, which is potentially deleterious to the organism.[78–80] In this regard, it is important to notice the report of Lee *et al.*,[81] which shows CR's selective suppression of glucocorticoid receptor expression in the hippocampus and cerebral cortex is supportive of such conclusion (more discussion below).

Resistance to Disease by CR

CR's ability to cause an organism's resistance to various stressors is perhaps best exemplified when animals were subjected to life-threatening toxic agents or radiation, as shown in the 1942 pioneering work of Tannenbaum[82] using mouse skin tumors induced by benzo (a) pyrene. A report by Chu *et al.* revealed further insights into how CR enhances resistance by inhibiting the interaction of DNA with the potent carcinogenic aflatox-in B (AFB).[83] These investigators found that AFB-induced hepatic tumors are reduced by more than 50% in CR rats. Moreover, CR reduced AFB-DNA adduct formation by as much as 71% in these rats, depending on the adduct type, compared to control rats. Furthermore, the authors found that *in vitro* nuclear DNA binding of AFB is 37% lower in CR rats than in controls, although exposure to activated AFB was the same for both groups. A similar resistance was shown in a DNA strand break experiment. A more stable double-stranded DNA was maintained following alkaline treatment in CR rats, whereas an approximate four-fold increase in damaged single-stranded DNA was found in *ad libitum*-fed rats.

A study by ThyagaRajan *et al.*[84] showed CR's ability to resist against hormone-inducible tumorigenesis.[18] This group investigated the mechanism by which CR suppresses carcinogen-induced mammary tumors in the rat and whether CR rats have the ability to blunt the action of tumor-promoting estrogen and/or prolactin. Their results show that when challenged with the tumor-promoting hormones, CR, indicating again a strong ability to resist tumor growth even under powerful hormonal stimulation by cancer-promoting estrogen and prolactin, significantly suppressed tumor progression.

An even more remarkable example of CR's ability to resist irradiation is shown in the work of Gross and Dreyfuss, who challenged mice with gamma irradiation to induce tumorigenesis.[85] They not only show a clear-cut tumor suppression by CR, but also that CR rats had a far stronger ability of resisting one of the most deleterious forms of stressors. While 43 out of 89 (48%) of the no irradiated control, *ad libitum*-fed rats showed tumors, none of 77 irradiated rats on the restricted diet developed tumors.

It would be important to know whether the suppression of the tumorigenesis by CR is related to the delay in the onset or the progression of the disease process.[86] Although there is no clear answer to this question at present, re-analyses of earlier works can shed some light on the CR effect against the incidence of leukemia in male Fischer-344 rats.[87] Correlative pathological data on neoplastic lesions and the survival data led to the conclusion that CR delays only the onset of the disease without modulating the progression or the course of leukemia in late life.[86] Additional information on the CR's differential action on the onset and progression of neoplastic lesions can be found by referring to Shimokawa and Higami.[87]

Stress Resistance by CR for Longevity

When exposed to a certain stressor, an organism's adaptive response undergoes three interrelated stages: alarm reaction, resistance, and exhaustion, according to Selye's definition.[88] Because stress itself is a most effective factor in eliciting adaptive responses, the organism's innate nature is to use some form of stress for its own benefit. Such a notion is in line with the basic premise of the proposed stress theory of aging[3] and adaptation hypothesis for longevity.[89]

Hardly an aspect of aging is more important than an organism's ability to withstand stress or to resist both internally and externally imposed insults. We know that as organisms lose their ability to resist these insults, aged organisms suffer more than the young. Therefore, the prime strategy for an organism's survival has been the evolutionarily adapted defense systems that guard against an insult, somewhat like to oxidative stress. Although terms like stress, resistance, and adaptability have long been used in biology, they remain mechanistically and quantitatively poorly defined. In a gerontological context, stress resistance or susceptibility is often discussed in association with an organism's vulnerability to disease and age-related damage.[90,91]

However, to date, there is no clear molecular delineation of the cellular and molecular mechanisms to define such complex biological phenomena. The life-prolonging action of CR seems to offer an excellent opportunity for investigating the interrelationship between stress and the aging process. As an omnipotent intervention, CR provides a unique opportunity to probe an organism's ability to withstand age-related stress as a survival strategy. In this context, the anti-aging action of CR can be viewed as "nutritional stress," because the organism's reduced caloric intake seems to be a stimulatory metabolic response for survivability. Recent gerontological research has provided sufficient experimental data supporting this anti-aging property of CR, of which several pertinent, key examples are discussed below.

Evidence of Stress Resistance by CR

One well-known, exemplified response to stress is the hormonal increase in adrenal corticosterone levels in plasma during aging, where increases in these levels appear to be proportional to the degree of stress. Aged animals appear to have a diminished ability to attenuate the increase, causing the aged to have continually elevated plasma levels of corticosterones. The authors[79] suggest that increased levels of corticosterone in aged rats result in hippocampal neuronal cell death, i.e., the stage of exhaustion. However, this scenario in the glucocorticoid cascade hypothesis is obviously not applicable in the case of the CR paradigm, because CR results in an increased life-span in spite of chronically elevated diurnal levels of serum corticosterone. This apparent contradiction makes the interrelation of glucocorticoid and aging far more complex than one might want to narrowly define it and needs other mechanistic explanations.[77,80,81]

Frame *et al.* viewed elevated glucocorticoid as the major adaptive response to nutrient stress.[91] Data clearly show that biologically active free corticosterone levels in CR rats are higher than those in *ad libitum*-fed rats throughout the animals' life-spans. Interesting and more important questions, however, are, how do chronically elevated, deleterious glucocorticoid levels cause no apparent harm to CR animals, and how do these animals use it to their own advantage. The answers can, in part, be found by CR's ability to maintain a stabilized neuronal membrane status.

Synergistic Effects of CR and Exercise

A consequence of the increased metabolic demand that physical exercise elicits is increased oxidative stress, as indicated by the increased production of oxidants in mitochondria, which seems paradoxical for what said about the anti-aging effects of exercise.[92] In fact, a report shows longer survival spans for exercised CR rats.[93] Exercised CR animals are shown to have an additional extended mean life-span of ~10% beyond their non-exercised CR counterparts.[21] An interesting question arises: How can this be possible if exercise promotes oxidative stress? The answer could come from the CR's unique ability to mobilize a series of adaptive defense mechanisms. For example, in addition to the enhanced anti-oxidative scavengers, the free radical generation of microsomes, in contrast to mitochondria, was shown to be significantly suppressed by exercise in CR rats, even at 20 months old.[92,93]

Another interesting manipulation by CR is its ability to resist exercise-induced oxidative stress on membrane integrity.[92] Data show that exercised CR rats, despite having the increased mitochondrial reactive species production, can maintain mitochondrial membrane fluidity as good as that of young sedentary rats.[21] The mechanisms for such remarkable resistance to stress are likely to be derived from a concerted network of stress-response elements, e.g., lowered membrane peroxidizability and antioxidant defenses. As a consequence, the membrane fluidity can be better preserved by exercise.[91]

CR as a Biological Hormesis

Numerous studies on CR have established its efficacy as the most effective life-prolonging intervention known today, extending both average and maximum life-spans.[94] Emerging views on extended longevity lead us to believe that CR's resistive action against stress is an evolutionary, adapted measure, which is characteristic of a hormetic response.[95]

The concept of hormesis may well offer a biological basis for the CR phenomenon.[95–97] The term *hormesis* by definition describes a beneficial, biological effect at low levels, as seen with reduced caloric intake, which at higher levels would cause deleterious effects. The evidence from genomic profiling data strongly suggests an organism's adaptive response to CR[68] turns on selective genes that are necessary to maintain high metabolically efficient state for the survival of the organism. The maintenance of homeostasis by adapting to stress during aging is likely the key determinant for extended longevity, as observed with the CR paradigm.[90]

Future Directions for CR Research

The important question one should ask is whether or not the anti-aging effects of CR observed in laboratory animals and nonhuman primates can be applicable to human aging; the answer could come the foreseeable future as two groups of investigators in the United States have been conducting CR studies on nonhuman primates[98] over 13 years. Their results so far are encouraging in that most of the data on monkey studies confirm the data collected on lower animals. With this experimental information, we will be one step closer to the applicability of this nutritional paradigm to humans.[99–101]

Some recent developments in CR research could turn out to be significant in revealing more information on the underlying molecular mechanisms of CR. For instance, a recent report by de Cabo *et al.*[102] showed this group's attempt to develop an *in vitro* model of caloric restriction model by treating cultured cells with sera obtained from CR rats or monkeys. Their results showed that cultured cells treated with CR sera had reduced cell proliferation, enhanced tolerance to oxidative stress, and increased heat stress-response gene expression, similar to those animals on CR. Another interesting development on CR is on the question of the relation between growth and life-span, as described in the introduction on MaCay's work. Shimokawa *et al.* reported the reduction of growth hormone-insulin-growth factor-1 in transgenic rat resulted in an extended life-span, implying that CR might mediate its life-extension ability partly through the reduction of the GH-IGF-1 axis.[103]

The CR model has been used as an excellent probe for uncovering a good deal of new information about aging and age-associated disease. By exploring CR's underlying mechanisms, researchers will learn even more about the aging process. Thus, based on past experiences, CR research in the future is expected.

References

1. Park S, Hwang ES, Kim HS, Park WY (eds.) *Healthy Aging for Functional Longevity*, vol. 928. New York Acad Sci, New York, USA.
2. Holliday R (1995) *Understanding Ageing*. Cambridge University Press, Cambridge, UK.
3. Merry BJ (2002) Molecular mechanisms linking calorie restriction and longevity. *Int J Biochem Cell Biol* **34**: 1340–1354.
4. Wei J (2000) *Aging Well*. John Wiley & Sons, Inc., New York, USA.
5. Kitani K, Aoba A, Goto S (eds.) (1996) *Pharmacological intervention in aging and age-associated disorders*, vol. 786. New York Acad Sci, New York, USA.
6. Yu BP (1999) Approaches to anti-aging intervention: the promises and the uncertainties. *Mech Ageing Dev* **111**: 73–87.
7. Yu BP, Kang CN, Han JS, Kim DS (1998) Can antioxidant supplementation slow the aging process? *Biofactors* 7: 93–101.
8. Yu BP (ed.) (1994) *Modulation of Aging Processes by Dietary Restriction*. CRC Press, Boca Raton, USA.
9. Weindruch R, Walford RL (1988) *The Retardation of Aging and Disease by Dietary Restriction*. Charles C Thomas, Publ. Springfield, USA.
10. Snyder DL (ed.) *Dietary Restriction and Aging*. Alan R. Liss, Inc., New York, USA.
11. Park DU, Lee JH, Kim HJ, Kim CH, Hong SE, Chang GY, Yu BP, Chung HY AgingDB: aging database for oxidative stress, calorie restriction in the aging study. *J Am Aging Asso*, in press.
12. McCay CM, Crowell MF, Maynard LA (1935) The effect of retard growth upon the length of life span upon the ultimate body size. *N Nutrition* **10**: 63–79.
13. Robertson TB, Ray LA (1920) On the growth of relatively long lived compared with that of relatively short lived animals. *J Biol Chem* **42**: 71–107.
14. MaCay CM, Maynard LA, Sperling G, Barnes LL (1939) Retarded growth, life span, ultimate body size and age changes in the albino rat after feeding diets restricted calories. *J Nutrition* **18**: 1–13.

15. Kristal BS, Yu BP (1994) Aging and its modulation by dietary restriction. In: Yu BP (ed.) *Modulation of Aging Processes by Dietary Restriction*, pp. 1–36. CRC Press, Boca Raton, USA.
16. Shimokawa I, Higami Y, Utsuyama M, Tsuchiya T, Otani HT, Komatsu T, Chiba T, Yamaza H (2003) Life span extension by reduction of the growth hormone-insulin-like growth factor-1 axis: relation to calorie restriction. *FASEB J* 02-0819fge.
17. Judge S, Judge A, Grune T, Leeuwenburgh C (2004) Shot-term CR decreases cardiac mitochondrial oxidant production and antioxidant enzyme activity. *Am J Physiol Regul Integr Com Physiol* **286**: R254–R259.
18. Cao SX, Dhahbi M, Mote PL, Spindler SR (2001) Genomic profiking of short- and long-term caloric restriction effects in the liver of aging mice. *Pro Natl Acad Sci* USA **98**: 10630–10635.
19. Masoro EJ (1995) Aging: current concepts. In: Masoro EJ (ed.) *Handbook of Physiology Series*, pp. 3–21. Oxford University Press, NY, USA.
20. Chung HY, Kim HJ, Kim WJ, Yu BP (2001) The inflammatory hypothesis of aging: molecular modulation by calorie restriction. In: Park S, Hwang ES, Kim HS, Park WY (eds.) *Healthy Aging for Functional Longevity*, pp. 327–335, vol. 928. New York Acad Sci, New York, USA.
21. Yu BP (ed.) (1993) *Free Radicals in Aging*. CRC Press, Boca Raton, USA.
22. Beckman KB, Ames BN (1998) The free radical theory of aging nature. *Physiol Rev* **78**: 547–581.
23. Warner H, Starke–Reed P (1997) Oxidative stress and aging. In: Clerch LB, Massaro DJ (eds.) *Oxygen, Gene Expression, and Cellular Function*, pp. 139–167. Marcel Dekker, Inc. New York, USA.
24. Yu BP (1994) Cellular defenses against damage from reactive oxygen species. *Physiol Rev* **74**: 139–162.
25. Herman D (1956) A theory based on free radical and radiation chemistry. *J Gerontol* **11**: 298–300.
26. Yu BP, Yang Y (1996) Critical evaluation of the free radical theory of aging. A proposal for the oxidative stress hypothesis. In: Kitani K, Aoba A, Goto S (eds.) *Pharmacological intervention in aging and age-associated disorders*, pp. 1–11, vol. 786. New York Acad Sci, New York, USA.
27. Yu BP (1996) Aging and oxidative stress: modulation by dietary restriction. *Free Rad Biol Med* **21**: 651–668.
28. Matsuo M (1993) Age-related alterations in antioxidant defense. In: Yu BP (ed.) *Free Radicals in Aging*, pp. 143–183. CRC Press, Boca Raton, USA.
29. Kim JW, No JK, Ikeno Y, Yu BP, Choi JS, Yokozawa T, Chung HY (2002) Age-related changes in redox status of rat serum. *Arch Geront Geriat* **34**: 9–17.
30. Sohal RS, Weindruch R (1996) Oxidative stress, caloric restriction and aging. *Science* **273**: 59–63.
31. Cho CG, Kim Hj, Chung SW, Jung KJ, Shim KH, Yu BP, Yodori J, Chung HY (2003) Modulation of glutathione and thioredoxin systems by calorie restriction during the aging process. *Exp Gerot* **38**: 539–548.
32. Lopez–Torres M, Gredilla R, Sanz A, Barja G (2002) Influence of aging and long-term caloric restriction on oxygen radical generation and oxidative DNA damage in rat liver mitochondria. *Free Rad Biol Med* **32**: 882–889.
33. Yu BP, Lee DY, Hwang EH, Lim BO (1999) Calorie restriction: a potent mechanistic solution to the oxygen paradox. In: Robine JM, Forette B, Franceschi C, Allard M (eds.) *The Paradoxes of Longevity*, pp. 93–102. Springer–Verlag, Berlin, Germany.

34. Sell DR, Lane MA, Obrenovich ME, Mattison JA, Handy A, Ingram D, Cutler RC, Roth G, Monnier VM (2003) The effect of caloric restriction on glycation and glycoxidation in skin collagen of nonhuman primates. *J Grentol* **58A**: 508–516.

35. Vereb G, Szollosi J, Matkko J, Nagy P, Farkas T, Vigh L, Matyus L, Waldmann TA, Damjanovich S (2003) Dynamic, yet structured: the cell membrane three decades after the Singer-Nicolson model. *Proc Natl Acad Sci USA* **100**: 8053–8058.

36. Zs–Nazy I (1994) The membrane hypothesis of aging. CRC Press, Boca Raton, USA.

37. Hulbert AJ (2003) Life, death and membrane bilayers. *J Exp Biol* **206**: 2303–2311.

38. Laganiere S, Yu BP (1993) Modulation of membrane phospholipid fatty acid composition by age and food restriction. *Gerontology* **39**: 7–18.

39. Matsuo M, Gomi F, Kuramoto K, Sagai M (1993) Food restriction suppresses an age-dependent increase in the exhalation rate of pentane from rats. *J Gerontol* **48**: B133–B138.

40. Chen JJ, Yu BP (1994) Alterations in mitochondrial membrane fluidity by lipid peroxidation products. *Free Rad Biol Med* **17**: 411–418.

41. Yu BP, Suescun EA, Yang SY (1992) Effect of age-related lipid peroxidation on membrane fluidity and phospholipase A2: modulation by dietary restriction. *Mech Age Dev* **65**: 17–33.

42. Pamplona R, Portero–Otin M, Ruiz C, Prat J, Bellmunt MI, Barja G (1998) Mitochondrial membrane peroxidizability index is inversely related to maximum life span. *J Lipid Res* **39**: 1989–1994.

43. Pieri C (1997) Membrane properties and lipid peroxidation in food restricted animals. *Age* **20**: 71–79.

44. Lee J, Yu BP, Herlihy JT (1999) Modulation of cardiac mitochondrial membrane fluidity by age and calorie intake. *Free Rad Biol Med* **26**: 260–265.

45. Choe M, Jackson C, Yu BP (1995) Lipid peroxidation contributes to age-related membrane rigidity. *Free Rad Biol Med* **18**: 977–984.

46. Yang S, Yu BP (1993) Age-related membrane alterations. In: Watson WW (ed.) *The Handbook of Nutrition in the Aged*, pp. 113–131. CRC Press, Boca Raton, USA.

47. Uchida K (2003) 4-Hydroxy-2-nonenal: a product and mediator of oxidative stress. *Progr Lipid Res* **42**: 318–343.

48. Chen JJ, Bertrand H, Yu BP (1995) Inhibition of adenine nucleotide translocator by lipid peroxidation products. *Free Rad Biol Med* **19**: 583–590.

49. Kristal BS, Chen JJ, Yu BP (1994) Sensitivity of mitochondrial transcription to free radicals and lipid peroxidation. *Free Rad Biol Med* **16**: 323–329.

50. Kristal BS, Park BJ, Yu BP (1994) Antioxidants reduce peroxyl-mediated inhibition of mitochondrial transcription. *Free Rad Biol Med* **16**: 653–660.

51. Kristal BS, Yu BP (1998) Dietary restriction augments resistance to oxidant-mediated inhibition of mitochondrial transcription. *Age* **21**: 1–6.

52. Kristal BS, Yu BP (1998) Dietary restriction augments protection against induction of the mitochondrial permeability transition. *Free Rad Biol Med* **24**: 1269–1277.

53. Kristal BS, Park BK, Yu BP (1996) 4-Hydroxyhexenal is a potent inducer of the mitochondrial permeability transition. *J Biol Chem* **271**: 6033–6038.

54. Lambert AJ, Merry BJ (2004) Effect of caloric restriction on mitochondrial reactive species production and bioenergetics: reversal by insulin. *Am J Physiol Regul Integr Com Physiol* **286**: R71–R79.

55. Lal SB, Ramsey JJ, Monemdjuo S, Weindruch R, Harper ME (2001) Effects of caloric restriction on skeletal muscle mitochondrial proton leak in aging rats. *J Gerontol* **56**: B116–B122.
56. Merry BJ (2002) Molecular mechanism linking calorie restriction and longevity. *Intl J biochem Cell Biol* **34**: 1340–1354.
57. Higami Y, Shimokawa I (2000) Apoptosis in the aging process. *Cell Tissue Res* **301**: 125–132.
58. Ando K, Higami Y, Tsuchiya T, Kanematsu T, Shimokawa I (2002) Impact of aging and lifelong calorie restriction on expression of apoptosis-related genes in male F344 rat liver. *Microsc Res Tech* **59**: 293–300.
59. Rajani R, Shelke J, Leeuwenburgh C (2003) Life-long calorie restriction (CR) increases expression of apoptosis repressor with a caspase recruitment domain (ARC) in the brain, *FASEB J* 02-0803fge.
60. Rao KS (2003) Dietary calorie restriction, DNA-repair and brain aging. *Mol Cell Biochem* **253**: 313–318.
61. Fraga CG, Shigenaga MK, Park JW, Degan P, Ames BN (1990) Oxidative damage to DNA during aging: 8-hydroxy-2-deoxyguanosine in rat organ DNA and urine. *Proc Natl Acad Sci USA* **87**: 4533–4537.
62. Richter CJ, Park JW, Ames BN (1988) Normal oxidative damage to mitochondrial and nuclear DNA is extensive. *Proc Natl Acad Sci USA* **85**: 6465–6467.
63. Chung MH, Kasai H, Nishimura S, Yu BP (1992) Protection of DNA damage by dietary restriction. *Free Rad Biol Med* **12**: 523–525.
64. Hruszkewycz A, Bergtold DS (1990) The 8-hydroxyguanine content of isolated mitochondria increases with lipid peroxidation. *Mutat Res* **244**: 123–128.
65. Kang CM, Kristal BS, Yu BP (1998) Age-related mitochondrial DNA deletions: effect of dietary restriction. *Free Rad Biol Med* **24**: 148–154.
66. Hoffmann S, Spitkovsky D, Radicella JP, Epe B, Wiesner R (2004) Reactive oxygen species derived from mitochondrial chain are not responsible for the basal levels of oxidative base modifications observes in nuclear DNA of mammalian cells. *Free Rad Biol Med* **36**: 765–773.
67. Pletcher SD, MacDonald SJ, Marguerie R, Certa U, Steams SC, Goldstein DB, Partridge L (2002) Genome-wide transcript profiles in aging and calorically restricted Drosophila melanogaster. *Curr Biol* **12**: 712–723.
68. Weindruch R, Kayo T, Lee CK, Prolla TA (2002) Gene expression profiling of aging using DNA microarrays. *Mech Ageing Dev* **123**: 177–193.
69. Lee CK, Klopp RG, Weindruch R, Prolla A (1999) Gene expression profile of aging and its retardation by calorie. *Science* **285**: 1390–1393.
70. Kim HJ, Yu BP, Chung HY (2002) Molecular exploration of age-related NFkB/IKK downregulation by calorie restriction in rat kidney. *Free Rad Biol Med* **32**: 991–1005.
71. Kim HJ, Jung KJ, Yu BP, Cho CG, Choi SJ, Chung HY (2002) Modulation of redox-sensitive transcription factors by calorie restriction during aging. *Mech Ageing Dev* **123**: 1589–1595.
72. Kim HJ, Jung KJ, Yu BP, Cho CG, Chung HY (2002) Influence of aging and calorie restriction on MAPK activity in rat kidney. *Exp Gerontol* **37**: 1041–1053.
73. Chung HY, Kim HJ, Jung KJ, Yoon JS, Yoo MA, Kim KW, Yu BP (2000) The inflammatory process in aging. *Rev Clin Gerontol* **10**: 207–222.
74. McGeer EG, McGeer PL (1999) Inflammation in Alzheimer disease and the therapeutic implications. *Curr Pharm Des* **5**: 821–836.

75. Colwell GA (1999) Inflammation and diabetic vascular complications. *Diabetes Car* **12**: 1927–1928.
76. Teunissen CE, van Boxtel MP, Bosma H, Bosmans E, Delanghe J, De Bruijn C, Wauters A, Maes M, Jolles J, Steinbusch HW, de Vente J (2003) Inflammation markers in relation to cognition in a healthy aging population. *J Neuroimmunol* **134**: 142–150.
77. Chung HY, Kim HJ, Kim KW, Choi JS, Yu BP (2002) Molecular inflammation hypothesis based on the anti-aging mechanism of calorie restriction. *Microscopy Res Techni* **59**: 264–272.
78. Sabatino F, Masoro EJ, McMahan CA, Kuhn RW (1991) Assessment of the role of the glucocorticoid system in aging process and in the action of food restriction. *J Gerontol* **46**: B171–B179.
79. Sapolsky RM, Kery LC, McEwen BS (1986) The neuroendocrinology of stress and aging: the glucocorticoid cascade hypothesis. *Endocr Rev* 7: 284–301.
80. Patel N, Finch CE (2002) The glucocorticoid paradox of calorie restriction in slowing brain aging. *Neurobiol Aging* **23**: 707–717.
81. Lee J, Herman JP, Mattson MP (2000) Dietary restriction selectively decreases glucocorticoid receptor expression in the hippocampus and cerebral cortex of rats. *Exp Neurol* **166**: 435–441.
82. Tannenbaun A, Silverstone H (1953) Effects of limited food intake on survival of mice bearing spontaneous mammary carcinoma and on the incidence of lung metastases. *Cancer Res* **13**: 532–536.
83. Chu MW, Pegram RA, Gao P, Allaben WT (1991) Effects of calorie restriction, aflatoxin B, metabolism and DNA modification in Fischer 344 rats. In: Fishbein L (ed.) *Biological effects of dietay restriction*, pp. 42–54. Springer–Verlag, Berlin, Germany.
84. ThygaRajan S, Meites J, Quadri SK (1993) Underfeeding-induced suppression of mammary tumors: counteraction by estrogen and haloperitol. *Proc Soc Exp Biol Med* **203**: 236–242.
85. Gross L, Dreyfuss Y (1990) Prevention of spontaneous and radiation-induced tumors in rats by reduction of food intake. *Proc Natl Acad Sci USA* **87**: 6795–6797.
86. Shimokawa I, Yu BP, Higami Y, Ikeda T, Masoro EJ (1993) Dietary restriction retards onset but not progression of leukemia in male F344 rats. *J Gerontol* **48**: B68–B73.
87. Shimokawa I, Higami Y (1994) Effect of dietary restriction on pathological processes. In: Yu BP (ed.) *Modulation of Aging Processes by Dietary Rrestriction*, pp. 247–265, CRC Press, Boca Ratan, USA.
88. Selye H, Tuchweber B (1976) Stress in relation to aging and disease. In: Everitt AV, Burgess JA (eds.) *Hypothalamus Pituitary and Aging*, pp. 553–569. Charles C Thomas, Publ. Springfield, USA.
89. Parson PA (1996) The limit to human longevity: an approach through a stress theory of ageing. *Mech Aging Dev* **87**: 211–218.
90. Yu BP, Chung HY (2001) Stress resistance by calorie restriction for longevity. In: Park S, Hwang ES, Kim HS, Park WY (eds.) *Healthy Aging for Functional Longevity*, pp. 39–47, 928 New York Acad Sci, New York, USA.
91. Frame LT, Hart RW, Leakey JEA (1998) Calorie restriction as a mechanism mediating resistance to environmental disease. *Environ Health Perspect* **106**: 313–324.
92. Kim JD, Yu BP, McCarter JM, Lee SY, Herlihy JT (1996) Exercise and diet modulate cardiac lipid peroxidation and anti-oxidant defenses. *Free Rad Biol Med* **20**: 83–88.

93. Kim JD, McCarter RJM, Yu BP (1996) Influence of age, exercise and dietary restriction on oxidative stress in rats. *Aging Clin Exp Res* **8**: 123–129.
94. Calabrese EJ, Baldwin IA (1998) Hormesis as a biological hypothesis. *Environ Health Perspect* **106**: 375–362.
95. Rattan SIS (2004) Aging, anti-aging, and hormesis. *Mech Ageing Dev* **125**: 285–289.
96. Hoffmann AA, Hercus MJ (2000) Environmental stress as an evolutionary force. *Bioscience* **50**: 217–226.
97. Neafsey PJ (1990) Longevity hormesis. A review. *Mech Ageing Dev* **51**: 1–31.
98. Mattison JA, Lane MA, Roth GS, Ingram DK (2003) Calorie restriction in rhesus monkeys. *Exp Gerontol* **38**: 35–46.
99. Walford RL, Mock D, Verdery R, MacCallum T (2002) Calorie restriction in biosphere 2: alterations in physiologic, hematologic, hormonal, and biochemical parameters in humans restricted for a 2-year period. *J Gerontol* **57**: B211–224.
100. Heibronn LK, Ravussin (2003) Calorie restriction and aging: review of the literature and implications for studies in humans. *Am J Clin Nutr* **78**: 361–369.
101. Hursting SD, Lavigne JA, Berrigan D, Perkins SN, Barrett JC (2003) Calorie restriction, aging, and cancer prevention: mechanisms of action and applicability to humans. *Annu Rev Med* **54**: 131–152.
102. de Cabo R, Furer-Galban S, Anson RM, Gilman C, Gorospe M, Lane MA (2003) An *in vitro* model of caloric restriction. *Exp Geront* **38**: 631–639.
103. Shimokawa I, Higami Y, Utsuyama M, Tuchiya T, Komatsu T, Chiba T, Yamaza H (2002) Life span extension by reduction in growth hormone-insulin-like growth factor-1 axis in a transgenic rat model. *Am J Pathol* **160**: 2259–2265.

11

Nutritional Interventions in Aging and Age-associated Disorders

Kenichi Kitani
Ex. Director General, National Institute for Longevity Sciences
36-3, Gengo, Moriokacho, Obu-shi, Aichi, Japan

INTRODUCTION

Nearly every country in the world is undergoing a rapid increase in the ratio of the elderly population to that of the younger, and the long-cherished human desire for longevity has now become the most urgent social and political issue in today's society. The only practical solution that gerontology has, at present, is to examine whether our eating and drinking habits can improve the health status of the elderly and reduce the risk of so-called age-associated disorders such as cancer, cardiovascular disease (CVD), osteoporosis and, hopefully, senile dementias including Alzheimer disease (AD).

This chapter deals with two main types of information. The first looks at experiments performed in the past on animals in an effort to increase their health status and extend their life spans. The second examines past information on food and drink of different kinds that have been reported to be beneficial for the health of the elderly, in an attempt to provide the scientific rationale underlying the effects of these substances, so that further improvements of these strategies can be explored. Although huge numbers of epidemiological studies related to this topic have been published, many of the results are at least partially contradictory and therefore inconclusive. The author has attempted here to report and emphasize some relatively new points so that this chapter is not simply repetitive but hopefully a complementary extension to other such chapters.

Nutritional Interventions in Aging

Since the first proposal of the "Free Radical Theory of Aging" (FRTA)[1] (for a more recent elaboration, see Beckman and Ames[2]), a considerable number of studies where various chemicals possessing some antioxidant properties have been administered to animals, in hopes that they would survive for a longer time. While a small but significant increase in longevity (average life span) was achieved in some of these studies, no study has ever convincingly demonstrated a dramatic increase in the longest survival by this type of approach, as reviewed by Harman[3] and more recently by Le Bourg.[4] A recent study[5] feeding mice with diets containing 5 different antioxidant chemicals also did not demonstrate any significant prolongation of life spans between treated animals and their control counterparts. These combined studies suggest that the administration of a single antioxidant or a combination of several does not dramatically modify the life spans of animals. However, our wish in general should not be how to live for a longer period but be how to live in better health and more independently in the last 20 to 30 years of our lives. For this purpose, the food and drink we consume every day appear to have a great impact, probably much greater than people usually think.

Nutritional Interventions in Age-associated Disorders: What Should We Eat and Drink?

Bad Foods Shorten the Life Span: An Example in Okinawa

Okinawa, the southern islands of Japan, had continuously enjoyed the longest longevity (average life span) among all forty-seven prefectures in Japan. The most recent survey on the public health trend taken by the Ministry of Health, Labor and Welfare of Japan has revealed, however, astonishing results. The longevity of male Okinawans is no longer at the top but is 26th (below the middle), although female Okinawans still remain at the top. Quite shockingly, while male Okinawans above 55 years of age still maintain lower death rates in every age group compared with corresponding average values in Japan, for younger male Okinawans below 55 years of age, the death incidences were all higher than corresponding average values of Japanese males of the same age. The end of World War II and the subsequent occupation by the U.S. army appear to have brought a drastic environmental impact to these islands, because Okinawans were suddenly exposed to a culture totally different from their own traditional one, including eating habits. A major change in their diet was the replacement of pork they had been eating for so many years with high calorie, high fat, high salt corned beef for preparing their traditional dishes. Corned beef may not be the only food to be "guilty". Okinawans also began eating "fast foods". Accordingly, the overall drastic changes in Okinawa's eating culture appear to have produced the outcome of the new 26th position, at least for males. The difference observed in longevity change between male and female Okinawans is an interesting but unanswered puzzle. It is possible that males who worked outside their homes had closer and more frequent contacts with American soldiers than females who tended to stay at their homes, so that the rapid change of eating habits started earlier and more intensely for males than females. Another possibility is that females are more resistant to adverse environmental changes than males. Both male and female Okinawans are now at the top in Japan in terms of obesity ratio. We need to observe how the longevity of female Okinawans will change in the future. Interestingly, those who emigrated from Okinawa to Hawaii in the USA and maintained their traditional cultures from Okinawa lived three years longer on average compared with their colleagues who lived in Okinawa. These facts tell us how important it is to select our foods so that we can live for a longer period in better physical and mental health conditions. A big question is *how*.

Green-yellow Vegetables and Various Fruits and Antioxidant Vitamins

Several decades after the proposal of the "FRTA", scientists and physicians began to suspect and attempted to prove that there might exist some positive or negative relationship between the foods and drinks people take and the incidences of several disorders, cancer and cardiovascular disease (CVD), etc. In the early eighties, Peto *et al.*[6] published a paper suggesting a beneficial effect of dietary β-carotene in reducing the risk of cancer. With growing awareness that these vegetables and fruits are the major sources of so-called antioxidant vitamins such as vitamins A, β-carotene, E and C, many (but not all) epidemiological studies have been published which demonstrated the associations between the intake from foods[6–15] and/or plasma levels[16–22] of these vitamins and the total mortality,[10,14,15,20] or incidence of specific disorders such as cancer[7–10,18] and CVD,[11,14,19–22] especially myocardial infarction (MI). In 1987, Gey *et al.*[22] demonstrated in 3 out of 6 different countries where the mortality due to ischemic heart disease (IHD) is high, people had much lower cholesterol standardized plasma vitamin E and C levels as well, compared with other regions where the mortality is low. They also suggested that these vitamins may reduce the risk of cancers of different organs such as lung, stomach and intestine, etc.[22] Numerous studies have reported that eating green-yellow vegetables and fruits may reduce the risk of CVD and cancers. One study reported that in healthy centenarians, plasma vitamin A and E levels were significantly higher than those in the younger three generations, implying that the high plasma concentration of these vitamins is a key factor for longevity.[23]

An important point which should not be missed from the majority of these past studies is, however, that a high consumption of foods containing vitamins A, E and C or high plasma levels of these vitamins do not necessarily mean that these vitamins *per se* resulted in the lower risk of these age-associated disorders. Foods containing these antioxidant vitamins also contain huge amounts and numbers of species of known and unknown components, mostly antioxidants such as carotenoids and flavonoids and some other micronutrients. Folate could be one example that is abundantly contained in Mediterranean green-yellow vegetables. One study reported that a higher intake of folate was significantly associated with the lower risk of MI.[24] Since elevated homocystein levels in plasma is a high risk factor for CVD and since it is known that folate can reduce homocystein levels, they concluded that the dietary folate intake may be one independent protective factor for MI, which sounds reasonable.[24]

Another example may be lycopene. Compared with vitamin A, α, β and γ-carotenes and vitamin E, studies on lycopene have begun rather recently. After a four-year follow-up of more than 50,000 male subjects who were initially free of any type of cancer, Giovannucci *et al.*[25] found a very strong negative association between the intake of carotenoids and the incidence of prostate cancer only for people who ate certain pre-pared tomato (tomato source) dishes but not raw tomatoes. Intakes of α, β and γ-carotenes, lutein and β-cryptoxanthin were not associated with the risk of prostate cancer. Eating both tomato source dishes and tomatoes or pizza showed weaker associations. This may be explained by the fact that the highly fat-soluble lycopene cannot be efficiently absorbed by simply eating raw tomatoes. Boiled (or fried) tomatoes with some oil in them is the best way to absorb lycopene. The highest negative association between the tomato source intake and the prostate cancer incidence suggests that lycopene is of paramount importance compared with other carotenoids as far as the prevention of prostate cancer is concerned. Very recently, a more direct study has reported that women with plasma lycopene levels above the medium level had a 34% reduction in the risk for MI compared with the lowest quartile.[26] They suggested that a lower lycopene plasma level may become a significant risk factor for MI.

Joseph and colleagues[27–29] have shown in a series of studies that dietary blueberries, strawberries or spinach supplementation can ameliorate age-related declines in neuronal, cognitive and motor function in old rats. Blueberries in particular have recently been shown to enhance memory-associated neuronal signaling molecules and to prevent the age related behavioral decline normally seen in a transgenic mouse model of AD.[29] Blueberries are known to be rich in antioxidants, in particular anthocyanine. They believe that some (but not all) of the mechanisms underlying these effects are due to antioxidants and partly some anti-inflammatory actions by substances contained in blueberries, strawberries, and spinach and other brightly colored fruits and vegetables.[27–30] However, they are not able to fully explain all of their observations as due to the combination of antioxidant and anti-inflammatory effects. They believe that a combination of both known and still unknown actions of unknown substances contained in these fruits and vegetables are responsible.

The beneficial effects of taking flavonoids from various sources (primarily fruits and vegetables) for reducing the risk of an age-related decline in cognitive function or various types of dementia are increasingly found in recent publications,[31] which fits into the observations made in old rats by Joseph and colleagues,[27–29] who emphasized the value of colorful fruits and vegetables.[30]

Recently, the concept of chemoprevention of cancer has been advanced. Many green vegetables, especially cabbage, broccoli and Brussels sprouts, were found to induce benzpyrene hydroxylase activities in the intestines and livers of rodents.[32,33] Benzpyrene (or other chemical carcinogen)-induced neoplasia can be prevented by these enhanced enzyme activities which hydroxylize these chemical carcinogens. Thus, this property of these vegetables may explain at least in part previous observations of reduced risks of gastrointestinal cancers in ingesters of these vegetables.[34,35] This may be only one example that vegetables work for the prevention of cancers by means of mechanism(s) other than antioxidant effects of vitamins.

Should Supplementation with Vitamins be Encouraged or Discouraged?

As has been reviewed above, high levels of antioxidant vitamins (A, E, C) in plasma or higher intake of foods containing these vitamins are associated with the lower risk of CVD and cancer. Given these findings, people and even doctors may be tempted to take vitamin supplements routinely with the hope that these vitamins will help improve their health and prevent these unwanted disorders. At least theoretically, however, all antioxidants can work as prooxidants in certain conditions. Accordingly, we must carefully examine whether vitamin supplementation is good or bad for our health. Although some studies suggest possible merits of supplementation (especially of vitamins E and C),[36,37] results of past studies are not convincing enough to encourage routine supplementation, since the results of many studies were negative in this regard.[38–40]

Further, one prospective double-blind study that administered a pharmacological dose of β-carotene in Finnish smokers found a significant increase in the incidence of lung cancer and a non-significant increase in cardiovascular mortality in the β-carotene treated group at the time of the termination of the study (5 ~ 8 years after the start).[41] There is another similar study reporting an adverse effect of a pharmacological dose of β-carotene in workers exposed to asbestos.[42] In an experimental study to follow up these results, ferrets which, unlike most other laboratory animals that do not absorb β-carotene, do absorb β-carotene as humans do, were exposed to cigarette smoking combined with a pharmacological dose of β-carotene and were found to have the greatest proliferative response in lung tissue and squamous metaplasia.[43] In this study, a higher accumulation of β-carotene in oxygen rich lung tissue caused a production of prooxidants from β-carotene. Interestingly, β-carotene treatment alone produced a higher incidence of proliferative response in the lung than in controls or even in ferrets exposed to smoking alone.[43]

Vitamin E supplementation was shown to enhance immune response in the elderly and was recommended for the prevention of infection (especially pneumonia) in the elderly.[44] However, several trials found no significant difference between control and supplemented elderly groups in terms of the incidence of infections.[45,46] The merit of vitamin E supplementation for infectious diseases in the elderly appears to be under debate.[47] The intake of vitamin A (and vitamin E also) increases the risk factor of osteoporosis and eventually bone fracture by several different mechanisms including the interaction with vitamin D absorption in the intestine.[48] Furthermore, in malnourished populations vitamin A impairs immune responses.[47]

Taking all the information together, there seems to exist no clear rationale to encourage vitamin supplementation. In particular, the routine intake of β-carotene and possibly vitamin A in high doses should be discouraged. Vitamin E, in a dose maintaining physiological levels in plasma, may not cause adverse effects, however, we should be cautious about a higher dose, since there is no clear evidence for its benefit and since a pharmacological dose was reported to drastically reduce vitamin C concentration in rat brain.[49] Whether supplementation should be recommended or not, however, remains controversial in the literature. Diplock[50] in a large review expressed a favorable view on vitamin E supplementation to reduce risk of CVD. However, there are some other reviews which convey a rather conservative view on this issue.[51] This author belongs to the latter group. Another demerit of supplementation is that vitamin supplemented people tend to think that they are sufficiently supplemented with all the necessary elements to maintain their health (which is certainly not true) and so tend to be lazy about eating enough foods containing these vitamins and many (mostly unknown) other important elements.

Meats; is Serum Cholesterol Related to Meat Intake?

Protein is an essential element of the body composition, and we need to eat proteins to make our own proteins. Hypoalbuminemia is a commonly observed sign of morbidities in the elderly, including protein calorie malnutrition.[52,53] A low intake of proteins is one big cause for hypoalbuminemia in the elderly which comes from different causes, such as poverty and poor food quality, loss of taste, loss of appetite, loneliness, depressive conditions and poor mastication and some gastrointestinal disorders.[52]

Hypoalbuminemia, in turn, causes multiple morbid conditions and/or worsens them.[52,53] It has been shown that the lower the serum albumin level, the shorter the life expectancy in the elderly.[53] Furthermore, albumin itself is a potent antioxidant. As a source of proteins, meat is as good as or probably better than other sources such as fish and beans, since the amino acid composition of meat is the closest to that of humans. Fish and beans have some merits over meat, not as protein sources but because they have additional components, as will be discussed later. Further, these three protein sources contain different vitamin Bs and thus should be mutually complementary. A number of epidemiological studies have shown a high positive correlation between serum cholesterol concentration and the incidence of CVD, mainly MI,[54] leading to a number of scientists to warn against an excess intake of proteins. However, less than one third of serum cholesterol comes from protein intake and the remaining two-thirds or more is synthesized in the liver in humans. Accordingly, the intake of proteins, especially from meat, is not likely to be responsible for hypercholesterolemia and hence for higher incidence of MI. Meat eating itself does not seem to elevate serum cholesterol concentration. In contrast, a cholesterol-lowering effect of eating meat has been reported, possibly because meat has a relatively large amount of polyunsaturated fatty acids (PUFAs), which can lower serum cholesterol. The risk of saturated fatty acids (SFAs) and unsaturated fatty acids (UFAs) will be discussed in the next section on fats.

As to the serum cholesterol levels, however, it is never "the lower, the better". A lack of association between cholesterol level and coronary heart disease morbidity and all-morbidity and all-mortality in persons older than 70 years of age was reported.[56] In the elderly population in the Netherlands, a low total cholesterol in serum was found to be a clear risk factor for their survival.[57] In this study, higher mortalities due to cancer and infection were demonstrated in the group with the lowest serum cholesterol level, while the mortality due to CVD was not associated with serum cholesterol level.[57] In agreement with this finding, a ten-year follow up study of residents over 40 years of age performed in Toda city, Japan, showed that the hypocholesterolemic groups had the highest mortality compared with normocholesterolemic or even hypercholesterolemic groups in both male and female subjects.[58] The low cholesterol level in serum was a significant risk factor for cancer in males, although in females the value did not reach statistical significance.[58]

In contrast with CVD, an inverse correlation between the serum cholesterol level and the mortality due to a stroke was reported.[59] A WHO-coordinated Cardiovascular Diseases and Alimentary Comparison (CARDIAC) study has also demonstrated an inverse correlation between the two.[54] Yamori and coworkers demonstrated that stroke prone spontaneously hypertensive rats (SHRSP) developed by this group could survive for a much longer time when given a protein and salt rich diet than those given salt only or a low protein and salt rich diet.[60] They have also shown that stroke and hypertension can also be prevented in humans by supplying appropriate amounts of protein. They have also shown a significant inverse association between arterial blood pressure and 24 hr-urinary excretion of methylhistidine (a marker of animal meat intake).[61]

Fats [Saturated (SFAs) vs. Unsaturated Fatty Acids (UFAs)]: a Myth or the Truth

It is a recent trend to discuss the idea that a higher intake of fats, particularly animal fats, leads to a higher risk of CVD (MI in particular) and cancer. Atherogenic diets containing fats with more SFAs mainly from animals are thought to operate by oxidizing serum low density lipoprotein (LDL), and a high oxidized LDL is thought to stimulate atherosclerosis leading to CVD. In contrast, UFAs, which are contained in high amounts in oils of seeds of various plants (*e.g.* olive oil, flax oil, safflower oil, sesame oil, etc.), are believed to counteract this change and are regarded as good oils.

Since Renaud and Lorgeri[62] demonstrated a highly positive significant correlation between the incidence of CVD mortality and the consumption of dairy fats in different countries in Europe, many data that go along with this idea of "bad fats from animals" have been published. However, many reports which more or less contradict this thesis can also be found in the literature.

Ravnskou[63] recently reexamined this issue by carefully collecting many data reported in the past. It is rather surprising that he found that only one cross-sectional study, three cohort studies, and one trial study were unequivocally supportive of the above idea, while all studies of the secular trends, more than 50 cross-sectional and cohort studies, and eight trial studies were contradictory.

He concluded that there is little evidence that SFAs as a group are harmful and that UFAs as a group are beneficial.[63] Ulbricht and Southgate already emphasized in 1991 that a simple comparison of SFAs vs. UFAs may be misleading, since there are differences in terms of thrombogenicity and atherogenicity among different SFAs and among different UFAs.[55] For example, while short-chain SFAs (C10 or below) do not raise serum cholesterol levels, longer-chain SFAs such as lauric acid (12:0), myristic acid (14:0) and palmitic acid (16:0) are potent raisers of cholesterol level (atherogenic) as well as thrombogenic. On the other hand, linoleic acid (18:3n−6) which was regarded as an efficient cholesterol lowering oil was found to reduce HDL as well. In contrast, oleic acid (18:1n−9) lowers total cholesterol as well as LDL, but not HDL.[55] They suggested that the thesis of SFAs vs. UFAs is much too simplistic and will certainly lead to much confusion in interpreting many epidemiological data. Even with this consideration in mind, however, discrepancies found in the past studies as pointed out by Ravnskou[63] do not appear to be clearly resolved. A 15 year-interventional study reported in 1991 from Finland showed that a group given an oil rich in linoleic acid had a 2.4-fold higher mortality due to CVD and a 1.4-fold higher general mortality than the control group.[64] In contrast, subjects who had acute MI and were given low linoleic and high α-linolenic (18:3*n*−3) and oleic acids for 5 years showed a 70% lower mortality due to CVD than did the control group,[65] and this was far less than the 30% lower mortality obtained by the use of simvastin for the same time period of 5 years.[66] These observations suggest that a higher intake of linoleic acid may be a risk factor for CVD and even cancer, as will be discussed later. Linoleic acid was shown to decrease serum cholesterol that was increased by dairy fat (butter)[62] and has been advocated and used for the prevention of CVD. However, its effect turned out to be transient, occurring only in the first few weeks of fat feeding, while on a long-term basis, there was no difference in serum cholesterol levels no matter which fat was fed due to a metabolic feed back mechanism (e.g. Iso *et al.*[59]). Okuyama[67] emphasizes that the *n*−6/*n*−3 ratio of dietary fatty acids rather than hypercholesterolemia is the major risk factor for coronary heart disease.

A recent study[68] has reported that higher intakes of fat, SFAs, in particular, increase the risk for dementia, especially of vascular nature, while fish consumption, an important source of *n*−3 polyunsaturated fatty acids (*n*−3 PUFA), was inversely associated with the incidence of dementia and in particular with that of AD. This subject will be discussed again in the following section for fish.

How about the cancer risk in terms of animal fat? Generally, the replacement of animal fat by vegetable oil is recommended to reduce the risk of cancer.[69] The primary support for this thesis is that countries with a low animal fat intake have had lower rates of cancers of different organs such as breast, colon, and prostate. However, much of the effect of dietary fat was suggested to be due to an increase in total energy intake,[55] which profoundly increased mammary tumor incidence in animals. The possibility that a high intake of linoleic acid or other PUFAs commonly found in vegetable oils may increase the risk for cancer has been suspected from animal studies and some population comparisons.[70] Zock and Katan[71] addressed this issue by the means of meta-analysis of about 50 studies reported in the past. They found no strong evidence indicating that a diet high in linoleic acid increases the risk of breast, colorectal, or prostate cancer.[71] They did, however, put forth the caution that even a small risk for these particular cancers can lead to an enormous number of cancer victims, suggesting that linoleic acid should be replaced by oleic acid, if the latter is proved as effective as the former for reducing the risk of CVD.

Olive oil is rich in oleic acid and is abundantly used in so-called Mediterranean dishes. Lipworth *et al.* concluded in their recent review on this issue that there is converging evidence for a protective effect of olive oil for breast cancer but no definitive evidence for other cancer types, and overall epidemiologic evidence, although promising, is quantitatively limited and qualitatively suboptimal.[72]

α-Linolenic acid, which is almost exclusively contained in perilla oil (from Perilla frutescens), linseed oil (amani-yu) or some nuts such as walnuts, is $n-3$ PUFA and is metabolized into docosahexaenoic acid (DHA; 22:6$n-3$) and eicosapentaenoic acid (EPA;20:5$n-3$) in the body. In contrast, linoleic acid is biotransformed into arachidonic acid, from which many bioactive substances called inflammatory mediators (or eicosanoids) are formed. Hirose *et al.* have shown that rats given safflower oil or soy oil and carcinogens showed much larger numbers of cancers than rats given perilla oil and the same carcinogens.[73] Further, Minoura *et al.* reported that the incidence of colonic cancer as well as prostaglandin E production were much higher in rats given linoleic acid rich oil than EPA rich oil.[74] Okuyama speculated that there must exist a heavy competition between α-linolenic acid-produced mediators ($n-3$ mediators) and ($n-6$) inflammatory mediators, which may explain the cancer promoting effect of linoleic acid and the anti-cancer effect of α-linolenic acid.[75] It appears that we need to reevaluate the whole aspect of fatty acids and health. It means that we may need to carefully reexamine the thesis that all plant oils are good for our health.

Apart from differences in fatty acid composition, there is another important difference between plant oils and animal fats, that is, their additional micronutrients, mostly antioxidants which are richly concentrated in seeds and eventually in oils from plants. For example, sesame oil was found to contain several potent antioxidants, which were shown to prevent a number of experimentally induced pathologies in animals.[76] More importantly, some chemicals are not radical scavengers in their original form in oil, but are biotransformed into potent radical scavengers.[76] Most other plant oils (especially olive oil which is obtained from whole fruits) contain similar but different antioxidants. In fact, Violio and Galli reported that nonglyceride components of olive oil, especially the polyphenols, and hydroxytyrosol in particular, are potent antioxidants which may contribute to the lower incidence of CVD in the Mediterranean area.[77]

As discussed earlier, the traditional Mediterranean dishes that are rich in olive oil have been shown to reduce the risk of CVD and cancers. In the future, epidemiological studies may also need to take into account this difference between animal fats and plants oils.

Fish, One Major Element of the Mediterranean Diet: *n*−3 Polyunsaturated Fatty Acids (*n*−3 PUFA) and Taurine

n−3 PUFA such as DHA and EPA are contained in certain fishes (e.g. sardines, mackerel, salmon, and other so called fatty fish) in large amounts. It is believed that these PUFAs can counteract the development of coronary thrombosis and reduce the risk of coronary arterial diseases by means of their anti-inflammatory, anti-arrhythmic and in particular anti-thrombotic effects by inhibiting platelet aggregation.[54,78] There is some evidence that these PUFAs are effective in preventing cancers.[79] The lower incidence of MI in Japan, which is less than one third that of the US, is attributed in part to the intake by the Japanese of fish containing *n*−3 PUFA. *n*−3 PUFA and taurine outputs in urine (markers for fish eating) is the highest in the Japanese among all other people in the world investigated by CARDIAC.[54]

A particular merit of fish protein which has not been recognized until recently is an amino acid, "taurine", which is rich in concentration in certain (relatively large) fishes like red snapper. Experimentally it has been shown that the feeding of taurine to SHRSP rats reduces the blood pressure as well as the death rate from apoplexy.[80] Furthermore, there are also some successful interventional studies in humans by feeding taurine or fish for reducing high blood pressure in a population.

The primary component of membrane phospholipids in the brain is DHA. The major source of preformed DHA is fish. Morris *et al.* have shown that those who consumed fish once a week or more had a 60% smaller risk of AD compared with those who rarely or never ate fish (RR, 0.4).[81] The total intake of *n*−3 PUFA and that of DHA were significantly associated with the reduced risk of AD.[81] In AD patients, plasma total *n*−3 PUFA levels were reported to be decreased to 60 to 70% of control levels.[82] There is increasing evidence in recent literature for a beneficial effect of eating fish for reducing the risk of dementia.[81,83–85] A highly significant inverse correlation was reported between the incidence of depression and the fish consumption among 9 different countries.[86] In the past, the benefits of the Mediterranean diet have been discussed with an emphasis on vegetables and fruits. Now, we need to recognize that fish is also a very important element of this diet as well as of traditional Okinawan one.

Soybeans and Other Beans

Categorically these should be included in the green-yellow vegetables. However, the importance of these beans can not be discussed simply in relation to antioxidant vitamins such as A, C and E. Soybeans, which contain a large amount of high quality protein (about 35%), can be a good protein source of vegetable origin. Some essential amino acids lacking in soybeans are abundant in rice protein, while certain essential amino acids rice protein lacks are rich in soy bean protein. Thus, the combination of both soybean and rice provides an ideal complementary protein source of plant origin for the human body. Conversely, an intake of protein from a single plant origin leads to a deficiency of essential amino acids.

Soybeans and other beans also contain a lot of vitamins, in particular vitamin E. Probably more importantly, these beans contain so many phytochemicals, primarily isoflavones[76,87] as well as lycopene and many more. Soybean isoflavones were suggested to have a weak estrogen-like hormonal activity which, once absorbed, may work to prevent the progression of osteoporosis, as well as to retard the cognitive function decline and importantly to reduce arterial blood pressure. Isoflavones are also considered to be effective in preventing breast cancer. Soy beans also contain many other micronutrients (lecithin, choline, vitamin E) and minerals (Ca^{2+}). Plant sterols such as sitosterol contained in soybean and other beans can reduce the cholesterol absorption from the intestine by means of mutual competition at the level of intestinal absorption.[88]

Salt

It has been established that the arterial blood pressure (and an incidence of cerebral stroke) and salt intake (or urinary output) are highly positively correlated.[54,61] In general, the intake of $6 \sim 7$ g/day is recommended for prevention of hypertension and apoplexy in Japan which is close to the traditional Okinawan intake. In the US, a much lower intake ($3 \sim 6$ g/day) is recommended. There exits a wide range of salt sensitivity in terms of developing hypertension among different ethnicities.

Seaweed, Wakame, Kelps and Fibers

One reason for the longevity of people living in Okinawa is believed to be due to their traditional habits of eating large amounts of seaweed, primarily kelp. Fibers contained in kelp adsorb salt in the intestine, reducing the salt absorption. Obata *et al.* have shown that a dietary fiber (psyllium) intake attenuated the intestinal absorption of Na^+, leading to the prevention of hypertension in SHRSP rats fed a high salt diet.[89] Also, water-soluble fibers in sea weeds such as arginic acid retard glucose and cholesterol absorption. Furthermore, seaweed contains many minerals and vitamins (Mg^{2+}, Ca^{2+} and vitamin E) and carotenoids (fucoxanthine). There is some evidence that dietary fibers reduce the risk of cancer of the colon by adsorbing carcinogens in the gut.

Wine and Other Alcohols

Ethyl alcohol itself is known for its anti-platelet aggregation activity.[90] It is effective for preventing arterial thrombosis but not arterial hemorrhage. An excess intake of heavy liquor leads to chronic alcoholism, however. Accordingly, the intake of alcoholic beverages has a U-shape effect suggesting that moderate drinking of alcoholic beverages may be good for the health.[91,92] For most alcoholic beverages, a sharp rise in the adverse effect with increasing amounts of alcohol intake is clearly documented.[91–93]

Different from almost all other alcohol beverages, red wine contains flavonoids, many of which are known to be potent antioxidants. It has been reported that moderate wine drinking (<500 mL), but not that of beer or spirits, has a beneficial effect on CVD.[94]

Highly antioxidant flavonoids contained in high amounts in red wine (the content is quite variable among different labels, vintages, etc.) are believed to be at least partially involved in this effect. Red wine (but not vodka) drinking prolonged the oxidation time of LDL in plasma, which may possibly prevent the progression of arteriosclerosis.[95,96] Maxwell *et al.* reported that in those who ingested 570 mL of red wine, the antioxidant activity in serum was increased more than two-fold.[97]

Recently, one study reported a possibly beneficial effect of moderate (22–32 g/day alcohol) wine drinking[98] which not only lowers the risk of overall mortality (RR 0.67, $P < 0.001$) and that by CVD (RR 0.55, $P < 0.003$) but that of cancer (RR 0.78, $P < 0.05$), while heavy drinking increased RR. In contrast, beer drinking did not demonstrate a lower risk for cancer. As a possible anti-cancer substance in red wine, they postulated that resveratrol, a phenolic compound reported to inhibit the 3 main stages of carcinogenesis,[99] in addition to its known antioxidant effects, might be responsible. RR values of the heaviest wine drinkers (>131 g/day) were 1.84 ($P < 0.001$) for all causes, 1.98 ($P < 0.001$) for cancer and 1.99 for other causes ($P < 0.002$), also suggesting a U-shape effect of wine drinking.

Orggozo *et al.*[100] in the 3-year follow up studies in 3777 community residents over 65 years of age have found that compared with non-wine drinkers, or mild wine drinkers (less than 250 mL/day), moderate drinkers (250–500 mL/day) had a significantly lower RR for dementia (RR 0.18, $P < 0.01$) and even for AD (RR 0.25, $P < 0.02$). Even heavy drinkers (>500 mL/day) also had lower RR values (0.24 for both dementia and 0.43 for AD), but these values were not significant ($P > 0.05$).[100] These authors were cautious enough in concluding that a moderate intake of red wine is recommended for the prevention of these disorders but suggested that there is no scientific rationale to recommend abstention of wine drinking for those who are used to drinking a moderate amount of wine. Since CVD, apoplexy, hypertension and diabetes are all risk factors for AD,[101] it appears reasonable to assume that wine flavonoids known to prevent arteriosclerosis are also of help in many ways for reducing the risk of AD.

Tea and its Polyphenols

Different kinds of teas contain different kinds of polyphenols. In green tea, catechins of different kinds possessing different antioxidant potencies are in abundance. Drinking tea was reported to reduce the risk of different cancers.[102] The anti-tumor promoting activity of (−)-epigallocatechin gallate (EGCG) (a major antioxidant catechin in green tea) was also reported.[103] A number of studies demonstrated that green tea drinking or catechin feeding can prevent experimentally-induced cancers in different organs in laboratory animals.[104] Fujiki *et al.* proposed a new concept of tumor promotion by tumor necrosis factor-α (TNF-α), which was inhibited by (−)EGCG and green tea.[104] Other mechanisms for the inhibitory effects of tea for tumor growth are also proposed.[105]

Tea catechins were also shown to reduce serum levels of lipids and inhibit peroxidation of LDL[106] suggesting that they may also be beneficial for preventing CVD.[107] Again results of epidemiological studies are inconsistent.[107–109] This is not surprising since the amount of tea intake by different people and quantities of efficient antioxidant catechins such as (−)EGCG contained in different teas are quite variable and since so many confounders are not easily controlled in these studies.

Again, the possibility that some micronutrients possessing antioxidant activities other than tea catechins might also be responsible for the beneficial effect of teas in preventing CVD and cancers should not be excluded. Mazur *et al.*[110] recently reported the presence of phytoestrogens and lignans in teas and to a lesser extent in coffee, hypothesizing that antioxidant lignan polyphenols may also contribute to the preventive effects of tea on CVD.

Herbs and Spices

Despite the failure to find any single element or even a clear group of elements which reduces the risk of diseases, people eating the so-called Mediterranean diets are reproducibly reported to be healthier.[15,65] One point that has not been studied carefully may be the abundant use of different spices in Mediterranean dishes. Almost all spices are high in content of antioxidants of different kinds, although few experiments have been done on these spices in animals or humans. However, as an example, a major antioxidant contained in turmeric for Indian curry was found to be curcumin, which is biotransformed during intestinal absorption into a more potent antioxidant, tetrahydrocurcumin.[76,111] Osawa and others have shown that feeding of curcumin or tetrahydrocurcumin to experimental animals prevented cancers in different organs, atherosclerosis etc.[76,111–113] The anti-tumorigenic effect of curcumin appears to be at least partially explained by its effect in inducing microsomal P-450 as well as cytosolic glutathione-S transferase (a Phase II potent detoxifying enzyme) activities in the liver,[114] which may detoxify chemical carcinogens causing "chemoprevention". So far, however, the author is not aware of any clinical or epidemiological study examining the effect of curcumin or turmeric. In India, China and Japan, turmeric has been used for thousands of years as a remedy for different disease conditions. In Okinawa, turmeric tea is a traditional drink routinely used even today. In the author's opinion, future studies examining the effects of spices used in Mediterranean dishes may uncover similar complex mechanisms of preventing cancer, CVD etc.

How Should We Prepare What We Eat?

So far, this chapter has focused on the issue of what we should eat. As was discussed with lycopene, however, another important issue seems to exist, how we should prepare what we eat. Not only fat-soluble lycopene, but also the advocated green-yellow vegetables in general, need to be heated before eating. This is particularly true for green vegetables, which are often eaten raw as salads. Unlike cows and goats, the human gastrointestinal system is unable to digest the hard cellular surface membrane of these green vegetables without breaking the membranes by (usually) heating, because it lacks enzymes for digesting cellulose. People often worry about the loss of vitamin C by heating. However, while considerable amounts of vitamin C remain active after boiling, these go into the boiling water. Thus, if vegetables are boiled, the boiled water must be consumed together with the vegetables. All other potent antioxidants residing inside the cells of vegetables can also be efficiently absorbed only after breaking their cellular walls. Not only antioxidants but other important elements such as potassium, for example, go into boiling water, so that the intake of this water is essential. This is also true for eating fish. If fish is boiled too long, most of the good elements such as $n-3$ USFA diffuse into the water. If only the heavily-boiled fish is eaten but the boiling water discarded, the most nutritious part of the fish may be missed.

In Okinawa, they first boil pork and then discard the pork fat which rises to the surface of boiling water. Thereafter, they put other vegetables and fish together in the boiling water and then consume the entire stew with the water. The replacement of pork with corned beef might have been a big failure since beef fat cannot be discarded this way. However, the most important aspect of Okinawan dishes is probably not the use of pork instead of beef, but is most likely their way of eating pork together with large amounts of vegetables, sea weeds and fish, which are boiled or fried (with pork fat). If the fatty acid compositions alone are compared between beef and pork fats, one must conclude that they are very similar.

The Mediterranean way of cooking beans simply by boiling appears to be a simple but a very good way. In Japan and other Far East Asian countries, they have other ways of eating soybean products. In addition to tofu and soy-milk, they have miso (a paste), soy sauce, and natto, which are all made of fermented soy beans.[115] Fermentation is a very good way to make foods digestible and can add important microorganisms for health (such as in yogurt). In miso, Esaki *et al.* found an antioxidant that is not present in soy beans.[115] That compound is also found as a biotransformed metabolite inside the body of animals that were given soybeans. In natto, there is an enzyme named nattokinase, which is similar to urokinase and is effective in preventing or even treating arterial thrombosis.[116]

Conclusions

As has been discussed, the subject of what we should eat and drink is a complex issue because foods and drinks contain so many elements, some of which may be good and some others may be bad for the health. Many epidemiological studies do not come to the same conclusions probably because so many confounders are not easily controlled. Despite these complexities, the recommendation of eating green-yellow vegetables and fresh fruits may still hold. Joseph *et al.*[30] emphasized the value of fruits and vegetables with dense colors which are usually evidence of a high content of different antioxidants. Our ways of eating vegetables, however, may need to be modified. This author discourages the eating of too many raw vegetables as salads, or at least we need to eat vegetables with a careful consideration of their digestibilities. Further, the beneficial effects of eating fish, which is one major element of the Mediterranean diet (and of traditional Okinawan diet as well) for prevention of CVD, cancer and presumably dementias of different causes including AD and age-induced decline in cognitive function, should be recognized. The issue of meat (for both proteins and fats) does not seem to have been clearly resolved at this time, however, it is the author's impression that the bad aspect of eating meat has been over-emphasized. The issue of pork vs. beef may be an oversimplification of reality. It appears that the drastically increased intake of total calories and a sharp decline in fish consumption by Okinawans after the last War could be major causes for the decline of male longevity from the 1st to the 26th position in Japan, although we need to analyze its cause in more detail. What should be studied further appears to be the value of many micronutrients (largely antioxidants) contained in the green-yellow vegetables, nuts, beans, fruits and spices, especially those of bright colors, which are all essential elements of Mediterranean dishes as well as of traditional Okinawan dishes.

Acknowledgments

The author deeply appreciates Drs. G. O. Ivy, T. Osawa and K. Uchida who kindly reviewed the manuscript and gave him many useful comments. Drs. H. Okuyama, T. Osawa and Y. Yamori generously provided valuable information based on their own research achievements. The efficient and dedicated secretarial work of Ms. Ohara is also gratefully acknowledged.

References

1. Harman D (1956) Aging: a theory based on free radical and radiation chemistry. *J Gerontol* **11**: 298–300.
2. Beckman KB, Ames BN (1998) The free radical theory of aging matures. *Physiol Rev* **78**: 547–581.
3. Harman D (1994) Free-radical theory of aging. Increasing the functional life span. In: Zs.-Nagy I, Harman D, Kitani K (eds.) *Pharmacology and Aging Process — Methods of Assessment and Potential Interventions*, vol. 71, pp. 1–15. Ann NY Acad Sci.
4. Le Bourg E (2003) Antioxidants as modulators. In: Rattan SIS (ed.) *Modulating Aging and Longevity*, pp. 183–203, Kluwer Academic Publishers, Dordrecht, The Netherlands.
5. Lipman RD, Bronson RT, Wu D, Smith DE, Prior R, Cao G, Han SN, Martin KR, Meydani SN, Meydani M (1998) Disease incidence and longevity are unaltered by dietary

antioxidant supplementation initiated during middle age in C57BL/6 mice. *Mech Ageing Dev* **103**: 269–284.

6. Peto R, Doll R, Buckley JD, Sporn MB (1981) Can dietary β-carotene materially affect human cancer rates? *Nature* **290**: 201–208.
7. Ziegler RG (1991) Vegetables, fruits, and carotenoids and the risk of cancer. *Am J Clin Nutr* **53**: 251S–259S.
8. Dartigues JF, Dabis F, Gros N, Bois MG, Salamon R, Dilhuydy J-M, Courty G (1990) Dietary vitamin A, beta-carotene and risk of cancer: a prospective study. *J Natl Cancer Inst* **79**: 443–448.
9. Enstrom JE, Kanim LE, Klein MA (1992) Vitamin C intake and mortality among a sample of the United States population. *Epidemiology* **3**: 194–202.
10. Fortes C, Forastiere F, Farchi S, Rapiti E, Pastori G, Perucci CA (2000) Diet and overall survival in a cohort of very elderly people. *Epidemiology* **11**: 440–445.
11. Gazziano JM, Manson JE, Branch LG, Colditz GA, Willet WC, Buring JE (1995) A prospective study of consumption of carotenoids in fruit and vegetables and decreased mortality in the elderly. *Ann Epidemiol* **5**: 255–260.
12. Malin AS, Qi D, Shu X-O, Gao Y-T, Friedmann JM, Jin F, Zheng W (2003) Intake of fruits, vegetables and selected micronutrients in relation to the risk of breast cancer. *Int J Cancer* **105**: 413–418.
13. Osler M, Schroll M (1997) Diet and mortality in a cohort of elderly people in a north European community. *Int J Epidemiol* **26**: 155–159.
14. Strandhagen E, Hansson P-O, Bosaeus I, Isaksson B, Eriksson H (2000) High fruit intake may reduce mortality among middle-aged and elderly men. The study of men born in 1913. *Eur J Clin Nutr* **54**: 337–341.
15. Trichopoulou A, Kouris-Blazos A, Wahlqvist MC, Gnardellis C, Lagiou P, Polychronopoulos E, Vasslakou T, Lipworth L, Trichopoulos D (1995) Diet and overall survival in elderly people. *Br Med J* **311**: 1457–1460.
16. Fletcher AE, Breeze E, Shetty PS (2003) Antioxidant vitamins and mortality in older persons: findings from the nutrition add-on study to the Medical Research Council Trial of Assessment and Management of Older People in the Community. *Am J Clin Nutr* **78**: 999–1010.
17. Gale CR, Ashurst HE, Powers HJ, Martyn CN (2001) Antioxidant vitamin status and carotid atherosclerosis in the elderly. *Am J Clin Nutr* **74**: 402–408.
18. Knekt P, Aromaa A, Maatela J, Aaran R-K, Nikkari T, Hakama M, Hakulinen T, Peto R, Teppo L (1991) Vitamin E and cancer prevention. *Am J Clin Nutr* **53**: 283S–286S.
19. Riemersma RA, Wood DA, Macintyre CCA, Elton RA, Gey KF, Oliver MF (1991) Risk of angina pectoris and plasma concentrations of vitamin A, C and E and carotene. *Lancet* **337**: 1–5.
20. Sahyoun NR, Jacques PF, Russell RM (1996) Carotenoids, vitamin C and E, and mortality in an elderly population. *Am J Epidemiol* **144**: 501–511.
21. Yokoyama T, Date C, Kokubo Y, Yoshiike N, Matsumura Y, Tanaka H (2000) Serum vitamin C concentration was inversely associated with subsequent 20-year incidence of stroke in a Japanese rural community. The Shibata study. *Stroke* **31**: 2287–2294.
22. Gey KF, Brubacher GB, Stähelin HB (1987) Plasma levels of antioxidant vitamins in relation to ishemic heart disease and cancer. *Am J Clin Nutr* **45**: 1368–1377.

23. Mecocci P, Polidori MC, Trolano L, Cherubini A, Cecchetti R, Pini G, Straatman M, Monti D, Stahl W, Sies H *et al.* (2000) Plasma antioxidants and longevity: A study on healthy centenarians. *Free Rad Biol Med* **28**: 1243–1248.
24. Hernandez-Diaz S, Marinez-Losa E, Fernandez-Jame E, Serrano-Martinez M, Martines-Gonzalez MA (2002) Dietary folate and the risk of nonfatal myocardial infarction. *Epidemiology* **13**: 700–706.
25. Giovannucci E, Ascherio A, Rimm EB, Stampfer MJ, Colditz C, Willett WC (1995) Intake of carotenoids and retinol in relation to risk of prostate cancer. *J Natl Cancer Inst* **87**: 1767–1776.
26. Sesso HD, Buring JE, Norkus EP, Gaziano JM (2004) Plasma lycopene, other carotenoids, and retinol and the risk of cardiovascular disease in women. *Am J Clin Nutr* **79**: 47–53.
27. Joseph JA, Shuhitt-Hale B, Denisova N, Prior RL, Cao G, Martin A, Taglialatela G, Bickford PC (1998) Long-term dietary strawberry, spinach or vitamin E, supplementation retards the onset of age-related neuronal signal-transduction and cognitive behavioral deficiting. *J Neurosci* **18**: 8047–8055.
28. Youdim KA, Shukitt-Hale B, Martin A, Wang H, Denisova N, Bickford PC, Joseph JA (2000) Short-term dietary supplementation of blueberry polyphenolics: Beneficial effects on aging brain performance and peripheral tissue function. *Nutr Neurosci* **3**: 383–397.
29. Joseph JA, Denisoza NA, Arendash G, Gordon M, Diamond D, Shukitt-Hale B, Morgan D (2003) Blueberry supplementation enhances signaling and prevents behavioral deficits in Alzheimer disease model. *Nutr Neurosci* **6**: 153–162.
30. Joseph JA, Nadeau DA, Underwood A (2003) The Color Code: A Revolutionary Eating Plan For Optimum Health. Hyperion, New York.
31. Commenges D, Scotet V, Renaud S, Jacqmin-Gadda, H, Barkerger-Gateau, Darigues J-F (2000) Intake of flavonoids and risk of dementia. *Eur J Epidemiol* **16**: 357–363.
32. Wattenberg LW (1971) Studies of polycyclic hydrocarbon hydroxylases of the intestine possibly related to cancer. Effect of diet on benzpyrene hydrosylase activity. *Cancer* **28**: 99–102.
33. Loub WD, Wattenberg IW, Davis PW (1975) Aryl-hydrocarbon hydroxylase induction in rat tissues by naturally occurring indoles of cruciferous plants. *J Natl Cancer Inst* **45**: 9985–988.
34. Graham S, Lilienfeld A, Tidings J (1967) Dietary and purgation factors in the epidemiology of gastric cancer. *Cancer* **20**: 2224–2234.
35. Graham S, Scholta W, Martino P (1972) Alimentary factors in the epidemiology of gastric cancer. *Cancer* **30**: 927–938.
36. Stephens NG, Parsons A, Schofield PM, Kelly F, Cheeseman K, Mitchinson MJ (1996) Randomised controlled trial of vitamin E in patients with coronary disease: Cambridge Heart Antioxidant Study (CHAOS). *Lancet* **347**: 781–786.
37. Osganian SK, Stempfer M, Rimm E, Spiegelman D, Hu FB, Manson JE, Willet WC (2003) Vitamin C and risk of coronary heart disease in women. *J Am Coll Cardiol* **42**: 246–252.
38. Albanes D, Heinonen OP, Huttunen J, Jaylov PR, Virtamo SV, Edwards BK, Haapakoski J, Rautahti M, Hartman AM (1996) Effects of α-tocopherol and β-carotene supplements on cancer incidence in the Alpha-Tocopherol Beta-carotene cancer prevention. *Am J Clin Nutr* **62**: 1427S–1430S.

39. Henneken CH, Buring JE, Manson JE, Stamper M, Rosner B, Cook NR, Belanger C, La Motte F, Bosner B, Cook NR *et al.* (1996) Lack of effect of long-term supplementation with beta carotene on the influence of neoplasms and cardiovascular disease. *N Engl J Med* **334**: 1145–1149.
40. Ward JA (1998) Should antioxidant vitamins be routinely recommended for older people? (1998) *Drugs & Aging* **12**: 169–175.
41. Alpha-Tocopherol, Beta Carotene Cancer Prevention Study Group (1994) The effect of vitamin E and beta carotene on the incidence of lung cancer and other cancers in male smokers. *N Engl J Med* **330**: 1029–1035.
42. Omenn GS, Goodman GE, Thornquist MD, Balmes J, Cullen MR, Glass A, Keogh JP, Meyskens FM, Valanis BW, Williams Jr, J *et al.* (1996) Effects of a combination of beta carotene and vitamin A on lung cancer and cardiovascular disease. *N Engl J Med* **334**: 1150–1155.
43. Wang X-D, Liu C, Bronson RT, Smith DE, Krinsky NI, Russell RM (1999) Retinoid signaling and activator protein-1. Expression in ferrets given β-carotene supplements and exposed to tobacco smoke. *J Natl Cancer Inst* **91**: 60–66.
44. Meydani SN, Meydani M, Blumberg JB, Leka LS, Siber G, Loszowski F, Thompson C, Pedrosa MC, Diamond RD, Stoller BD (1997) Vitamin E supplementtion and *in vivo* immune response in healthy elderly subjects, a randomized controlled trial. *JAMA* **277**: 1380–1386.
45. Graat JM, Schouten EG, Kok F (2002) Effect of daily vitamin E and multivitamin-mineral supplementation on acute respiratory tract infections in elderly persons. A randomized controlled trial. *JAMA* **288**: 715–721.
46. Merchant AT, Curhan G, Bendich A, Singh VN, Villett AC, Fawzi WW (2004) Vitamin intake is not associated with community-acquired pneumonia in U.S. men. *J Nutr* **134**: 439–444.
47. Chandra RK (2004) Impact of nutritional status and nutrient supplements on immune responses and incidence of infection in older individuals. *Ageing Res Rev* **3**: 91–104.
48. Chandra RK (2003) Serum retinol levels and fracture risk. *N Engl J Med* **348**: 1927–1928.
49. Martin A, Prior R, Shukitt-Hale B, Cao G , Joseph JA (2000) Effect of fruits, vegetables, or vitamin E-rich diet on vitamins E and C distribution in peripheral and brain tissues; Implications for brain function. *J Gerontol* **55A**: B144–B151.
50. Diplock AT (1997) Will the "good fairies" please prove to us that vitamin E lessens human degenerative disease? *Free Radic Res* **27**: 511–532.
51. McCall MR, Frei B (1999) Can antioxidant vitamins materially reduce oxidative damage in human? *Free Rad Biol Med* **26**: 1034–1053.
52. Lehmann AB (1989) Review: Undernutrition in elderly people. *Age and Ageing* **18**: 339–353.
53. Corti M-C, Gurelnik JM, Salive ME, Sorkin JD (1994) Serum albumin level and physical disability as predictors of mortality in older persons. *JAMA* **272**: 1036–1042.
54. Yamori Y, Ikeda K, Tagami M, Yamagata K, Nara Y (2001) Nutritional pathogenesis and prevention of stroke. In: Fernstrom JD, Uauy R, Arroyo P (eds.) Nitrition And Brain, pp. 231–246. Karger A G, Basel.
55. Ulbricht TLV, Southgate DAT (1991) Coronary heart disease, seven dietary factors. *Lancet* **338**: 985–992.

56. Krumholz HM, Seerman TE, Merrill SS, De Leon CFM, Vaccarino V, Silverman DI, Tsukahara R, Ostfeld AM, Bedeman LF (1994) Lack of association between cholesterol and coronary heart disease mortality and morbidity and all-cause mortality in persons older than 70 years. *JAMA* **272**: 1335–1340.
57. Weverling-Rijnsburger AWE, Blauw GJ, Lagaay AM, Knook D, Meinders AE (1997) Total cholesterol and risk of mortality in the oldest old. *Lancet* **350**: 1119–1123.
58. Iso H, Naito Y, Kitamura A, Sato S, Kiyama M, Takayama Y, Lida M, Shimamoto T, Sankai T, Kodama Y (1994) Serum cholesterol and mortality in a Japanese population. *J Clin Epidemiol* **5**: 78–94.
59. Iso H, Jacobs DR, Wentworth D, Neaton JD, Cohen JD (1989) Serum cholesterol levels and six-year mortality from stroke in 350,977 men screened for the multiple risk factor intervention trial. *N Eng J Med* **320**: 904–910.
60. Yamori Y, Lovenberg W, Freis ED (eds.) (1979) Prophylactic Approach To Hypertensive Diseases. Raven Press, New York.
61. Liu LJ, Ikeda K, Yamori Y (2000) Twenty-four hour urinary sodium and 3-methylhistidine excretion in relation to blood pressure in Chinese: results from the China-Japan cooperative research for the WHO-CARDIAC Study. *Hypertens Res* **23**: 151–157.
62. Renaud S, De Lorgeril M (1992) Wine, alcohol, platelets and French paradox for coronary heart disease. *Lancet* **339**: 1523–1526.
63. Ravnskov U (1998) The questionable role of saturated and polyunsaturated fatty acids in cardiovascular disease. *J Clin Epidemiol* **6**: 443–460.
64. Strandberg TE, Salomaa VV, Naukkarinen VA, Vanhaanen HT, Sarna ST, Miettinen TA (1991) Long term mortality after 5-year multifactorial primary prevention of cardiovascular diseases in middle-aged men. *JAMA* **266**: 1225–1229.
65. de Lorgeril M, Renaud S, Mamelle N, Salen P, Maetin J-P, Mojjaud I, Gudollet J, Tonboul P, DeLaye J (1994) Mediterranean alph-linolenic acid-rich diet in secondary prevention of coronary heart disease. *Lancet* **343**: 1454–1459.
66. Scandinavian Simvastatin Survival Study Group (1994) Randomized trial of cholesterol lowering in 4444 patients with coronary heart disease: the Scandinavian Simvastin Survival Study (4S). *Lancet* **344**: 1384–1389.
67. Okuyama H (2001) High *n*-6 to *n*-3 ratio of dietary fatty acids rather than serum cholesterol as a major risk factor for coronary heart disease. *Eur J Lipid Sci Technol* **103**: 418–422.
68. Kalmijn S, Launer LJ, Ott A, Witteman JCM, Hofman A, Breteler MMB (1997) Dietary fat intake and the risk of incident dementia in the Rotterdam study. *Ann Neurol* **42**: 776–782.
69. Willett WC (1994) Diet and health: what should we eat? *Science* **264**: 532–537.
70. Cantrill RC, Yung-Sheng H (1998) Fatty acids and cancer. *Nutrition* **14**: 235–236.
71. Zock P, Katan MB (1998) Linoleic intake and cancer risk: a review and meta-analysis. *Am J Clin Nutr* **68**: 142–153.
72. Lipworth L, Martinez ME, Angell J, Hsieh C-C, Trichopoulos D (1997) Olive oil and human cancer: an assessment of the evidence. *Prev Med* **26**: 181–190.
73. Hirose M, Masuda A, Ito N, Kamano K, Okuyama H (1990) Effects of dietary perilla oil, soybean oil and safflower oil on 7,12-dimethylbenz[a]anthracene (DMBA) and 1,2-dimethylhydrazine (DMH)-induced mammary gland and colon carcinogenesis in female SD rats. *Carcinogeneis* **11**: 731–735.

74. Minoura T, Tahata T, Sakaguchi M, Takada H, Yamamura M, Hioki K, Yamamoto M (1988) Effect of dietary eicosapentaenoic acid on azoxymethane-induced colon carcinogenesis in rats. *Cancer Res* **48**: 4790–4794.
75. Okuyama H (20000) Prevention of excessive linoleic acid syndrome. *Lipid Tech Newslett* **6**: 128–132.
76. Osawa T (2000) Protective effect of dietary antioxidants in oxidative stress. In: Yoshikawa T, Toyokuni Y, Yamamoto Y, Naito Y (eds.) Free Radicals In Chemistry, Biology And Medicine. Oika International, London.
77. Visioli F, Galli C (1988) The effect of minor constituents of olive oil on cardiovascular disease: new findings. *Nutr Rev* **56**: 142–147.
78. Lagarde M (1990) Metabolism of *n*-3/*n*-6 fatty acids in blood and vascular cells. *Biochem Soc Trans* **18**: 770–772.
79. Karmali RA, Marsh J, Fuchs C (1984) Effect of omega-3 fatty acids on growth of a rat mammary tumor. *J Natl Cancer Inst* **73**: 457.
80. Nara Y, Yamori Y, Lovenberg W (1978) Effect of dietary taurine on blood pressure in spontaneously hypertensive rats. *Biochem Pharmacol* **27**: 2689–2692.
81. Morris MC, Evans DA, Bienlas JL, Evans DA, Bientas JL, Tangney CC, Bennett DA, Wilson RS, Aggarwal N, Schneider J (2003) Consumption of fish and n-3 fatty acids and risk of incident Alzheimer disease. *Arch Neurol* **60**: 940–946.
82. Conquer JA, Tierney MC, Zecevic J, Better WJ (2000) Fatty acid analysis of blood plasma of patients with Alzheimer's disease, other types of dementia, and cognitive impairment. *Lipids* **35**: 1305–1312.
83. Barberger G, Letenner L, Deschamps PO Peres K, Dartguea J-F, Renaud S (2002) Fish, meat and risk of dementia: cohort study. *Br Med J* **325**: 932–933.
84. Kalmijn S, Launer LJ, Ott A, Willerman JCM, Hofman A, Breteler MB (1997) Dietary fat intake and the risk of incident dementia in the Rotterdam study. *Ann Neurol* **42**: 776–782.
85. Friedland RP (2003) Fish consumption and the risk of Alzheimer disease. Is it time to make dietary recommendations? *Arch Neurol* **60**: 923–924.
86. Hibbbeln JR (1998) Fish consumption and major depression. *Lancet* **351**: 1213.
87. Watanabe S, Adlercreutz (1998) Pharmacokinetics of soy phytoestrogens in humans. In: Shibamoto T, Terao J, Osawa T (eds.) *Functional Foods For Disease Prevention II*, pp. 198–208, American Chemical Society, Washington.
88. Potter SM (1995) Overview of proposed mechanisms for hypocholesterolemic effects of soy. *J Nutr* **125**: 606S–611S.
89. Obata K, Ikeda K, Yamasaki M, Yamori Y (1998) Dietary fieber, psyllium, attenuates salt accelerated hypertension in stroke prone spontaneously hypertensive rats. *J Hypertens* **12**: 1959–1964.
90. Renaud S, Ruf JC (1996) Effect of alcohol on platelet functions. *Clin Chim Acta* **246**: 77–89.
91. Doll R, Peto R, Hall E, Wheatley K, Gray R (1994) Mortality in relation to consumption of alcohol: 13 years' observations on male British doctors. *Brit Med J* **309**: 911–918.
92. Friedmann GD, Klatsky AL (1993) Is alcohol good for your health? *N Engl J Med* **329**: 1189–1193.

93. Rimm EB, Klatsky A, Grobbee D, Stampfer MJ (1996) Review of moderate alcohol consumption and reduced risk of coronary heart disease: Is the effect due to beer, wine or spirits? *Br Med J* **312**: 731–736.
94. Gronbaek M, Deis A, Sorensen TIA, Becker U, Schnoh R, Jensen G (1995) Mortality associated with moderte intake of wine, beer, or spirits. *Br Med J* **310**: 1165–1169.
95. Frankel EN, Kanner J, Germann JB, Parks E, Kinsella JE (1993) Inhibition of oxidation of human low-density lipoprotein by phenolic substances in red wine. *Lancet* **341**: 454–457.
96. Kondo K, Matsumoto A, Kurata H, Tanahashi H, Koda H, Amachi T, Itakura, H (1994) Inhibition of oxidation of low-density lipoprotein with red wine. *Lancet* **344**: 1152.
97. Maxwell S, Cruickshank A, Thorpe G (1994) Red wine and antioxidant activity in serum. *Lancet* **344**: 193–194.
98. Renaud SC, Guéguen R, Siest G, Salamon R (1999) Wine, beer, and mortality in middle-aged men from eastern France. *Arch Intern Med* **159**: 1865–1870.
99. Jang M, Cai L, Udeani GO, Slowing KV, Thomas CF, Beecher CWW, Fong HHS, Farnsworth NR, Kinghorn D, Mehta RG (1999) Cancer chemopreventive activity of resveratol, a natural product derived from grapes. *Science* **727**: 218–220.
100. Orgogozo J-M, Dartigues J-F, Lafont S, Letenneur L, Commenges D, Salamon R, Renaud S, Breteler MB (1997) Wine consumption and dementia in the elderly: a prospective community study in the Bordeaux area. *Rev Neurol* (Paris) **153**: 185–192.
101. Sparks DL, Martin TA, Gross DR, Hunsaker 3rd, JC (2000) Link between heart disease, cholesterol and Alzheimer's disease, a review. *Microsci Res Tech* **50**: 287–290.
102. Zheng W, Doyle TJ, Kushi LH, Sellers TA, Hong C-P, Folsom AR (1996) Tea consumption and cancer incidence in a prospective cohort study of postmenopausal women. *Am J Epidemiol* **144**: 175–182.5.
103. Yoshizawa S, Horiuchi T, Fujiki H, Yoshida T, Okuda T, Sugimura T (1987) Antitumor promoting activity of (−)-epigallocatechin gallate, the main constituent of "tannin" in green tea. *Phytother Res* **1**: 44–47.
194. Fujiki H, Suganuma M, Okabe S, Sueoka E, Suga K, Imai K, Nakachi K (2000) A new concept of tumor promotion by tumor necrosis factor-alpha, and cancer preventive agents, (−)-epigallo-catechin gallate and green tea — a review. *Cancer Detect Prev* **24**: 91–99.
105. Meydani M (2001) Nutrition interventions in aging and age-associated disease. *Ann NY Acad Sci* **928**: 226–235.
106. Yokozawa T, Dong E (1997) Influence of green tea and its three major components upon low-density lipoprotein oxidation. *Exp Toxicol Pathol* **49**: 329–335.
107. Arts ICW, Hollman PCH, Feskens EJM, De Mesquita HBB, Kromhout D (2001) Catechin intake might explain the inverse relation between tea consumption and ischemic heart disease; the Zutphen elderly study. *Am J Clin Nutr* **74**: 227–232.
108. Hollman PCH, Feskens EJM, Katan MB (1999) Tea flavonols in cardiovascular disease and cancer epidemiology. *Proc Soc Exp Biol Med* **220**: 198–202.
109. Peters U, Poole C, Arab L (2001) Does tea affect cardiovascular disease? A meta-analysis. *Am J Epidemiol* **154**: 495–503.
110. Mazur WM, Wahala K, Rasku S, Salakla A, Hase T, Adlercreutz H (1998) Lignan and isoflavonoid concentrations in tea and coffee. *Br J Nutr* **79**: 37–45.

111. Osawa T, Sugiyama Y, Inayoshi M, Kawakishi S (1995) Antioxidative activity of tetrahydrocurcuminoids. *Biosci Biotech Biochem* **59**: 1609–1612.
112. Kim JM, Araki S, Kim DJ, Park CB, Takasuka N, Baba-Toriyama H, Ota T, Nir Z, Khachik F, Shimidzu N *et al.* (1998) Chemopreventive effects of carotenoids and curcumins on mouse colon carcinogenesis after 1,2-dimethylhydrazine initiation. *Carcinogenesis* **19**: 81–85.
113. Huang M-T, Lou Y-R, Ma W, Newmark HL, Reuhl KR, Conney AH (1994) Inhibitory effects of dietary curcumin on forestomach, duodenal, and colon carcinogenesis in mice. *Cancer Res* **54**: 5841–5847.
114. Oetari S, Sudibyo M, Commandeur JNM, Samhoedi R, Vermeulen NPE (1995) Effects of curcumin on cytochrome P450 and glutathione S-transferase activities in rat liver. *Biochem Pharmacol* **51**: 39–45.
115. Esaki H, Onozaki H, Osawa T (1994) Antioxidant activity of fermented soybean products. In: Functional Foods For Disease Prevention II, Shibamoto T, Terao J, Osawa T (eds.), pp. 355–360, ACS, Washington.
116. Sumi H, Hamada H, Nakanishi K, Hiratani H (1990) Enhancement of the fibrinolytic activity in plasma by oral administration of nattokinase. *Acta Haematol* **84**: 139–143.

12

Telomere- and Telomerase-based Therapies

Maria A. Cerone,[1,3,a] Ryan Ward[2,3,a] and Chantal Autexier[1,2,3,*]

Department of Anatomy and Cell Biology[1]

Department of Medicine[2]

McGill University, Montréal (Québec), Canada, H3A 2B4

and

Bloomfield Centre for Research in Aging[3]

Lady Davis Institute for Medical Research

Sir Mortimer B. Davis Jewish General Hospital

Montréal (Québec), Canada, H3T 1E2

TELOMERES AND TELOMERASE

The German cytologist Hermann J. Muller was one of the first to propose the idea that the ends of chromosomes were in some way different than the rest.[1] In 1938, Muller spoke of two types of genes: those that were unipolar with only one neighboring gene due to their location at the end of the chromosome, and those that were bipolar with neighboring genes on both sides and located throughout the chromosome. These unipolar genes he called telomeres (telos meaning end, and meros meaning part) and attributed to them the special function of sealing the chromosomes, a function that could not be reproduced by simply tearing two bipolar genes apart. In the late 1920's, Muller had produced the first experimentally-induced mutations by X-Ray irradiating *Drosophila*. His observations, for which he was later awarded the

[a]These two authors contributed equally to this work.

*Corresponding author.

Tel.: 514-340-8260, Fax: 514-340-8295, E-mail: chantal.autexier@mcgill.ca

Nobel Prize, allowed him to note that many mutations were associated with chromosomal rearrangements that involved large translocations between non-homologous chromosomes or inversions in the same chromosome. Muller hypothesized that these rearrangements were the consequence of two independent breaks and the exchange or inversion of the broken ends. Initially these observations were made by genetic analysis of irradiated *Drosophila*, however the rediscovery of the large polytene chromosome allowed Muller to directly observe this phenomenon. All together, Muller's experiments enabled him to deduce that the rearrangements were the consequence of joining two newly broken chromosomes, and not the attachment of a broken end to an originally free end (the telomere).

Around the same time, Barbara McClintock's work in maize closely supported Muller's observations by demonstrating that, independent of how the break in the chromosome occurred, broken chromosome ends were extremely "sticky" and would readily join to other broken ends.[1] In the system that she established, a breakage-fusion-bridge cycle occurred where dicentric chromosomes (containing two centromeres) were pulled to both sides of the mitotic spindle creating a chromosome bridge. Though this bridge was broken during anaphase or telophase to create broken chromosomal ends, McClintock noted that the broken ends of the sister chromatids fused to one another before the next mitosis and thus generated a new dicentric chromosome to further continue the breakage-fusion-bridge cycle. From this work, she went on to other studies that led to her theory that spontaneous chromosomal breaks could sometimes be generated by the existence of transposable elements, work for which she was eventually awarded the Nobel Prize. Thus, both Muller and McClintock recognized that broken chromosomal ends were highly unstable, preferring to join with other broken ends as compared to normal telomeres, which were stable and in some way sealed.

In this first section of the chapter, we will further discuss the role of telomeres, and provide an introduction of the function of telomerase, and telomere-binding proteins in the maintenance of chromosomal stability and cell viability. The second and third sections of the chapter will focus on a discussion of telomere- and telomerase-based anti-cancer and anti-aging therapies respectively.

The End Replication Problem

Though the importance of telomeres was understood from the work described above, their role was not fully appreciated until the elucidation of DNA structure and the mechanisms by which it is replicated. It was recognized that the terminal ends of the chromosome would be incompletely replicated at each round of cellular division due to the 5′ to 3′ directionality of the conventional DNA replication machinery, and the use of short RNA primers.[1] Leading strand synthesis is continuous from the origin of replication to the end of the chromosome. The discontinuous nature of lagging strand synthesis and the use of RNA primers for the initiation of replication generate a region of unreplicated DNA at the most 5′ end of the lagging strand. With successive rounds of cell division, this unreplicated region would become larger and eventually erode the ends of the chromosomes potentially leading to genomic instability.

Telomere Structure and Function

Telomeres are most simply defined as the terminal ends of linear chromosomes. In humans this DNA sequence usually ranges from 2 to 15 kilobase pairs (kb) in length;[2] however great length variation exists between organisms, for example the telomeres of the ciliate *Euplotes aediculatus* are less than 50 base pairs versus that of the mouse which can be 60 kb.[2] Structurally, telomeres are stretches of non-coding, repetitive DNA that serve to protect the ends of the chromosomes from being recognized by the cellular machinery as broken or damaged DNA, from non-homologous end joining, end-to-end fusions or endogenous exonucleases.[3] In humans, telomeres consist of the hexanucleotide repeat TTAGGG on the 3′ strand (termed the G-rich strand, or G-strand) and AATCCC on the 5′ strand (termed the C-rich strand, or C-strand). Though the telomeric sequence can differ between organisms, in general telomeres are G/C rich structures that terminate in a single-stranded 3′ G-strand overhang.[2] It is now thought that this overhang provides a structure by which chromosomes and telomeres are protected or "capped". Electron microscopy of telomeric DNA and purified de-proteinated telomeres from a variety of organisms has revealed that the 3′ overhang can loop and invade the double-stranded telomere.[4] This generates the telomere-loop (t-loop), and the displacement-loop (d-loop) at the region of invasion.[4] These loops can form in the absence of interacting proteins or cross-linking agents however their formation is greatly increased in the presence of telomere binding proteins (discussion to follow).[5]

Further characterization of the G-rich DNA tracks suggests that telomeric DNA may exist in higher structural forms. NMR spectroscopy and X-ray crystallography of telomeric DNA sequences demonstrated that the guanine-rich sequences can form G-quartet structures in which multiple guanines hydrogen bond to form stacking structures in either parallel or anti-parallel conformation.[6] Primarily observed *in vitro*, there is now some evidence that G-quartet structures can form *in vivo*[6] and data from recent *in vitro* studies suggest that these structures may involve the telomere binding proteins TRF1 and TRF2.[5] Though direct evidence of the existence of G-quartet structures in human cells is still lacking, as well as their exact orientation and structure, they do provide a model whereby the single-stranded G-rich overhang might be further protected and stabilized.

Telomere Binding Proteins

A number of proteins have been shown to localize to telomere, either directly by binding telomeric DNA, or indirectly by associating with proteins found on the telomeres. The best characterized telomere binding proteins in humans are TRF1, TRF2 and POT1.[7] TRF1 and TRF2 both localize to the double-stranded telomeric DNA, with TRF2 preferentially found at the junction between the double-stranded telomere and the invasion of the single-stranded 3′ overhang, while POT1 is found at single-stranded telomeric DNA.[8] All appear to bind the telomeres as dimers (and in the case of TRF1 and TRF2 perhaps amalgamated tetramers), and participate in the stabilization of the t-loop structure,[5] aid in the loading of other telomere binding proteins,[9] and play roles in telomere length maintenance.[10] Overexpression of both TRF1 and TRF2 have negative regulatory effects on telomere length, suggesting that when bound by these proteins, telomeres are stabilized and their elongation is prevented.[7] TRF1 null cells exhibit a growth defect with an extended population doubling time, chromosomal instability in the form of end-to-end sister chromatid fusions, and decreased association of

TRF2 and TIN2 with the telomeres.[9] Expression of a mutant TRF2 protein that is defective in its ability to interact with telomeric DNA induces rapid onset of apoptosis or senescence (depending on the particular cell line used) in an ATM/p53- or p16/RB-dependent fashion, end-to-end fusions and chromosomal instability, and erosion of the 3′ overhang.[7] Our understanding of TRF1 and TRF2 suggests that the two telomere binding proteins act to stabilize the structure of telomeric DNA preventing it from being recognized as damaged DNA, from joining to other chromosomal ends, and to sequester the 3′ end from the active telomerase enzyme.[7,9] In contrast, Pot1 has been shown to bind the single-stranded G-rich sequence, but not the C-rich sequence, consistent with the notion that it preferentially binds the single-stranded 3′ overhang portion of the telomere.[8] Overexpression of Pot1 induces telomere lengthening in a telomerase-dependent way, and accordingly its inhibition by antisense oligonucleotides induces telomere shortening,[8] reduces 3′ overhang signal, and increases the frequency of anaphase bridges.[11] Our current view of Pot1 function is that it recruits telomerase to the ends of the telomeres and allows their elongation.[8]

The disruption of telomere binding proteins causes telomere dysfunction in a length-independent way. Uncapped telomeres have now been associated with DNA damage response factors such as 53BP-1, γ-H2AX, Rad17, ATM and Mre11.[12] The binding proteins cap the telomere in dynamic ways to prevent genomic instability and to regulate the activity of telomerase. Uncapping of the telomeres, either by critical shortening or inhibition of binding proteins induces a DNA damage response.[7]

As mentioned above, at each round of cellular replication, DNA lagging strand synthesis remains incomplete. Telomeres cap the ends of linear chromosomes with non-coding sequences, providing a strategy to ensure that coding sequences are not lost, but they do not resolve the issue of incomplete DNA replication. Accordingly, small amounts of telomeric DNA are lost at each round of cellular division, and thus telomeres shorten with every round of replication.[1]

This continual and gradual loss of telomeric DNA has been proposed to be a mechanism by which cells can "count" their number of divisions: a putative "Replicometer". This idea was first proposed by Leonard Hayflick who is credited with the observation that normal human cells have a finite replicative lifespan. As a student, Hayflick demonstrated that cells, when explanted from tissues and grown in cell culture conditions did not (as widely thought at the time) have infinite growth capacity.[1] Rather the cells in his experiments grew for a reproducible number of population doublings before they stopped dividing. Even cryogenically frozen cells that were thawed retained their limited capacity for the same number of cell divisions, suggesting that their limited growth potential was not determined by the simple lapse of time. We now call the point at which normal cells cease proliferating the Hayflick limit, and associate it with a cellular state called senescence. Primary human cells will proliferate in cell culture conditions for a certain number of population doublings, losing increasing amounts of telomeric DNA.[1] Eventually, telomeres shorten to a point where they are recognized as damaged DNA and cells stop proliferating. This proliferative arrest is characterized by the stabilization of p53, the activation of the DNA damage response mechanisms, and the hyperphosphorylation of the cell cycle control protein Rb.[13] Inactivation of cell cycle control checkpoints allows further proliferation beyond the Hayflick limit and generates dysfunctional telomeres which, as well as being recognized as DNA damage, are prone to G-overhang loss, end-to-end chromosomal fusions and general genomic instability.[14] Therefore, without the presence or activation of a mechanism to either maintain or elongate telomeres, cells cannot grow indefinitely. Interestingly, there does exist a highly conserved mechanism by which cells can maintain their telomeres, that being the enzyme telomerase. As discussed in the next section, telomerase has the ability to elongate telomeres and can, in fact, impart cellular immortality when transfected into certain cell types.[15]

Telomerase

The enzyme that maintains telomeres was first discovered in the ciliate *Tetrahymena* by Carol Greider and Elizabeth Blackburn.[1] In subsequent studies the enzyme initially termed "telomere terminal transferase" but now known as telomerase, was identified as a ribonucleoprotein consisting of a reverse transcriptase protein component (denoted as TERT, *te*lomerase *r*everse *t*ranscriptase, hTERT in humans) and an RNA template component (denoted as TR, *t*elomerase *R*NA, hTR in humans).[16]

Within a few years, telomerase enzymes were discovered in numerous other organisms including humans.[1]

The principle function of telomerase is to catalyze *de novo* telomeric repeat synthesis at the 3′ end of DNA.[1] Telomerase elongates the G-strand overhang by reverse transcribing the nucleotide template sequence encoded by its TR.[1] In humans, telomerase synthesizes multiple TTAGGG sequence repeats in a reiterative way: binding the 3′ end of the DNA substrate, elongating the DNA, translocating to the newly synthesized 3′ end, and repeating the process.[1] Traditional DNA polymerases can then copy the G-stand in the 5′ to 3′ direction and it now appears likely that further processing of the telomeres occurs to generate the structurally important 3′ overhangs.[17,18]

Telomerase activity can be measured in the laboratory using a polymerase chain reaction-based assay called the Telomere Repeat Amplification Protocol (TRAP).[19] Using this sensitive assay, limited numbers of cells and amounts of tissue can be screened for the presence of telomerase activity.[19] What was exciting to discover was that most normal human tissues do not display telomerase activity, whereas a large percentage of all cancer tissues and cell lines tested demonstrate activity.[19] Accordingly, much interest has been generated in using telomerase as an anti-cancer target in humans, or telomerase activity as a marker for the early detection of cancers (as further discussed below).

Two recent studies have shown that telomerase mediates elongation of the shortest telomeres. In the first study, late generation TR (mTR) knockout mice that display short telomeres were crossed with a mTR heterozygous strain with long wild-type telomeres.[20] The results indicated that telomerase was targeted to elongate the shortest telomeres. As such, the reintroduction of an active telomerase enzyme did not induce global telomere elongation but rather a reduction in the total number of telomeres that lacked telomeric signals. In the second study, telomerase RNA knockout yeasts with short telomeres were mated to yeasts expressing an active telomerase enzyme.[21] The direct analysis of individual telomere lengths demonstrated that telomerase does not act on every telomere during a particular cell cycle, but rather acts preferentially to elongate the shortest telomeres. Further, the frequency of elongation steadily increases as the length of the telomere decreases. When the same experiments were performed with yeasts double deficient for telomere regulatory proteins and TR, telomeres of normal length were also elongated indicating a role for telomere binding proteins in modulating the accessibility of the telomerase enzyme to the telomeric substrate. Taken together, a model now exists whereby telomeres exist in one of two conformational states: telomerase extendible and non-extendible depending on telomere length. When a telomere is of sufficient length, regulatory proteins and protective structures such as the t-loop and/or G-quartets inhibit the association of telomerase with the end of the telomere. In contrast, when telomeres are short they may become accessible to the telomerase enzyme due to loss of the above mentioned structures or dis-association of telomere binding proteins or their regulators.[21]

The active *in vivo* telomerase complex has long been suspected to be a multimer of telomerase enzymes and associated proteins. Initial observations that affinity-purified catalytically active human telomerase sedimented at a molecular weight of ~550 kilodaltons (kDa) showed a striking difference to telomerase sedimented from nuclear extracts of HeLa cells which displayed a molecular weight of ~1000 kDa.[22] Coupled to the observation that hTERT and hTR, with a minimal molecular weight calculated at ~280 kDa, are sufficient to reconstitute telomerase activity, argued strongly that *in vivo* the telomerase complex consists of multimerized hTERTs and hTRs and other associated proteins.[22] Based on recent data we now have a working model whereby two hTERT-hTR complexes multimerize to generate an active telomerase complex.[22] Other data indicate that hTERT and hTR can associate with numerous other proteins, some specific for telomerase and others common binding partners of many ribonucleoproteins. Disruption of these protein interactions with hTERT or hTR could have the consequences of improper enzyme assembly, telomerase mislocalization, and telomere length changes.[22]

Though the vast majority of human cells that do maintain their telomeres do so by the expression of telomerase, it is not the only mechanism by which cells can solve the end replication problem. At least three other mechanisms exist: The first being the elimination of chromosome ends as exemplified by the circular chromosomes of bacteria. The second is the use of retrotransposable elements that mobilize from more centromeric DNA sequences and translocate to the telomeres to template the addition of telomeric repeats.[23] Amazingly, this phenomenon has thus far only been observed in *Drosophila*, the model organism used by Muller in his original work to differentiate between newly broken DNA ends and the relatively stable telomeres. Finally, a third mechanism has been identified in the yeast *S. cerevisiae*. Yeast cells normally use telomerase to maintain telomeres. Upon inactivation of telomerase most cells die but rare survivors that are able to maintain telomeres through recombination arise.[24] These survivors can be distinguished into type I and type II based on their telomere structure. Type I survivors have short telomeres that are maintained by recombination in the telomere-associated Y' element, while type II survivors have very long and heterogeneous telomeres that are likely maintained by inter-telomeric recombination.[24] As previously said, human cells mostly rely on telomerase for telomere maintenance. However, 20–30% of human immortalized cell lines and 10–15% of human tumors do not express telomerase, but use an alternative mechanism to maintain telomeres, called ALT or *a*lternative *l*engthening of *t*elomeres.[25] Human ALT cells are reminiscent of type II survivors in yeast. They are characterized by very long and heterogeneous telomeres (up to 50 kb) that undergo cycles of rapid shortening and elongation and also by the presence of nuclear structures, called APBs or ALT-associated promyelocytic leukemia (PML) bodies, which contain telomeric DNA, telomere-binding proteins such as TRF1 and TRF2, proteins involved in recombination, and the PML protein.[25] There are convincing evidences in the literature that, as in yeast, ALT in human cells relies on homologous recombination and copy switching.[25] However, it is not clear whether recombination occurs between telomeres on different chromosomes or between the double-stranded and the single-stranded regions of the same telomere.

Telomere and Telomerase Therapies against Cancer

The most important difference between normal human cells and tumor cells is their proliferative behavior: normal cells can divide only a limited number of times before entering replicative senescence, whereas tumor cells are immortal, and can divide indefinitely. This proliferative behavior correlates with the fact that normal cells undergo telomere shortening with cell divisions, whereas immortal cells are able to maintain stable telomeres.[1] Therefore, telomere maintenance is an essential requisite for unlimited cell proliferation and transformation. Almost all human tumors possess some mechanism for telomere maintenance, with over 85% up-regulating telomerase.[19,26] In normal cells, telomerase activity has been detected only in germ line and stem cells, and at very low level in some somatic cells.[16,27]

The observation that telomerase up-regulation appears to be almost universal in human tumors, while it is mostly absent in somatic cells, makes telomerase a useful and promising target for cancer diagnosis as well as for the development of anti-cancer therapies. Telomerase inhibition in telomerase-positive tumor cells would result in erosion of telomeric DNA, genomic instability and ultimately growth arrest and/or cell death due to critically short and dysfunctional telomeres. The effects of telomerase inhibition would be relatively specific to cancer cells with minor side-effects for somatic cells and stem cells, which have longer telomeres and slower duplication time.

In this section of the chapter we will review the data available in the literature regarding the possibility of interfering with telomere maintenance and/or structure in human cancer cells as an anti-cancer therapeutic approach and highlight the advantages and disadvantages of these strategies.

Telomerase Inhibition as an Anti-cancer Approach

The telomerase protein complex contains multiple sites for potential inhibition. Both hTERT and hTR have been targeted to inhibit telomerase activity. In particular, catalytically inactive variants of hTERT (hTERT-DN) confer dominant-negative phenotypes in different telomerase-positive cancer cell lines, resulting in telomere loss, chromosome aberrations and apoptotic cell death.[28,29] Moreover, reverse transcriptase inhibitors have been widely tested for anti-telomerase activity. Nucleotide analogs, which bind to the dNTP-binding site on reverse transcriptase enzymes, and non-nucleotide analogs, which bind to a hydrophobic pocket near the catalytic site, have been used. However, most were ineffective or weak inhibitors of telomerase.[30] A more promising family of molecules are compounds that interfere with the processivity of the enzyme, i.e. the ability of telomerase to reiteratively reverse transcribe the RNA template to synthesize multiple telomeric repeats, a characteristic that is unique to telomerase.[31] Treatment of cancer cells with this class of molecules resulted in highly selective inhibition of the telomerase enzyme, progressive telomere shortening followed by growth arrest and cell senescence.[32]

The RNA subunit of telomerase has also been targeted by several approaches. Short anti-sense oligodeoxynucleotides (ODN) directed against either the template region or non-template regions of hTR have been tested in cultured cancer cells.[30] However, in many cases, continued treatment with these ODN, though reducing telomerase activity *in vitro*, did not result in telomere shortening. To improve the efficacy, the specificity and the delivery of these compounds, different chemical modifications have been made. The attachment of a 2′,5′-oligoadenilate moiety (2-5A) to an ODN directed against a non-template region of hTR increased its anti-telomerase activity and resulted in apoptosis of different cancer cells.[33,34] Similarly, treatment of tumor cells with other modified ODN directed against the template region of hTR resulted in growth inhibition and apoptosis due to telomere shortening.[35,36] Examples of modified ODN that were tested are peptide nucleic acids (PNAs) oligonucleotides, which are very resistant to cellular exonucleases and endonucleases, and 2′-O-methylRNA and 2′-O-methoxyethoxyRNA oligonucleotides, which have an increased specificity for the telomerase RNA.[35,36]

All these approaches for anti-telomerase therapy rely on the loss of telomerase activity and, therefore, telomere shortening to inhibit cancer cell growth. The major limitation of such inhibition is the time necessary for the telomeres to shorten sufficiently to engage a proliferative arrest. The delay prior to the anti-proliferative effects of telomerase inhibition implies that this strategy could be effective only in cancer cells with short telomeres, whereas cells with longer telomeres would not respond to this treatment.[30] The same delay in the effects of telomerase inhibition has been observed in mice deficient of either telomerase subunit.[37,38] Initially these mice did not show any overt phenotype due to the lack of telomerase, most likely because mouse telomeres are much longer than human telomeres. However, knockout mice in successive generations showed severe proliferative defects and no survivor pups could be obtained. This reduction in proliferative capacity correlated with dramatic loss of telomeric DNA and increased genomic instability, due to critically short and dysfunctional telomeres.[37,39,40] Though these mice had a slightly higher incidence of spontaneous tumors than wild-type mice, they showed significantly fewer carcinogen-induced skin tumors even in early generations and enhanced mortality in response to γ-irradiation.[41]

A possible strategy to circumvent the lag phase prior to the anti-proliferative effects of telomerase inhibition is to cause uncapping of the telomeres independently of telomere shortening. This approach is based on reprogramming the telomerase enzyme by introducing mutant template RNAs into the cells. Two studies reported that reconstitution of a mutant holoenzyme in telomerase-positive cancer cells resulted in synthesis of mutant telomeres and consequently decreased cell proliferation and viability, and increased apoptosis.[42,43] More interestingly, these effects were rapid and did not require telomere shortening, suggesting that they may be caused by alteration of the telomere structure likely due to the formation of aberrant DNA-protein complexes. However, the presence of the wild-type telomerase mitigated the deleterious effects of the mutant hTR in these cells, probably because telomere function could be restored by addition of wild-type sequences to terminal mutant sequences.[42] Therefore the use of mutant template RNA, although extremely effective in cells that do not express wild-type hTR,[44] would only be partially effective in telomerase-positive cancer cells and human tumors.

Although very promising, the available data indicate that all telomerase-based strategies for tumor inhibition are only partially effective in cancer cells and most of them do not result in immediate cell killing, making them unsuitable for clinical application. A more effective anti-cancer strategy could be a combination approach based on both telomerase inhibition and more conventional clinical treatments, such as chemotherapy, radiotherapy or angiogenesis inhibition. Several reports have shown, in fact, that inhibition of telomerase by expression of hTERT-DN increased the sensitivity of both lung cancer and melanoma cells to a variety of DNA damaging drugs, such as cisplatin, etoposide and temozolomide.[45,46] Similarly, glioblastoma and breast cancer cells treated with 2-5A oligonucleotides, anti-sense vectors or ribozymes against hTR resulted in enhanced susceptibility to doxorubicin, other topoisomerase II inhibitors, and cisplatin.[47–50] As with any strategy, however, the responsiveness to such treatments depends on both cell-and drug-type.[51]

Use of the hTERT Promoter and hTERT-based Immunotherapy as Anti-cancer Strategies

Recently, exciting avenues of research have been pursued in the application of gene therapy to telomerase. One of the approaches is based on the use of the hTERT promoter. hTERT is the limiting factor for telomerase activity and it has been shown that its promoter has high transcriptional activity in tumor cells whereas its activity is very low or absent in non-tumor cells.[15,52] Therefore, its selective activity could be used to drive the expression of cytotoxic or apoptotic genes specifically in tumor cells. Using this strategy, different laboratories have shown that the expression of apoptotic genes such as caspases, FADD, Bax or TRAIL under the control of the hTERT promoter resulted in cell death by apoptosis *in vitro* and suppression of tumor growth in nude mice[53–56] with minimal effects on bone marrow mesenchymal stem cells and liver cells.[56]

Another intriguing alternative for cancer treatment is hTERT-based immunotherapy. This strategy is based on the fact that hTERT represents a good candidate as a universal tumor-specific antigen for the development of vaccines directed against cells that express hTERT. This approach has the advantage of targeting tumor cells independently of telomere shortening, thereby eliminating the lag phase typical of telomerase inhibition. Tumor cells exhibiting strong telomerase activity and expressing hTERT can present peptides derived from hTERT in the context of cell surface class I MHC molecules, which can be recognized by cytotoxic T cells.[57] Vonderheide and colleagues identified a hTERT-derived peptide that binds to the most frequently expressed HLA allele in humans. They showed that cytotoxic T lymphocytes specific for this peptide recognized and killed a wide range of telomerase-positive cancer cell lines, but did not recognize telomerase-negative cells.[58] These results have been confirmed in mouse models, where vaccination with dendritic cells transduced with murine TERT induced an anti-TERT immune response and resulted in protection from tumor initiation.[59] Similarly, hTERT-expressing dendritic cells from human patients have been used to generate *ex vivo* hTERT-specific cytotoxic T cells, which are able to kill primary human tumors in a MHC-dependent fashion.[59] The major concern regarding hTERT immunotherapy is the possibility of killing normal telomerase-positive cells, due to their recognition by cytotoxic

T cells. None of the mentioned studies reported any effects on hematopoietic progenitors or activated T-lymphocytes,[60,61] most likely due to relatively low protein levels or inefficient peptide processing in normal cells. Whether other stem cell populations or telomerase-positive somatic cells may be affected by this treatment still remains to be evaluated.

Telomeres as Anti-cancer Targets

Inhibition of telomerase has been considered a tumor-specific anti-proliferative strategy due to its widespread presence in human tumors. Though telomerase is the most common mechanism for telomere maintenance, a small percentage of human tumors (10–15%) use an alternative mechanism to maintain telomere length, called ALT, which is independent of telomerase and is based on recombination.[25,62] It is expected that these kinds of tumors would not be responsive to anti-telomerase therapy. Moreover, inhibition of telomerase activity in cancer cells may result in strong selective pressure for survival, activation of ALT and resistance to therapy.[63,64]

Because telomere stability and maintenance is essential for the growth of any tumor cell, a possible anti-cancer approach independent of the telomere maintenance mechanism would be the targeting of the telomeres themselves. Both the telomeric sequence and the unique structure of the T-loop at the very end of the chromosomes have been exploited as suitable targets for anti-cancer drug development. The G-rich 3′-overhang at the end of the telomeres can form G-quadruplexes *in vitro*.[6] Molecules that can bind and stabilize these structures may have anti-proliferative effects due to the sequestration of the 3′-end, which would block telomerase access, resulting in telomere shortening.[65] A number of such compounds, such as cationic porphyrins and anthraquinones derivatives have been designed.[65,66] However, only recently have there been reports showing that new generation G-quadruplex-interacting agents specifically can induce both anti-proliferative and pro-apoptotic effects associated with telomere shortening in lung adenocarcinoma and mieloma cells.[67,68] Destabilization of telomeric structure has also been achieved by using PNA oligonucleotides that are complementary to the single-stranded telomeric overhang.[68] In this situation, cells stopped growing and died because the presence of the PNA oligonucleotide specifically blocked telomerase access to the telomeres causing telomere erosion and/or disruption of telomere structure.[68]

Conclusion

In the last decade, enormous progress has been made in the telomerase and telomere research field. These research efforts have expanded the development of possible approaches for the treatment of human tumors. The different telomerase- and telomere-based strategies described above, although very promising for future clinical applications, require a more thorough evaluation of any benefits as well as any detrimental side-effects for the patients. More specifically, it is necessary to evaluate the toxicity and the deleterious effects of these treatments on normal telomerase-positive cells. Studies in mice have reported that viable, late generation telomerase knockout mice show deficiencies in hematopoietic stem and liver cell functions, and in wound healing, indicating that blocking telomerase may compromise the function of normal fast renewing stem cells.[41] More recently, it has been shown that telomerase is present at very low levels in some normal proliferating somatic cells. Interfering with its functions induces premature senescence, suggesting that telomerase is necessary for the maintenance of telomere structure and cellular lifespan in normal cells.[27] From a therapeutic viewpoint, it is important to consider that tumor cells proliferate at a much higher rate than stem cells, and in normal somatic cells, telomerase activity is only transiently present during the cell cycle. Thus, the effects of telomerase inhibition and telomere destabilization may be much greater for tumor cells with only mild effects on normal and stem cells. As with all therapies, the clinical applicability of anti-telomerase/telomere strategies should be based on a careful balance of benefits and risks for the patients.

Telomere and Telomerase Therapies against Aging

The mechanisms underlying the aging process in humans are incompletely defined and likely to be complex and multifactorial. At a cellular level, many changes identified during aging are also features of certain age-related diseases. The molecular basis for some of these changes include oxidative stress, DNA damage, genomic instability, telomere reduction and regulation, and cell death or apoptosis.[69]

Cellular senescence is a natural barrier to cellular proliferation, and may be an exploitable anti-aging target, at least at a cellular level. However there are a number of mechanisms controlling cellular senescence in cell culture, where this aspect of cell behavior was first described.[1] Moreover, the regulation or impact of cellular senescence in the context of an aging organism remains unknown.

In this final section of the chapter, we will review the different mechanisms regulating cellular senescence; provide a short perspective regarding the relation between cellular senescence in cell culture to aging in the organism; describe some recent data implicating telomerase in cell survival independent of its role in telomere maintenance; and discuss how modulating mechanisms implicated in maintaining genomic integrity, such as telomere stability and survival-promoting proteins like telomerase could ultimately affect cellular survival.

Telomere-dependent and Independent Cellular Senescence

Senescence in cell culture, regardless of the initiating signal, is characterized by withdrawal from the cell cycle followed by growth arrest and morphological changes that include an increase in cell volume and a flattened cytoplasm. Senescent cells also exhibit alterations in gene expression, nuclear structure, protein processing and metabolism, and chromosomal instabilities, yet remain metabolically viable.[70,71] A commonly used marker for senescence is the appearance of senescence-associated β-galactosidase; this assay allowed for the first time the detection of senescent cells *in vivo*.[70]

Until recently, cellular senescence was equated with replicative, telomere-dependent senescence. Besides telomere shortening, other mechanisms also appear to control cellular senescence. Specifically in mouse and other rodent fibroblast cells that have long telomeres and are telomerase-positive, telomere-independent senescence mechanisms operate.[70] Initial suggestions that cellular senescence of mouse embryonic fibroblasts might be due to the culture environment were recently supported by experiments describing the lack of replicative senescence in normal rodent cells grown under specific culture conditions, for example low, 3%, physiological oxygen.[70] Other differences between cellular senescence in human and mice have been extensively reviewed elsewhere, and will not be the focus of this chapter.[70]

A number of regulators of cellular senescence in human cells have also been reported. These include, but are not limited to, telomere shortening, suboptimal culture conditions, overexpression of oncogenes, such as RAS, DNA-damaging stresses such as X- and UV irradiation, or oxidative stress and chromatin remodelling.[70–74] For several years, distinctions were made between telomere-independent senescence, which typically cannot be prevented by the forced expression of hTERT, and telomere-dependent senescence.[75,76] Interestingly, overexpression of telomerase, similarly to oncogenic expression of Ras, Raf and E2F1, can also lead to cellular senescence.[77] The picture that is now emerging is that despite differences in the initiating events, effectors leading to cellular senescence typically activate the tumor suppressor proteins p53 or pRb, supporting the view of cellular senescence as a tumor suppressor mechanism in human cells.[71] Upregulation of p16, an activator of pRb, occurs during stress-induced senescence, while telomere shortening, through telomere dysfunction or uncapping leads to activation of p53.[71,73] Interestingly, Satyanarayana and colleagues have described that senescence via activation of the DNA damage response through telomere shortening is dependent on mitogen stimulation.[78] Additionally, though stress-induced premature senescence is generally telomere independent, rapid telomere shortening can be associated with oxidative stress, suggesting that strictly stress-independent, telomere-dependent senescence may be the exception rather than the rule.[70,74]

Relating Cellular Senescence *in vitro* to Aging *in vivo*

Despite our incomplete understanding of cellular senescence *in vitro*, we face an even more daunting challenge: that of determining if cellular senescence occurs *in vivo*. If it does, is the regulation of cellular senescence similar *in vitro* and *in vivo*, and does cellular senescence contribute to the aging phenotype?[74] The appearance of senescence-associated β-galactosidase activity has been used extensively to document the existence of senescent cells *in vivo*, during aging, for example in skin, retina and liver, and in chronic diseases associated with aging, such as hyperplastic prostate and atherosclerotic lesions.[70] Senescent cells *in vivo* have also been recently documented in mouse models.[71] As one example, proliferation-incompetent hepatocytes from telomerase-deficient mice with critically short telomeres are unable to regenerate the liver and subsequently undergo senescence.[79] Satyanarayana and colleagues recently demonstrated that the regenerative defects of the hepatocytes from these mice depend on both telomere length and mitogen stimulation, mirroring the results they obtained using human fibroblasts.[78] In their report these authors highlight the importance of analyzing cell cycle activity *in vivo* in order to assess true levels of senescent cells in organs and tissues. Despite increased evidence for the existence of senescent cells *in vivo*, we can only begin to suggest that such cells may contribute to the age-related decline of tissues and organs or the development of age-related diseases. The impact of cellular senescence in relation to the ill-characterized role of other mechanisms such as apoptosis in aging remains to be investigated.

Telomerase and Cell Survival

The primary function of telomerase is the synthesis of telomeric DNA. Recently however, a number of additional, as yet incompletely characterized roles, some independent of its catalytic activity or telomere maintenance function, have been documented for telomerase.[80] The authors highlight five possible ways by which telomerase may promote growth and survival including enhancement of DNA repair, healing of broken chromosome ends, mitogenic signaling, regulation of tumor suppressors and activation of DNA methyltransferase I transcription. Below, we will focus on a discussion of the role of telomerase in resistance to growth inhibition and/or cell death, tumorigenesis, and tumor dissemination, and cell survival and proliferation.

Early evidences of a possible role for telomerase in mediating protection or resistance to cell death came in part from studies in human cancer cells and rodent neuronal cells.[81,82] A recent analysis revealed that hTERT suppresses etoposide- and staurosporine-mediated apoptosis of telomerase-positive and ALT cells at a pre-mitochondrial step and requires catalytic activity and binding to 14-3-3 protein.[83] The authors suggest that telomerase may suppress a DNA damage response by preventing or repairing damage to the telomeres. Ectopic expression of TERT has also been recently reported to protect against brain injury resulting from ischemia and NMDA-induced neurotoxicity.[84]

A telomere length-independent role for telomerase in tumorigenesis has been described primarily using mouse models. Transgenic mice that overexpress TERT under the control of a keratin-5 or constitutive β-actin promoter have an increased incidence of developing spontaneous tumors.[85–87] These observations have been recently expanded using a new mouse model with constitutive expression of TERT in thymocytes and peripheral T cells (Lck-TERT mice).[88] These mice not only had an increased incidence of T-cell lymphomas, but the lymphomas were more disseminated than in control mice, affecting nonlymphoid organs. As TERT expression in primary thymocytes resulted in greater chromosomal instability (upon gamma irradiation), the authors suggest that TERT may affect the DNA damage response, or alternatively, TERT may increase the survival of cells with chromosomal damage.

A telomere maintenance-independent role for telomerase in tumorigenesis has also been documented in human cells. Stewart and colleagues elegantly demonstrated that ALT GM847 cells are expressing oncogenic H-ras tumorigenic in immunodeficient mice in the presence of hTERT expression.[89] The same phenotype was observed when a mutant hTERT, incapable of maintaining telomere length, was expressed in these cells. Another role for telomerase in carcinogenesis has been suggested based on the evidence that $p16^{INK4A(-)}$ human mammary epithelial cells ectopically expressing hTERT are resistant to transforming growth factor β growth inhibition, a key characteristic of epithelial cell-derived malignancies.[90] This phenotype was however dependent on the expression of a catalytically active telomerase able to maintain telomeres.

Several reports have implicated telomerase in cell survival and proliferation, only a subset of which are discussed below.[80] DNA microarray studies identified a set of growth-controlling genes whose expression was modified by the ectopic expression of hTERT in human mammary epithelial cells.[91] Specifically, the expression of genes implicated in a cellular mitogenic program, such as EGRF and FGF was upregulated and the expression of suppressors of cell growth was downregulated, resulting in enhancement of cell proliferation. A role for telomerase in regulating cell survival, independent of its catalytic activity was also recently reported.[92] In this study, expression of a catalytically inactive mutant was able to rescue human breast cancer cells expressing hTERT antisense RNA from cell death.

Telomere and Telomerase-based Anti-aging Therapy

A large number of studies have investigated and validated telomerase as an anti-cancer therapy. Fewer studies have addressed the effects of telomerase overexpression and investigated a putative telomerase-based anti-aging therapy or therapy for age-related diseases. While forced expression of telomerase in human cells does not impart a tumorigenic phenotype, it does extend the lifespan of some human primary cells and rescue the chromosomal instability and premature aging phenotypes of $mTR^{-/-}$ mice.[14,15,40,93] The inhibition of liver cirrhosis in mice by telomerase gene delivery has also been reported.[94] These experiments validate the possibility of using telomerase-based therapies to augment the cellular lifespan of cells destined to senescence or death, for instance during neurodegeneration, or in diseases with premature aging phenotypes, such as dyskeratosis congenita.[14,82] Moreover, telomerase, though imparting immortality to various human somatic cell types, does not impart uncontrolled cell growth, a feature typically attributed to oncogenes.[14] The distinction between telomerase and oncogenes is further evident when considering that telomerase is present in stem cells, germline cells, and many mouse somatic cell types without features of uncontrolled cell growth.

Nonetheless, telomerase expression in mice can, in certain contexts, promote telomere length-independent tumorigenesis, highlighting the need to cautiously evaluate the use of telomerase expression in therapy.[85–87] To fully evaluate the effects of using telomerase-based therapy, it will be essential to completely understand telomerase functions that are independent of telomere lengthening or catalytic activity. Studies may also reveal fundamental differences of telomerase-based therapies in cell culture and between human and rodents. Besides the forced expression of telomerase, it is possible to envision other indirect mechanisms to alter telomerase expression, for example, by targeting the hTERT promoter. Stabilizing telomere length or integrity by modulation of telomere-binding proteins or telomerase-associated proteins could also be investigated as alternative strategies to increase telomere and genomic stability, and cell survival. Lastly, strategies to increase cellular resistance to oxidative stress, to which telomeres are particularly susceptible, may also favor cell survival.

Conclusion

As our understanding of telomerase function in cell survival and tumorigenesis is incomplete, and provided in large part by studies using human cell culture and mouse models, it is with diligence that we must continue to design telomere- and telomerase-based anti-cancer and anti-aging strategies to inhibit cancer cell growth and promote cellular survival respectively. Modulating levels of inhibition or activation (transient, conditional) and duration and location of treatment (cell-type specific) may ultimately provide mechanisms by which we can optimize the benefits of telomere- and telomerase-based strategies while minimizing the risks.[14]

Acknowledgments

Work in the laboratory of C. Autexier has been funded by the Canadian Institutes of Health Research (CIHR, Institute of Aging, Institute of Cancer Research), Cancer Research Society Inc., Valorisation Recherche Quèbec, and the Fonds de Recherches en Santè du Quèbec (FRSQ). M.A. Cerone is supported by an U.S. Army Department of Defense Breast Cancer Research Program Award. R. Ward was supported by a CIHR Cancer Consortium Training Grant Studentship Award. C. Autexier is a Chercheur-Boursier of the FRSQ and the recipient of a Boehringer–Ingelheim (Canada) Inc. Young Investigator Award.

References

1. Blackburn EH, Greider CW (1995) *Telomeres*, Cold Spring Harbor, Cold Spring Harbor Laboratory Press.
2. Chakhparonian M, Wellinger RJ (2003) Telomere maintenance and DNA replication: how closely are these two connected? *Trends Genet* **19**: 439–446.
3. Blackburn EH (2001) Switching and signaling at the telomere. *Cell* **106**: 661–673.
4. Griffith JD, Comeau L, Rosenfield S, Stansel RM, Bianchi A, Moss H, de Lange T (1999) Mammalian telomeres end in a large duplex loop. *Cell* **97**: 503–514.

5. Yoshimura SH, Maruyama H, Ishikawa F, Ohki R, Takeyasu K (2004) Molecular mechanisms of DNA end-loop formation by TRF2. *Genes Cells* **9**: 205–218.
6. Neidle S, Parkinson GN (2003) The structure of telomeric DNA. *Curr Opin Struct Biol* **13**: 275–283.
7. de Lange T (2002) Protection of mammalian telomeres. *Oncogene* **21**: 532–540.
8. Wei C, Price M (2003) Protecting the terminus: t-loops and telomere end-binding proteins. *Cell Mol Life Sci* **60**: 2283–2294.
9. Iwano T, Tachibana M, Reth M, Shinkai Y (2004) Importance of TRF1 for functional telomere structure. *J Biol Chem* **279**: 1442–1448.
10. Loayza D, De Lange T (2003) POT1 as a terminal transducer of TRF1 telomere length control. *Nature* **423**: 1013–1018.
11. Kondo T, Oue N, Yoshida K, Mitani Y, Naka K, Nakayama H, Yasui W (2004) Expression of POT1 is associated with tumor stage and telomere length in gastric carcinoma. *Cancer Res* **64**: 523–529.
12. Takai H, Smogorzewska A, de Lange T (2003) DNA damage foci at dysfunctional telomeres. *Curr Biol* **13**: 1549–1556.
13. Sherr CJ, McCormick F (2002) The RB and p53 pathways in cancer. *Cancer Cell* **2**: 103–112.
14. Harley CB (2002) Telomerase is not an oncogene. *Oncogene* **21**: 494–502.
15. Bodnar AG, Ouellette M, Frolkis M, Holt SE, Chiu C-P, Morin GB, Harley CB, Shay JW, Lichtsteiner S, Wright WE (1998) Extension of life-span by introduction of telomerase into normal human cells. *Science* **279**: 349–352.
16. Collins K (2000) Mammalian telomeres and telomerase. *Curr Opin Cell Biol* **12**: 378–383.
17. Jacob NK, Kirk KE, Price CM (2003) Generation of telomeric G strand overhangs involves both G and C strand cleavage. *Mol Cell* **11**: 1021–1032.
18. Zhu XD, Niedernhofer L, Kuster B, Mann M, Hoeijmakers JH, de Lange T (2003) ERCC1/XPF removes the 3′ overhang from uncapped telomeres and represses formation of telomeric DNA-containing double minute chromosomes. *Mol Cell* **12**: 1489–1498.
19. Kim NW, Piatyszek MA, Prowse KR, Harley CB, West MD, Ho PLC, Coviello GM, Wright WE, Weinrich SL, Shay JW (1994) Specific association of human telomerase activity with immortal cells and cancer. *Science* **266**: 2011–2015.
20. Hemann MT, Strong MA, Hao LY, Greider CW (2001) The shortest telomere, not average telomere length, is critical for cell viability and chromosome stability. *Cell* **107**: 67–77.
21. Teixeira MT, Arneric M, Sperisen P, Lingner J (2004) Telomere length homeostasis is achieved via a switch between telomerase-extendible and -nonextendible states. *Cell* **117**: 323–335.
22. Harrington L (2003) Biochemical aspects of telomerase function. *Canc Lett* **194**: 139–154.
23. Biessmann H, Mason JM (2003) Telomerase-independent mechanisms of telomere elongation. *Cell Mol Life Sci* **60**: 2325–2333.
24. Chen Q, Ijpma A, Greider CW (2001) Two survivor pathways that allow growth in the absence of telomerase are generated by distinct telomere recombination events. *Mol Cell Biol* **21**: 1819–1827.
25. Henson JD, Neumann AA, Yeager TR, Reddel RR (2002) Alternative lengthening of telomeres in mammalian cells. *Oncogene* **21**: 598–610.

26. Shay JW, Bacchetti S (1997) A survey of telomerase activity in human cancer. *Eur J Cancer* **33**: 787–791.
27. Masutomi K, Yu EY, Khurts S, Ben-Porath I, Currier JL, Metz GB, Brooks MW, Kaneko S, Murakami S, DeCaprio JA, Weinberg RA, Stewart SA, Hahn WC (2003) Telomerase maintains telomere structure in normal human cells. *Cell* **114**: 241–253.
28. Hahn WC, Stewart SA, Brooks MW, York SG, Eaton E, Kurachi A, Beijersbergen RL, Knoll JHM, Meyerson M, Weinberg RA (1999) Inhibition of telomerase limits the growth of human cancer cells. *Nature Med* **5**: 1164–1170.
29. Zhang X, Mar V, Zhou W, Harrington L, Robinson MO (1999) Telomere shortening and apoptosis in telomerase-inhibited human tumor cells. *Genes Devel* **13**: 2388–2399.
30. Huard S, Autexier C (2002) Targeting human telomerase in cancer therapy. *Curr Med Chem Anti* **2**: 577–587.
31. Pascolo E, Wenz C, Lingner J, Hauel N, Priepke H, Kauffmann I, Garin-Chesa P, Rettig WJ, Damm K, Schnapp A (2002) Mechanism of human telomerase inhibition by BIBR1532, a synthetic, non-nucleosidic drug candidate. *J Biol Chem in press. February 19, 2002.*
32. Damm K, Hemmann U, Garin-Chesa P, Hauel N, Kauffmann I, Priepke H, Niestroj C, Daiber C, Enenkel B, Guilliard B, Lauritsch I, Muller E, Pascolo E, Sauter G, Pantic M, Martens UM, Wenz C, Lingner J, Kraut N, Rettig WJ, Schnapp A (2001) A highly selective telomerase inhibitor limiting human cancer cell proliferation. *EMBO J* **20**: 6958–6968.
33. Kondo S, Kondo Y, Li G, Silverman RH, Cowell JK (1998) Targeted therapy of human malignant glioma in a mouse model by 2-5A antisense directed against telomerase RNA. *Oncogene* **16**: 3323–3330.
34. Kushner DM, Paranjape JM, Bandyopadhyay B, Cramer H, Leaman DW, Kennedy AW, Silverman RH, Cowell JK (2000) 2-5A antisense directed against telomerase RNA produces apoptosis in ovarian cancer cells. *Gynecol Oncol* **76**: 183–192.
35. Shammas MA, Simmons CG, Corey DR, Reis RJ (1999) Telomerase inhibition by peptide nucleic acids reverses 'immortality' of transformed cells. *Oncogene* **18**: 6191–6200.
36. Herbert B-S, Pitts AE, Baker SI, Hamilton SE, Wright WE, Shay JW, Corey DR (1999) Inhibition of human telomerase in immortal human cells leads to progressive telomere shortening and cell death. *Proc Natl Acad Sci USA* **96**: 14276–14281.
37. Blasco MA, Lee H-W, Hande MP, Samper E, Lansdorp PM, DePinho RA, Greider CW (1997) Telomere shortening and tumor formation by mouse cells lacking telomerase *RNA Cell* **91**: 25–34.
38. Liu Y, Snow BE, Hande MP, Yeung D, Erdmann NJ, Wakeham A, Itie A, Siderovski DP, Lansdorp PM, Robinson MO, Harrington L (2000) The telomerase reverse transcriptase is limiting and necessary for telomerase function *in vivo*. *Curr Biol* **10**: 1459–1462.
39. Hemann MT, Rudolph KL, Strong MA, DePinho RA, Chin L, Greider CW (2001) Telomere dysfunction triggers developmentally regulated germ cell apoptosis. *Mol Biol Cell* **12**: 2023–2030.
40. Samper E, Flores JM, Blasco M (2001) Restoration of telomerase activity rescues chromosomal instability and premature aging in *Terc*$^{-/-}$ mice with short telomeres. *EMBO Rep* **2**: 1–8.
41. Goytisolo FA, Blasco MA (2002) Many ways to telomere dysfunction: *in vivo* studies using mouse models. *Oncogene* **21**: 584–591.

42. Marusic L, Anton M, Tidy A, Wang P, Villeponteau B, Bacchetti S (1997) Reprogramming of telomerase by expression of mutant telomerase RNA template in human cells leads to altered telomeres that correlate with reduced cell viability. *Mol Cell Biol* **17**: 6394–6401.
43. Kim MM, Rivera MA, Botchkina IL, Shalaby R, Thor AD, Blackburn EH (2001) A low threshold level of expression of mutant-template telomerase RNA inhibits human tumor cell proliferation. *Proc Natl Acad Sci USA* **98**: 7982–7987.
44. Guiducci C, Cerone MA, Bacchetti S (2001) Expression of mutant telomerase in immortal telomerase-negative human cells results in cell cycle deregulation, nuclear and chromosomal abnormalities and rapid loss of viability. *Oncogene* **20**: 714–725.
45. Misawa M, Tauchi T, Sashida G, Nakajima A, Abe K, Ohyashiki JH, Ohyashiki K (2002) Inhibition of human telomerase enhances the effect of chemotherapeutic agents in lung cancer cells. *Int J Oncol* **21**: 1087–1092.
46. Tentori L, Portarena I, Barbarino M, Balduzzi A, Levati L, Vergati M, Biroccio A, Gold B, Lombardi ML, Graziani G (2003) Inhibition of telomerase increases resistance of melanoma cells to temozolomide, but not to temozolomide combined with poly (ADP-ribose) polymerase inhibitor. *Mol Pharm* **63**: 192–202.
47. Kondo Y, Kondo S, Tanaka Y, Haqqi T, Barna BP, Cowell JK (1998) Inhibition of telomerase increases the susceptibility of human malignant glioblastoma cells to cisplatin-induced apoptosis. *Oncogene* **16**: 2243–2248.
48. Ludwig A, Saretzki G, Holm PS, Tiemann F, Lorenz M, Emrich T, Harley CB, von Zglinicki T (2001) Ribozyme cleavage of telomerase mRNA sensitizes breast epithelial cells to inhibitors of topoisomerase. *Cancer Res* **61**: 3053–3061.
49. Kondo Y, Komata T, Kondo S (2001) Combination therapy of 2-5A antisense against telomerase RNA and cisplatin for malignant gliomas. *Int J Oncol* **18**: 1287–1292.
50. Chen Z, Koeneman KS, Corey DR (2003) Consequences of telomerase inhibition and combination treatments for the proliferation of cancer cells. *Cancer Res* **63**: 5917–5925.
51. Folini M, De Marco C, Orlandi L, Daidone MG, Zaffaroni N (2000) Attenuation of telomerase activity does not increase sensitivity of human melanoma cells to anticancer agents. *Eur J Cancer* **36**: 2137–2145.
52. Meyerson M, Counter CM, Eaton EN, Ellisen LW, Steiner P, Dickinson SC, Ziaugra L, Beijersbergen RL, Davidoff MJ, Liu Q, Bacchetti S, Haber DA, Weinberg RA (1997) *hEST2*, the putative human telomerase catalytic subunit gene, is up-regulated in tumor cells and during immortalization. *Cell* **90**: 785–795.
53. Gu J, Kagawa S, Takakura M, Kyo S, Inoue M, Roth JA, Fang B (2000) Tumor-specific transgene expression from the human telomerase reverse transcriptase promoter enables targeting of the therapeutic effects of the Bax gene to cancers. *Cancer Res* **60**: 5359–5364.
54. Koga S, Hirohata S, Kondo Y, Komata T, Takakura M, Inoue M, Kyo S, Kondo S (2000) A novel telomerase-specific gene therapy: gene transfer of caspase-8 utilizing the human telomerase catalytic subunit gene promoter. *Hum Gene Ther* **11**: 1397–1406.
55. Komata T, Kondo Y, Kanzawa T, Hirohata S, Koga S, Sumiyoshi H, Srinivasula SM, Barna BP, Germano IM, Takakura M, Inoue M, Alnemri ES, Shay JW, Kyo S, Kondo S (2001) Treatment of malignant glioma cells with the transfer of constitutively active caspase-6 using the human

telomerase catalytic subunit (human telomerase reverse transcriptase) gene promoter. *Cancer Res* **61**: 5796–5802.
56. Jacob D, Davis J, Zhu H, Zhang L, Teraishi F, Wu S, Marini FC 3rd, Fang B (2004) Suppressing orthotopic pancreatic tumor growth with a fiber-modified adenovector expressing the TRAIL gene from the human telomerase reverse transcriptase promoter. *Clin Cancer Res* **10**: 3535–3541.
57. Vonderheide RH (2002) Telomerase as a universal tumor-associated antigen for cancer immunotherapy. *Oncogene* **21**: 674–679.
58. Vonderheide RH, Hahn WC, Schultze JL, Nadler LM (1999) The telomerase catalytic subunit is a widely expressed tumor-associated antigen recognized by cytotoxic T lymphocytes. *Immunity* **10**: 673–639.
59. Nair SK, Heiser A, Boczkowski D, Majumdar A, Naoe M, Lebkowski JS, Vieweg J, Gilboa E (2000) Induction of cytotoxic T cell responses and tumor immunity against unrelated tumors using telomerase reverse transcriptase RNA transfected dendritic cells. *Nat Med* **6**: 1011–1017.
60. Minev B, Hipp J, Firat H, Schmidt JD, Langlade-Demoyen P, Zanetti M (2000) Cytotoxic T cell immunity against telomerase reverse transcriptase in humans. *Proc Natl Acad Sci USA* **97**: 4796–4801.
61. Vonderheide RH, Anderson KS, Hahn WC, Butler MO, Schultze JL, Nadler LM (2001) Characterization of HLA-A3-restricted cytotoxic T lymphocytes reactive against the widely expressed tumor antigen telomerase. *Clin Cancer Res* **7**: 3343–3348.
62. Dunham MA, Neumann AA, Fasching CL, Reddel RR (2000) Telomere maintenance by recombination in human cells. *Nat Genet* **26**: 447–450.
63. Hande MP, Samper E, Lansdorp P, Blasco MA (1999) Telomere length dynamics and chromosomal instability in cells derived from telomerase null mice. *J Cell Biol* **144**: 589–601.
64. Bechter OE, Zou Y, Walker W, Wright WE, Shay JW (2004) Telomeric recombination in mismatch repair deficient human colon cancer cells after telomerase inhibition. *Cancer Res* **64**: 3444–3451.
65. Bearss DJ, Hurley LH, Von Hoff DD (2000) Telomere maintenance mechanisms as a target for drug development. *Oncogene* **19**: 6632–6641.
66. Hurley LH, Wheelhouse RT, Sun D, Kerwin SM, Salazar M, Fedoroff OY, Han FX, Han H, Izbicka E, Von Hoff DD (2000) G-quadruplexes as targets for drug design. *Pharmacol Ther* **85**: 141–158.
67. Riou JF, Guittat L, Mailliet P, Laoui A, Renou E, Petitgenet O, Megnin-Chanet F, Helene C, Mergny JL (2002) Cell senescence and telomere shortening induced by a new series of specific G-quadruplex DNA ligands. *Proc Nat Acad Sci USA* **99**: 2672–2677.
68. Shammas MA, Liu X, Gavory G, Raney KD, Balasubramanian S, Shmookler Reis RJ (2004) Targeting the single-strand G-rich overhang of telomeres with PNA inhibits cell growth and induces apoptosis of human immortal cells. *Exp Cell Res* **295**: 204–214.
69. Troen BR (2003) The biology of aging. *Mount Sinai J Med* **70**: 3–22.
70. Itahana K, Campisi J, Dimri GP (2004) Mechanisms of cellular senescence in human and mouse cells. *Biogerontology* **5**: 1–10.
71. Ben-Porath I, Weinberg RA (2004) When cells get stressed: an integrative view of cellular senescence. *J Clin Invest* **113**: 8–13.

72. Ramirez RD, Morales CP, Herbert BS, Rohde JM, Passons C, Shay JW, Wright WE (2001) Putative telomere-independent mechanisms of replicative aging reflect inadequate growth conditions. *Genes Dev* **15**: 398–403.
73. Munro J, Barr NI, Ireland H, Morrison V, Parkinson EK (2004) Histone deacetylase inhibitors induce a senescence-like state in human cells by a p16-dependent mechanism that is independent of a mitotic clock. *Exp Cell Res* **295**: 525–538.
74. Von Zglinicki T (2003) Replicative senescence and the art of counting. *Exp Gerontol* **38**: 1259–1264.
75. Wei S, Sedivy JM (1999) Expression of catalytically active telomerase does not prevent premature senescence caused by overexpression of oncogenic Ha-Ras in normal human fibroblasts. *Cancer Res* **59**: 1539–1543.
76. Gorbunova V, Seluanov A, Pereira–Smith OM (2002) Expression of human telomerase (hTERT) does not prevent stress-induced senescence in normal human fibroblasts but protects the cells from stress-induced apoptosis and necrosis. *J Biol Chem* **277**: 38540–38549.
77. Gorbunova V, Seluanov A, Pereira–Smith OM (2003) Evidence that high telomerase activity may induce a senescent-like growth arrest in human fibroblasts. *J Biol Chem* **278**: 7692–7698.
78. Satyanarayana A, Greenberg RA, Schaetzlein S, Buer J, Masutomi K, Hahn WC, Zimmermann S, Martens U, Manns MP, Rudolph KL (2004) Mitogen stimulation cooperates with telomere shortening to activate DNA damage responses and senescence signaling. *Mol Cell Biol* **24**: 5459–5474.
79. Satyanarayana A, Wiemann SU, Buer J, Lauber J, Dittmar KE, Wustefeld T, Blasco MA, Manns MP, Rudolph KL (2003) Telomere shortening impairs organ regeneration by inhibiting cell cycle re-entry of a subpopulation of cells. *Embo J* **22**: 4003–4013.
80. Gorbunova V, Seluanov A (2003) Telomerase as a growth-promoting factor. *Cell Cycle* **2**: 534–537.
81. Holt SE, Glinsky VV, Ivanova AB, Glinsky GV (1999) Resistance to Apoptosis in Human Cells Conferred by Telomerase Function and Telomere Stability. *Mol Carcin* **25**: 241–248.
82. Mattson MP, Klapper W (2001) Emerging roles for telomerase in neuronal development and apoptosis. *J Neuro Res* **63**: 1–9.
83. Zhang P, Chan SL, Fu W, Mendoza M, Mattson MP (2003) TERT suppresses apoptotis at a premitochondrial step by a mechanism requiring reverse transcriptase activity and 14-3-3 protein-binding ability. *Faseb J* **17**: 767–769.
84. Kang HJ, Choi YS, Hong SB, Kim KW, Woo RS, Won SJ, Kim EJ, Jeon HK, Jo SY, Kim TK, Bachoo R, Reynolds IJ, Gwag BJ, Lee HW (2004) Ectopic expression of the catalytic subunit of telomerase protects against brain injury resulting from ischemia and NMDA-induced neurotoxicity. *J Neurosci* **24**: 1280–1287.
85. Gonzalez-Suarez E, Samper E, Ramirez A, Flores JM, Martin-Caballero J, Jorcano JL, Blasco MA (2001) Increased epidermal tumors and increased skin wound healing in transgenic mice overexpressing the catalytic subunit of telomerase, mTERT, in basal keratinocytes. *EMBO J* **20**: 2619–2630.
86. Gonzalez-Suarez E, Flores JM, Blasco MA (2002) Cooperation between p53 mutation and high telomerase transgenic expression in spontaneous cancer development. *Mol Cell Biol* **22**: 7291–7301.

87. Artandi SE, Alson S, Tietze MK, Sharpless NE, Ye S, Greenberg RA, Castrillon DH, Horner JW, Weiler SR, Carrasco RD, DePinho RA (2002) Constitutive telomerase expression promotes mammary carcinomas in aging mice. *Proc Natl Acad Sci USA* **99**: 8191–8196.
88. Canela A, Martin-Caballero J, Flores JM, Blasco MA (2004) Constitutive expression of tert in thymocytes leads to increased incidence and dissemination of T-cell lymphoma in Lck-Tert mice. *Mol Cell Biol* **24**: 4275–4293.
89. Stewart SA, Hahn WC, O'Connor BF, Banner EN, Lundberg AS, Modha P, Mizuno H, Brooks MW, Fleming M, Zimonjic DB, Popescu N, Weinberg RA (2002) Telomerase contributes to tumorigenesis by a telomere length-independent mechanism. *Proc Natl Acad Sci USA*.
90. Stampfer MR, Garbe J, Levine G, Lichtsteiner S, Vasserot AP, Yaswen P (2001) Expression of the telomerase catalytic subunit, hTERT, induces resistance to transforming growth factor beta growth inhibition in p16INK4A(-) human mammary epithelial cells. *Proc Natl Acad Sci USA* **98**: 4498–4503.
91. Smith LL, Coller HA, Roberts JM (2003) Telomerase modulates expression of growth-controlling genes and enhances cell proliferation. *Nat Cell Biol* **5**: 474–479.
92. Cao Y, Li H, Deb S, Liu JP (2002) TERT regulates cell survival independent of telomerase enzymatic activity. *Oncogene* **21**: 3130–3138.
93. Jiang X-R, Jimenez G, Chang E, Frolkis M, Kusler B, Sage M, Beeche M, Bodnar AG, Wahl GM, Tisty TD, Chiu C-P (1999) Telomerase expression in human somatic cells does not induce changes associated with a transformed phenotype. *Nat Genet* **21**: 111–114.
94. Rudolph KL, Chang S, Millard M, Schreiber-Agus N, DePinho RA (2000) Inhibition of experimental liver cirrhosis in mice by telomerase gene delivery. *Science* **287**: 1253–1258.

13

Clinical Perspective of the Present Status of Treatment for Major Age-related Diseases

Ashit Syngle

Healing Touch City Clinic; Fortis Heart Institute & Multispeciality Hospital

#547, Sector 16-D Chandigarh -160015, India

Email: ashitsyngle@yahoo.com

Although aging is a universal biological phenomenon and not a disease *per se*, the fact is that we progressively become more prone to a variety of diseases with aging. Hundreds of biogerontologists are committed to understanding and unraveling the molecular mechanisms of aging, and significant advances have been made in this respect. These advances will surely lead to the development of new and effective means of prevention and treatment of age-related diseases. One does not know which of these therapies will become available and when. But, the need for the treatment of diseases exists now and here, and the patients must be treated with the best possible approach. Geriatric medicine is quite an advanced field, and there is a whole range of clinical perspectives in the treatment of age-related diseases. This chapter is devoted to management principles of major age-related diseases, as seen from the perspective of a clinical practitioner.

Cardiovascular Disease

Cardiovascular diseases — coronary artery disease (CAD), hypertension, heart failure and stroke — reach epidemic proportions in the elderly. A common pathologic substrate for these diseases is atherosclerosis.

There has been a definite change in the traditional approach of symptomatic treatment of these diseases to treating the damaged vascular bed with combination medical therapy, including a statin (regardless of lipid levels), aspirin, a beta blocker and an angiotensin-converting enzyme inhibitor (ACE inhibitor). Any patient who presents with clinically evident atherosclerosis is at life long risk and life long treatment is recommended, so long as it is well tolerated. Overwhelming scientific evidence demonstrates that treatment targeting the atherosclerotic disease process alters the natural history of this disease, improves clinical outcomes and prolongs survival. Diabetics have similar cardiovascular risk as patients with established atherosclerosis and should also be targeted for treatment.[1] The guiding principle should be: *target the disease not the symptom or the lumen alone.* For this a holistic approach — lifestyle modification, patient education and empowerment, treatment of risk factors (dyslipidemia, diabetes, hypertension, smoking and physical inactivity), appropriate pharmacotherapy and procedures — needs to be adopted.

Who Should be Treated?

All patients with known atherosclerosis should be treated, whether they have coronary, peripheral or cerebral vascular disease unless contraindications exist or treatment is not tolerated. It does not matter how the diagnosis was made, whether the patients have symptoms, or whether they have undergone revascularization. Since diabetes is a CAD risk equivalent, it is recommended that diabetics be treated as if they have known CAD with the combination cardioprotective therapy.

What Should be the Treatment?

There is unequivocal evidence that the following therapies are of benefit:

- Antiplatelet therapy with aspirin (81–325 mg/d), clopidogrel (75 mg/d) or both.[2]
- Neurohumoral inhibition with beta blockers and ACE inhibitors in all patients without contraindications, including patients with normal blood pressure and normal ejection fraction.[3,4]
- Antiatherogenic therapy with statins regardless of LDL concentrations, in all patients with known cardiovascular disease.[5–9] Target lipid

levels in coronary, other vascular disease, or diabetes: LDL ≤ 100 mg/dL (ideally < 70mg/dL), HDL < 45 mg/dL (if starting with baseline LDL < 100, target is 70 mg/dL), TG < 150 mg/dL.

It is important to realize that the benefits of combination therapy outweigh those of any individual therapy.[10,11] The Heart Outcomes Prevention Evaluation Study[11] (HOPE) demonstrated that in patients with CAD, CVD, PVD or diabetes the use of an ACE inhibitor (ramipril) was associated with a reduction in cardiovascular events, cardiovascular mortality and all cause mortality. This benefit was seen in patients without hypertension and with normal left ventricular ejection fractions. The benefits were observed among patients who were already taking a number of effective treatments such as aspirin, beta-blockers and lipid-lowering agents. Another landmark study EUROPA[4] involving a large number of CAD patients commonly seen in clinical practice (as opposed to high risk patients in HOPE) demonstrated a 20% risk reduction in cardiovascular events and cardiovascular mortality with another ACE inhibitor Perindopril.

Furthermore, it is important to urge smoking cessation for the patient and family. The patient should also undergo cardiac rehabilitation. An aerobic exercise program consisting of moderate intensity activity 30–60 minutes a minimum of 5–7 times a week is recommended. Dietary counseling should be provided. Weight goal should be BMI 21–25 kg/m^2. The patient and his/her family member should be instructed regarding the use of medications. The purpose, dose and major side effects of each medication prescribed should be explained. The warning signs of a heart attack should be discussed with each patient and their immediate plan of action reviewed. A good glycaemic control with HbA1c < 7 is recommended in diabetics.

Individualization of therapy depending on tolerance of combination therapy and other medical issues and risk of side effects may be appropriate in certain circumstances.

Chronic Ischemic Heart Disease

The treatment of angina in older and younger patients is similar. Pharmacotherapy includes an antiplatelet agent, anti-ischemic therapy with nitrates, ACE inhibitor, statin, beta-blocker and/or calcium antagonist.

However, older individuals may be more susceptible to symptoms related to any hypotensive effects of some anti-ischemics because of decreased sympathetic responsiveness. Reversible precipitating factors need to be identified and treated. Of these, anemia, hyperthyroidism, congestive cardiac failure and non-compliance with medication may all be more common and more difficult to treat in the elderly. Though atherosclerosis is a progressive disease and risk factor reduction may appear less important in the elderly, both successful treatment of hypertension and smoking cessation decrease cardiovascular mortality in the elderly.

If medical management fails to control symptoms adequately, percutaneous transluminal coronary angioplasty (PTCA) should be considered in patients with appropriate anatomy. In the more elderly individual whose life expectancy is limited regardless of therapy and whose activity is restricted by other disease, treatment goals should be directed toward symptom alleviation and continuation of an independent life style. Although immediate success and complication rates in those 65 to 80 years of age are often similar to those in the younger age groups when angioplasty is performed by experienced operators, it is associated with a lower clinical success rate and higher vascular and cardiac complication rates in those over 80 years. This probably relates to more difficult access because of increased likelihood of peripheral vascular disease, coronary calcification, multivessel disease, renal impairment and cerebrovascular complications.

Acute Myocardial Infarction (AMI)

The two major strategies for managing AMI are: thrombolytic therapy and primary PTCA. There are many limitations to thrombolytic therapy in the elderly. More older individuals have absolute or relative contraindications to these agents, particularly hypertension, history of stroke, and gastrointestinal bleeding. Thrombolysis is unsuccessful in 20% of patients; less than 60% achieve brisk flow in infarct related artery and there is increased risk of intracranial bleeding. On the other hand, there may be survival advantage in those over 70 years with primary PTCA. However, PTCA requires not only a cath lab but one that can be galvanized into action at a short notice.

Hypertension

Elderly are at an increased risk of developing hypertension. There is a 90% residual lifetime risk of developing hypertension for individuals 55 years and older with normal blood pressures.[12] It is important to treat hypertension as it is a precursor of many devastating conditions, including coronary heart disease, stroke, end-stage renal disease and peripheral arterial disease. Among the elderly in the general population 27% of CVD in women and 32% in men is attributable to hypertension. High blood pressure causes 1 in every 8 deaths worldwide making hypertension the third leading killer in the world.[13] In people older than 50 years of age, elevated systolic (SBP) is a much more important risk for cardiovascular disease than diastolic blood pressure (DBP)[14] making SBP control the focus of treatment. Even modest increases in marginally elevated blood pressure within the high normal range significantly increases the risk of CVD.[14] Framingham Study investigation of the CVD risk associated with high normal BP (130–139/85–89 mm Hg), compared with optimal BP (<120/80 mm Hg), found that it imposed a 2.5-fold hazard ratio in women and 1.6 in men. Hence it is important to treat even minimal elevation of SBP.

Concerns in the Elderly

Treatment of hypertension in older persons requires consideration of polypharmacy, altered drug metabolism, changes in physiological characteristics and co-morbidities.

Target Blood Pressures

Joint National Committee **(JNC 7)** established SBP/DBP goals of <140/90 mm Hg to decrease cardiovascular disease complication. In hypertensive patients with diabetes or renal insufficiency, target BP measurements are <130/80 mm Hg.

Treatment Goals

Individuals, especially those over age 65 years, with BP level ≥ 140/90 mm Hg or on treatment for hypertension, are prone to angina pectoris, MI and sudden death at a rate that is 2 to 3 times greater than their normotensive counterparts. The major objectives of antihypertensive therapy are to reduce the risk of cardiovascular injury, cerebrovascular events, heart failure, kidney disease and other target organ damage. Although there are no guarantees, the most appropriate correlation in hypertension therapy is the simplest: the higher the patients BP the higher the risk for morbidity and mortality; as the patient's BP appears normal, the rate falls.

Nonpharmacologic Treatment

Lifestyle modification is the initial and concurrent therapy for all patients. Yet with lifestyle modifications alone, only about 20% patients will achieve target BP. The majority will require drug therapy. Table 1[14] provides an algorithm for management of hypertension and Table 2[14] summarizes various nonpharmacologic approaches and their benefits in reducing blood pressure.

Table 1. Algorithm for treatment of hypertension.

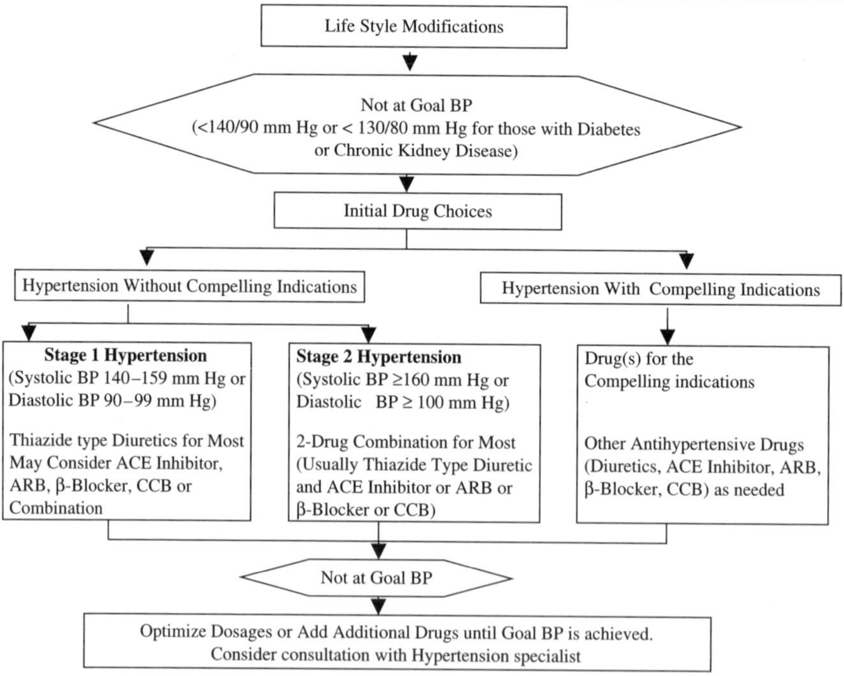

Table 2. Lifestyle modifications to manage hypertension.*

Modification	Recommendation	Approximate systolic BP reduction, range
Weight reduction	Maintain normal body weight (BMI, 18.5–24.9)	5–20 mm Hg/10 kg weight loss
Adopt DASH eating plan	Consume a diet rich in fruits, vegetables, and low-fat dairy products with a reduced content of saturated and total fat	8–14 mm Hg
Dietary sodium reduction	Reduce dietary sodium intake to no more than 100 mEq/L (2.4 g sodium or 6 g sodium chloride)	2–8 mm Hg
Physical activity	Engage in regular aerobic physical activity such as brisk walking (at least 30 minutes per day, most days of the week)	4–9 mm Hg
Moderation of alcohol consumption	Limit consumption to no more than 2 drinks per day (1 oz or 30 ml ethanol [e.g. 24 oz beer, 10 oz wine, or 3 oz 80-proof whiskey]) in most men and no more than 1 drink per day in women and lighter-weight persons	2–4 mm Hg

Abbreviations: BMI, body mass index calculated as weight in kilograms divided by the square of height in meters; BP, blood pressure; DASH, Dietary Approaches to Stop Hypertension.
*For overall cardiovascular risk reduction, stop smoking. The effects of implementing these modifications are dose and time dependent and could be higher for some individuals.

Pharmacologic Treatment

How long should a patient be maintained on lifestyle interventions before initiating medication? In a patient with stage 1 hypertension (SBP 140–159 or DBP 90–99) and no other risk factors it might be appropriate to try life style modification alone for 3–6 months. However, in the presence of risk factors medication is recommended either initially or after confirming the elevated BP and 3–4 weeks of nonpharmacologic therapy. Patients with diabetes should be started on medication at the same time as lifestyle interventions. Thiazide diuretics are recommended as initial drug therapy for most hypertensives.[14,15] Thiazide diuretic usage in long term reduces stroke (by 40–50%), heart failure (by 40–50%) and CAD (by 16–25%). However, BP will not be controlled with monotherapy in more than 50–60% of patients. Often a second or a third medication is necessary to reduce BP to goal level. In patients with stage 2 hypertension consideration should be given to initiating therapy with two agents, one of which should be a diuretic. In addition, multiple drugs as initial therapy are indicated in patients with stage 2 hypertension if they have diabetes, renal disease or coronary heart disease. Combination therapy makes physiologic sense, has compliance advantage and is time tested.

There are certain compelling indications for the use of specific medication. For example, in patients with heart failure and hypertension, use of an ACE inhibitor, ARB, beta blocker or aldosterone antagonist along with a diuretic is indicated. In patients with diabetic nephropathy, an ACE inhibitor or an ARB are medications of choice, but it is often necessary in these patients to use a diuretic to achieve goal BP.

Use of an ACE inhibitor or an ARB in a treatment protocol may prevent some cases of new onset diabetes especially in the patient with the metabolic syndrome.

Finally a dedicated physician implementation of the treatment guidelines esp. in the elderly is of utmost importance. Practitioner empathy for the patient contributes to trust, motivation and hypertension control.

Management of Acute Ischemic Stroke[16]

The management of patients with acute ischemic stroke is multifaceted. Outcomes after stroke can be improved and death or disability from stroke can be reduced with appropriate treatment. Main recommendations are:

1) Stroke should be approached as the life-threatening emergency it is. Patients with acute ischemic stroke should be evaluated and treated immediately. A regional or local organized program to expedite stroke care is recommended.
2) Urgent evaluation is aimed primarily at determining that ischemic stroke is the likely cause of the patient's symptoms and whether the patient can be treated with intravenous rtPA.
3) Urgent treatment should include measures that protect the airway, breathing, and circulation (life support), especially among seriously ill or comatose patients. An elevated blood pressure should be

lowered cautiously. Antihypertensive agents should be avoided unless the SBP > 220 mm Hg or DBP > 120 mm Hg. Agents that have a short duration of action and little effect on cerebral blood vessels are preferred. Situations that require urgent antihypertensive therapy include hypertensive encephalopathy, aortic dissection, acute renal failure, acute pulmonary edema or acute myocardial infarction.

4) Supportive care should include treatment of sources of fever (if any) and the use of antipyretics to control body temperature and control of hypoglycemia or hyperglycemia.

5) Intravenous administration of recombinant tissue plasminogen activator (rtPA) (0.9 mg/kg; maximum 90 mg) is strongly recommended for treatment of carefully selected patients who can receive the medication within 3 hours of onset of stroke. Safe use of rtPA requires adherence to National Institute of Neurological Disorders and Stroke (NINDS) selection criteria, close observation, and careful ancillary care. Intravenous administration of streptokinase or other thrombolytic agents cannot be substituted safely for rtPA. The intra-arterial administration of thrombolytic agents is being given to an increasing number of patients. While intra-arterial thrombolysis holds promise for treating patients at time periods longer than 3 hours after the onset of stroke, the patient selection criteria and effectiveness of this form of therapy have not been fully established.

6) Urgent administration of anticoagulants has not yet been associated with lessening of the risk of early recurrent stroke or improving outcomes after stroke. Because urgent anticoagulation can increase the risk of brain hemorrhage, especially among patients with moderately severe strokes, the routine use of this therapy cannot be recommended. Aspirin can be administered within the first 48 hours because of reasonable safety and a small benefit.

7) Surgical interventions — carotid endarterectomy, extracranial-intracranial arterial bypass, endovascular treatment — are not recommended for most patients with acute ischemic stroke.

8) Corticosteroids are not recommended for the management of cerebral edema and increased intracranial pressure following ischemic stroke. Osmotherapy and hyperventilation are recommended for patients whose condition is deteriorating secondary to increased intracranial pressure, including those with herniation syndromes.

9) Surgical interventions, including drainage of cerebrospinal fluid, can be used to treat increased intracranial pressure secondary to hydrocephalus. Surgical decompression and evacuation of large cerebellar infarctions that are leading to brain stem compression and hydrocephalus is recommended. Surgical decompression and evacuation of a large infarction of the cerebral hemisphere can be a life-saving measure, but survivors have severe residual neurological impairments. Recurrent seizures should be treated as with any other acute neurological condition. Prophylactic administration of anticonvulsants to patients who have had stroke but not seizures is not recommended.
10) No medication with putative neuroprotective effects has yet been shown to be useful for treatment of patients with acute ischemic stroke. However in authors experience intravenous administration of piracetam has been helpful in recovery from an acute ischemic stroke particularly if started early.
11) Comprehensive stroke unit care, including comprehensive rehabilitation, can be given to a broad spectrum of patients.
12) Subsequent treatment in the hospital should include measures to prevent or treat medical or neurological complications of stroke. An evaluation to determine the most likely cause of the patient's stroke should lead to institution of medical or surgical therapies to lessen the risk of recurrent stroke.
13) Rehabilitation and plans for care after hospitalization also are important components of acute management of patients with stroke.

Cancer

Cancer is the second leading cause of death in older adults; it will strike 1 in 3 adults and accounts for about 22% of all deaths in people 65 and older. Among men, the most common cancer is prostate cancer, followed by lung and then colorectal cancer. Among women, the most common cancer is breast cancer, followed by lung and then colorectal cancer. About 58% of all cancers occur among the elderly who have disproportionately higher rates of lung, colon, rectum, urinary bladder, stomach, and pancreatic cancers. Older adults are also more likely to die of cancer than younger people, they account for two-thirds of all cancer deaths. Overall, the mortality rate is highest for lung cancer.

The management of a patient with cancer depends on several factors — age of the patient, performance status, co-morbid conditions, life expectancy and the type and stage of the cancer.

Lung Cancer

It is the leading cause of cancer death in both men and women and in all races. The incidence of lung cancer peaks between ages 55 and 65 years. Lung cancer accounts for 31% of all cancer deaths in men and 25% in women. Management of lung cancer depends on the histology — small cell (SCLC) versus non-small cell lung cancer (NSCLC; Table 3).

Long-term survival is seen in those patients with NSCLC who undergo surgery which is the mainstay of treatment in early stages and attempted in advanced stage after chemoradiotherapy.

Management of SCLC

It is a systemic disease and the simplest way to stage is: Localized Disease (confined to one hemithorax) and Extensive Disease (beyond hemithorax). Chemotherapy is the main modality with radiotherapy/surgery in select cases.

Table 3. Management of NSCLC.

Stage	Treatment options	5 year survival (%)
I	Surgical Resection. If surgery is contraindicated, definitive radiotherapy.	38–61
II	Surgical Resection. If surgery is contraindicated, definitive radiotherapy.	22–34
IIIA	Neoadjuvant therapy followed by surgery or definitive chemoradiation.	9–13
IIIB (without pleural effusion)	Inoperable. Chemoradiation.	<5
IIIB (with malign, pleural effusion)	Chemotherapy. Best supportive care for those with poor performance status.	<5
IV	Chemotherapy. Best supportive care for those with poor performance status.	<5

Prostatic Cancer

Table 4. Treatment options for prostatic cancer.

Stage	Treatment options	5 year survival (%)
I	Watchful waiting/Radical Prostatectomy/ Radical Radiotherapy	83–87
II	Watchful waiting/Radical Prostatectomy/ Radical Radiotherapy	73–78
III	Watchful waiting/Radical Prostatectomy/ Radical Radiotherapy	61–65
IV	Castration (Medical/Surgical) Hormonal Therapy Hormonal Therapy Failures-Chemotherapy	29

Prostatic cancer is the second most common cancer diagnosis and the second leading cause of death in men. The treatment options for prostatic cancer are summarized in Table 4. The patient most likely to benefit from either radical prostatectomy or radiation should have a relatively long life expectancy and no significant surgical or radiation toxicity risk factors, whereas those most likely to benefit from surveillance are those with a shorter life expectancy ($\leq$10 years) and/or a low grade tumor histology.

Breast Cancer

Worldwide, breast cancer is a major public health problem; there are approximately 1 million new cases annually. During the last two decades significant advances have been made in the development of early detection methods, less radical and morbid primary therapy, the use of radiation therapy in primary therapy (permitting breast conservation for many women), the use of systemic adjuvant therapy, and improved systemic treatments and supportive care for metastatic disease. The principles of managing breast cancer are summarized in the Table 5.[17]

Colorectal Cancer

This type of cancer occurs generally in individuals $\geq$50 years of age. It comprises 18% of all cancer deaths and is the third leading cause of cancer death in western world. Annual occult foecal blood testing above

Table 5. The principles of managing breast cancer.

Stage	0	I	II	III	IV
Presentation[a]	12.4%	41.8%	33.1%	8.0%	4.7%
Treatment[b]	S (R)	S (C, R)	S, C (R)	C, S, R	C, R
Goal	Cure	Cure	Cure	Cure	Palliation
Survival[c]	98%	90%	70%	50%	15%

[a]Listed under 'presentation' is the percent of women with a given disease stage at the time of diagnosis.
[b]Main modality of treatment is listed first.
[c]Overall survival at 5 years.
Abbreviations S, surgery; C chemotherapy; R, radiation therapy.

Table 6. Treatment options in various stages of colorectal cancer.

Stage	Treatment option	Approx. 5-year survival (%)
I	Surgery alone	>90
II*	Surgery ± Chemotherapy	70–80
III*	Surgery + Chemotherapy	35–65
IV	Chemotherapy ± Surgery	5

*Radiotherapy for rectal cancer.

50 years of age is an inexpensive method for early detection of this cancer. The goal of the treatment is to cure in early stages (I & II), to prolong survival in stage III; and to palliate in stage IV. Main modality of treatment is surgery in early stages; surgery + adjuvant chemotherapy in stage III; observation vs chemotherapy in stage IV. Various treatment options in various stages and the 5-year survival rates are summarized in the Table 6.

Parkinson's Disease (PD)

PD is the second most common neurodegenerative disorder after Alzheimer's disease with a prevalence of 0.5 to 1% in the age-group 65 to 69 years, rising to 1 to 3% among those 80 years of age or older. Age is the single most consistent risk factor. Although, the starting point for symptomatic treatment is still a matter of some debate, most specialists opt for treatment when patient begins to experience functional impairment.

Neuroprotection (with disease-modifying compounds) should be considered as soon as diagnosis of PD is made. It may be provided (though not proven) by selective monoamine oxidase B (MAO-B) inhibitor inhibitor selegiline and dopamine agonists — pramipexole, ropinirole and cabergoline. *Anticholinergics* used are trihexyphenidyl 2–15 mg daily and benztropine 1–8 mg daily. *Amantadine* provides modest symptomatic improvement in tremor, rigidity and bradykinesia and is also effective for dyskinesia. Selegiline is preferably used with levodopa at doses 5–10 mg daily. It reduces the metabolism of dopamine and therefore increases the "on" time and reduces motor fluctuations, when used as an adjunct to levodopa.

Dopamine agonists[18] (DAs) as the initial drug treatment may postpone the use of levodopa and may delay or prevent motor complications and disease progression. DAs are now being increasingly used for initiation of symptomatic treatment of PD. A number of factors; such as age, drug cost, practice protocols etc; need to be taken into account when choosing DAs as initial monotherapy. Levodopa is still preferable in PD patients with cognitive impairment, in the elderly who have a decreased tendency to develop motor complications and those with atypical parkinsonism. Most physicians feel comfortable treating patients 65 years of age or younger with DAs, both as initial monotherapy and as adjuvant therapy, with the caveat that every drug regimen must be individualized in accordance with patient requirements. DAs may trigger CNS side-effects, such as cognitive dysfunction in older (70–75 years) PD patients and should probably be avoided in this subgroup. Life expectancy after 70 to 75 years of age is usually short enough that the development of motor complications related to long term levodopa use becomes of lesser importance.

Levodopa is the most effective therapy for treatment of PD. It is combined with carbidopa to prevent peripheral conversion of levodopa to dopamine. Virtually all patients with PD experience a dramatic and sustained response to levodopa, and almost all patients with PD require levodopa at some point during the disease course. Postural instability, freezing, speech abnormalities, depression, dementia, constipation, sexual dysfunction, urinary problems and sweating and sensory phenomenon like tingling, numbness and pain do not respond to levodopa.[19] When levodopa is initiated, patients show a global improvement in parkinsonian symptoms. As the disease progresses, the dopaminergic neurons lose the buffering capacity and the symptoms improve when plasma levodopa levels rise and become worse when levels dip. These motor fluctuations are known as "on" and "off" periods. On average, 50% patients develop some involuntary movements known as dyskinesia and/or levodopa induced motor fluctuations within 5 years of treatment. Risk of above side effects increases with doses higher than 300 mg/day and with early onset PD (<50 years). Another long term complication of levodopa therapy is mental status changes such as confusion, hallucinations and psychosis.[18]

Cathecol-O-methyltransferase (COMT) inhibitors have no role as monotherapy but are indicated for patients with PD who have end-of-dose wearing off. Currently, 200 mg entacapone with each dose of levodopa upto a total of 8 doses daily or tolcapone 100 mg every 6 hours for total of 3 doses daily can be given. Extensive liver function monitoring for patients on tolcapone is recommended.

Treatment of "Wearing-Off" Symptoms

For patients receiving levodopa monotherapy, dopamine agonists can be added and vice versa. The frequency and/or dose of levodopa can be increased. Another option would be to add a COMT inhibitor or substitute or add sustained release levodopa/carbidopa. If these medication changes are not helpful, surgical treatment such as deep brain stimulation (DBS) of the subthalamic nucleus (STN) or globus pallidus may be helpful.

Treatment of Dyskinesia

Dyskinesia most commonly takes 2 forms: peak dose, also known as improvement-dyskinesia-improvement (IDI); and diphasic dyskinesia, also known as dyskinesia-improvement-dyskinesia (DID). The treatment of peak-dose dyskinesia includes substitution of immediate with sustained-release levodopa/carbidopa, discontinuation of selegiline, lowering individual doses of levodopa/carbidopa, concomitantly reducing levodopa/carbidopa, and adding or increasing doses of dopamine agonists. For diphasic dyskinesia, more frequent dosing of levodopa/carbidopa, addition or increase in dopamine agonists, and restriction of levodopa/carbidopa to early or midday doses may be attempted. For both peak-dose and diphasic dyskinesia, if these medical strategies are not beneficial, surgical therapy such as DBS of the subthalamus or globus pallidus may be helpful. Treatment of psychosis and hallucinations requires withdrawal or reduction in dosage of antiparkinsonian drugs. Newer antipsychotic drugs like Quitiapine may be useful.

Surgical Treatment of Parkinson's Disease[20]

The last 10–15 years has seen an increase in the use of surgical treatments for PD due to limitations of current drug therapies, improvements and advances in neurosurgical and imaging techniques, and a better understanding of the functional circuitry of the basal ganglia. Surgical treatment for PD includes ablative surgery, DBS, and transplantation. In ablative surgery for PD, a lesion is created in the thalamus, globus pallidus, or subthalamus. Thalamotomy and pallidotomy are rarely used today and the safety and efficacy of subthalamotomy is yet to be determined. DBS involves high-frequency, pulsatile, electrical stimulation delivered with placement of an electrode tip in the targeted brain region, namely the thalamus, globus pallidus, or subthalamus. A high-frequency pulsatile stimulation is then used to mimic the effects of ablative surgery. The stimulation can be turned on or off by the patient with a hand-held magnet or an access control device. In transplantation, stem cells are placed in the striatum in an attempt to mimic the function of the substantia nigra cells which have been lost in PD. Transplantation is currently experimental and is being performed at only a limited number of centers.

Management of Osteoarthritis (OA)

OA has afflicted man and other vertebrates from prehistoric times and it has accompanied man throughout his evolutionary history. It is the most common joint disorder in the world and is the leading cause of disability and pain in the elderly. Management of OA needs to be individualized, holistic, patient centered and situational. The optimal management of OA requires a combination of non-pharmacological and pharmacological treatment modalities. Various treatment modalities are summarized in Table 7 and treatment algorithm is given in Table 8.[22]

Table 7. Treatment modalities available for management of OA.

Non pharmacologic treatment
Corner stone of OA management and is recommended in every OA patient to start with.

Patient Education and social support
Self management programs (e.g. arthritis self-help course)
Health professional social support via telephone contact.

Exercises
Physical therapy
- Range of motion exercises
- Strengthening exercises (quadriceps exercises in knee)
- Assistive devices for ambulation (canes, crutches, orthotic shoes, knee braces)

Occupational therapy
- Joint protection and energy conservation
- Assistive devices for activities of daily living (ADLs) and instrumental activities of daily living (IADLs)

Low-impact aerobic exercise programs (aquatic for hip)

Other non pharmacologic measures
Weight reduction in obese
Acupuncture
Transcutaneous nerve stimulation
Pulsed electromagnetic field therapy
Dietary additions including glucosamine, chondroitin, vitamins C and D, ginger extracts, avocado, and soybean derivatives

Pharmacologic measures
Analgesics: Acetaminophen (ACET) 1 gm. 3–4 times/d (max 4 gm/d)
Tramadol and opioid analgesics
Antidepressants for analgesia (and for depression)
Systemic nonsteroidal anti-inflammatory drugs (NSAIDs) including coxibs for those not relieved with ACET.
Topical agents including capsaicin and NSAID creams and gels.
Intra-articular (IA) injections including steroids and hyaluronan (HA)
Colchicine for demonstrable CPPD crystals in joint fluid

DMOAD
Glucosamine sulphate, chondroitin sulphate, avocado/soybean unsaponifiables (ASU), diacerein, hyaluronic acid have symptomatic effects and may modify structure.

Surgical measures
Tidal irrigation (washout) of the joint (in knee OA)
Arthroscopic debridement
Cartilage transplantation and tissue engineering techniques
Osteotomy
Partial or complete joint replacement

Complementary and alternative therapies
Almost every known type of complementary and alternative medicine has been used in attempts to help people who have OA.

Table 8. Algorithm of OA management.

Nonpharmacological treatment
Education, support, exercise, weight loss, joint protection

plus

Acetaminophen up to 4 g/d for control of pain and other symptoms, as well as before activity

↓

Add topical capsaicin cream 4 times daily, if needed, to painful joints

↓

If joint effusion is present, consider aspiration and intra-articular injection of a corticosteriod as initial therapy

↓

If more pain and symptom control is needed, add an NSAID in low dosage or a nonacetylated slicylate such as choline magnesium trisalicylate or salsalate. Tramadol may be useful as adjunctive therapy

↓

If more pain or symptom control is needed, use full dosage of an NSAID, plus misoprostol or a proton pump inhibitor if the patient is at risk for upper gastrointestinal tract bleeding or ulcer disease, or subsitute a COX-2-specific inhibitor for the NSAID. Some patients may benefit from intra-articular injections of a hyalronic acid product.

↓

If response is inadequate, consider referring the patient for joint lavage, arthroscopic debridement, osteotomy, or joint replacement.

Disease-modifying OA Drugs

Better understanding of the pathogenetic mechanisms underlying the breakdown of articular cartilage in OA and particularly of the mediators involved in tissue breakdown has led the pharmaceutical industry on a search for the "holy grail": a disease modifying osteoarthritis drug (DMOAD). There are several candidates: chemically modified tetracyclines, diacerein, glucosamine and chondroitin sulfate. Glucosamine and chondroitin sulfate have achieved striking popularity for treatment of OA recently. They are widely sold as nutraceuticals although they are not

FDA approved. Several studies have shown glucosamine to be superior to placebo and comparable to NSAIDS with respect to efficacy in patients who have knee OA and it has a better safety profile than NSAIDS. However, the efficacy of glucosamine and chondroitin have not been examined in large, well-designed, placebo-controlled trials. Much of the beneficial effects are believed to be overestimated in trials most of which were industry sponsored. There is growing evidence to support the use of two of these agents for their symptomatic effects, but for the others the evidence is weak or absent.[21]

Results of two recent virtually identical randomized controlled trials[23,24] have led to the suggestion that glucosamine not only improves joint pain in patients who have knee OA but are also chondroprotective. A NIH supported multicenter study currently in progress is comparing glucosamine with chondroitin sulfate, the combination of the two, celecoxib and placebo in patients who have knee OA.

Osteoporosis

Osteoporosis is a skeletal disorder characterized by compromised bone strength that predisposes one to an increased risk of fracture. Bone strength in this definition reflects the integration of 2 main features: bone density and bone quality. Although all bones are affected, the most common fractures caused by osteoporosis occur in the spine, wrist, and hip. The fracture risk increases with advancing age at comparable bone mineral density. Age is the most important risk factor for osteoporotic fractures. Worldwide, the lifetime risk for a woman to have an osteoporotic fracture is 30–40%; in men the risk is about 13%. A woman's lifetime risk of hip fracture equals the combined risk of breast, uterine and ovarian cancer, and the risk of dying from hip fracture equals mortality from breast cancer. Men account for nearly one-third of all hip fractures that occur, and one-third of these men do not survive more than a year.

Management of Osteoporosis

The past few decades have seen a change in the approach to management of osteoporosis; from passively accepting osteoporosis as an inevitable consequence of aging to active aggressive management with rewarding results. The goal of treatment is to reduce fracture.

Universal Measures to Reduce Risk Factors

This includes risk factors for osteoporosis, risk factors for fracture, or risk factors of falling. Optimize nutrition to make sure that patients get enough calcium and vitamin D. Patients should be mobile and exercising. In terms of US Food and Drug Administration (FDA)-approved osteoporosis therapies, raloxifene, alendronate, and risedronate are approved for both prevention and treatment. Calcitonin is not approved for prevention, but is approved for treatment.

Calcium Supplementation

The recommended daily calcium intake is 1500 mg for postmenopausal women and 1000 mg for pre-menopausal women. Calcium supplementation prevents bone loss in postmenopausal women and combination with vitamin D3 prevents hip and vertebral fracture in the elderly.

Vitamin D Supplementation

The role of two potent analogues of vitamin D (1-alpha-(OH) and 1,25 $(OH)_2D_3$) in the treatment of osteoporosis is controversial. However, vitamin D supplementation (400–800 IU daily) is particularly useful in vitamin deficient elderly.

Phytoestrogens

These are plant derived molecules that have estrogenic actions. The best source is soya products. Ipriflavone has been shown to increase bone mass in established osteoporosis.

Pharmacotherapy — Antiresorptive Medications

The bisphosphonates (alendronate and risedronate), calcitonin, and raloxifene affect the bone remodeling cycle by reducing the number and/or activity of bone-resorbing osteoclasts but do not slow the bone-forming portion of the bone cycle. As a result, new formation continues at a greater rate than bone resorption, and bone density may increase over time. However, they do not significantly increase bone volume and have minimal capacity to improve skeletal architecture. Bone volume, bone composition, and bone architecture are important determinants of bone strength.

Bisphosphonates

Alendronate sodium is approved for both the prevention (5 mg per day or 35 mg once a week) and treatment (10 mg per day or 70 mg once a week) of postmenopausal osteoporosis. It is also approved for treatment of glucocorticoid-induced osteoporosis in men and women as a result of long-term use of these medications (i.e., prednisone and cortisone) and for the treatment of osteoporosis in men. It reduces bone loss, increases bone density and reduces the risk of spine, wrist and hip fractures. *Risedronate sodium* is approved for the prevention and treatment of postmenopausal osteoporosis. It is also approved for use by men and women to prevent and/or treat glucocorticoid-induced osteoporosis. Taken daily (5 mg dose) or weekly (35 mg dose), risedronate slows bone loss, increases bone density and reduces the risk of spine and non-spine fractures. The prolonged use of risedronate in women with established osteoporosis has been linked with a statistically significant increase in the occurrence of pulmonary cancer, a finding which needs to be confirmed. Alendronate and risedronate must be taken on an empty stomach, first thing in the morning, with eight ounces of water (no other liquid), at least 30 minutes before eating or drinking. Patients must remain upright during this 30-minute period. *Ibandronate sodium* (2.5 mg once daily orally) is a newer bisphosphonate approved for the treatment and prevention of osteoporosis in postmenopausal women. It is also used for cancer-related bone complications.

Calcitonin

This is a naturally occurring hormone involved in calcium regulation and bone metabolism. In women who are more than 5 years beyond menopause, calcitonin slows bone loss, increases spinal bone density, and may relieve the pain associated with bone fractures. Calcitonin reduces the risk of spinal fractures but has not been shown to decrease the risk of non-spine fractures. Because calcitonin is a protein, it cannot be taken orally as it would be digested before it could work. Calcitonin is available as an injection (50–100 IU daily) or nasal spray (200 IU daily).

Tibolone

A steroidal, synthetic compound, it exerts an estrogenic, progestogenic or androgenic effect depending on the tissue. It is has been evaluated for prevention and treatment of postmenopausal osteoporosis with results on bone mineral density comparable to those obtained with traditional hormone replacement therapy.

Raloxifene

It is approved for the prevention and treatment of postmenopausal osteoporosis (60 mg a day). It is from a class of drugs called Selective Estrogen Receptor Modulators (SERMs) that have been developed to provide the beneficial effects of estrogens without their potential disadvantages.

Bone Forming Medications — Parathyroid Hormone (PTH)

Teriparatide,[25] recombinant human parathyroid hormone 1–34, is approved for the treatment of osteoporosis in postmenopausal women who are at high risk for fracture and to increase the bone mass in men with primary or hypogonadal osteoporosis who are at high risk for fracture. This medication increases the number and action of osteoblasts to stimulate new bone formation. It significantly increases bone mineral density, also increases bone volume and improves skeletal architecture by increasing connectivity through modeling effects, and through effects on periosteum, increase cortical bone diameter. It reduces vertebral and nonvertebral fracture risk. Teriparatide is self-administered as a daily injection up to 24 months.

In a meta-analysis of therapies for postmenopausal osteoporosis[26] (including calcium, vitamin D, Etidronate, Alendronate, Raloxifene,

Calcitonin, Risedronate, HRT, Fluoride), it was concluded that vitamin D, calcitonin, HRT, raloxifene, and the bisphosphonates etidronate, risedronate, and alendronate reduce vertebral fractures and that there is convincing evidence for nonvertebral fracture reduction for only two agents, risedronate and alendronate. The choice of an agent for osteoporosis treatment is influenced by several factors — strength of the evidence, emerging newer therapies, additional benefits, risks, adverse effects, inconveniences and price associated with different medications all weigh heavily in treatment decisions.

Combination Therapies

There are a number of published reports of using combination therapy in women with established osteoporosis. These include alendronate plus oestrogen, alendronate plus raloxifene, risedronate plus oestrogen, calcitonin plus oestrogen, and PTH plus oestrogen. In general, combination therapy has resulted in slightly greater increases in BMD than were seen when either drug was used alone. However, these increments in BMD were not always statistically significant and there has been no documentation of any synergism when two drugs have been used in combination. Additionally, there are no fracture data to support the use of combination therapy. There has been a considerable interest in the potential combined or sequential use of PTH and antiresorptive therapy. Animal studies suggest that antiresorptive therapy maintains effects produced by PTH, but few studies have been done in human subjects to confirm the work in animals. If confirmed, the issue of whether antiresorptive therapy should be administered before, during, or after PTH treatment will need to be addressed.

Agents in Development

Agents in development for the treatment of established osteoporosis include intravenous zoledronate, intravenous ibandronate, and oral strontium ranelate. Zoledronate, a third generation bisphosphonate approved for the treatment of bone metastases in cancer patients, is effective in preventing and treating osteoporosis and has the potential of emerging as an attractive alternative for patients who cannot tolerate oral bisphosphonates or who have compliance issues as an annual or biannual dose. Strontium ranelate substitutes for calcium in hydroxyapatite crystals increasing bone density and may emerge as a new effective and safe treatment of vertebral and nonvertebral osteoporosis.

Age-related Cataract

Cataracts begin to form in everyone over age 30 but they progress at different rates in individuals. The prevalence of cataracts increases with age. People with cataracts experience a painless, progressive loss of vision that may involve one or both eyes. There is currently no proven intervention to prevent cataract and surgery is the only form of treatment. Cataract surgery is one of the most common procedures performed. It is safe and effective. Cataracts can be removed through either of the following procedures:

- *Phacoemulsification*: A small incision is made on the side of the cornea. A probe is inserted into the eye and ultrasound waves are emitted to soften and break up the cloudy center of the lens so it can be removed by suction.
- *Extracapsular surgery*: In this procedure, a slightly longer incision is made on the side of the cornea and removes the hard center of the lens. The remainder of the lens is removed by suction.

Usually an intraocular lens (IOL), a clear, artificial lens, is inserted and becomes a permanent part of the eye. Those with eye problems who are not candidates for IOL implantation need to use a soft contact lens or strong magnifying glasses. A recent review of surgical interventions for age-related cataract[27] provides evidence from one randomized controlled trial that phacoemulsification gives a better visual outcome than extracapsular extraction with sutures.

IOL Opacification[28]

Ophthalmologists in the UK have found that several patients treated with an acrylic hydrophilic intraocular lens (IOL) implant after phacoemulsification experience severe vision loss within 2 years. The vision loss is due to lenses becoming totally opaque, but in two cases, patients were treated for posterior capsular opacification before the correct diagnosis was made. Opacification was due to an interaction between the silicone sleeve used to hold the IOL in the vial and the acrylic material of the IOL itself. Apparently all these IOLs have been withdrawn and a modified version without the silicone sleeve has been introduced.

Age-related Macular Degeneration (ARMD)

ARMD is the leading cause of untreatable legal blindness for older people. It involves a deterioration of the retinal cells that causes a loss of the central field of vision. The macula is the part of the retina that provides clear, sharp central vision. While their central vision is impaired, people with ARMD rarely become totally blind. The prevalence of ARMD rises with age. The exact cause of macular degeneration is not known.

Dry macular degeneration is the major type of ARMD and accounts for 80–90% of the cases. Only 10% of people with dry macular degeneration have vision loss severe enough to qualify as legal blindness. Wet macular degeneration is more likely to result in severe visual impairment. Among older adults who are legally blind, 80–90% have this form of the condition.

Currently there is no treatment for dry AMD, but there is some exciting news. The progression of dry macular degeneration to wet macular degeneration is inhibited by zinc and antioxidants.[29] Laser treatment is not effective in dry AMD.[30] Wet AMD may be treated with laser surgery. The laser treatment aims a high energy beam of light directly onto the leaking blood vessels and can help stop further vision loss. After surgery, frequent eye examinations are needed to detect any recurrence of leaking blood vessels. People who smoke and are hypertensive have a greater risk of recurrence than those who don't. Our present arsenal for AMD includes photodynamic therapy (PDT) for patients who have predominantly classic neovascularization and transpupillary thermotherapy (TTP) for occult neovascularization. Changing the parameters of (PDT) may make it more effective, and this is being investigated. Considering the size of the lesion rather than the various components of the lesion (e.g., classic vs occult) appears to show that smaller lesions benefit more from PDT than large ones. Corticosteroids have limited role in treating ARMD.[31]

Anti-VEGF Therapy

Another approach to the vascular hyperpermeability and the neovascularization seen with AMD and diabetic retinopathy is the use of anti-VEGF therapy. Vascular endothelial growth factor (VEGF) is a mediator for angiogenesis and is also a potent vascular permeability factor. Agents that inhibit VEGF include an aptamer and an antibody fragment. Aptamer blocks the VEGF receptor when given intravitreally and is presently in phase 2–3 clinical trials in patients with AMD. An antibody fragment, called rhuFab V2, is given intravitreally, and is presently in phase 1 trials for treatment of patients with subfoveal choroidal neovascularization. Some Day 98 results suggested a beneficial effect. It is likely that future treatments will include the combination of PDT with corticosteroids, angiostatic steroids, or the antiangiogenic, antipermeability drugs. The possibility of transplanting healthy cells into a diseased retina is also being explored but is at a very early stage and still experimental.

Finally, in taking care of the elderly a gentle human approach with empathy is important, which sometimes is lost in this era of evidence-based medicine. To be felt loved and wanted is a universal human need but perhaps more important at this phase of life than at any other time. A desire and purpose to live builds hope, and hope is everything in the twilight of life!

References

1. Adult Treatment Panel III (2001) Executive summary of the third report of the National Cholesterol Program (NCEP) Expert Panel on Detection, Evaluation, and Treatment of High Blood Cholesterol in Adults. *JAMA* **285**: 2486–2497.
2. Tan WA, Moliterno DJ (1999) Aspirin, ticlopidine and clopidogrel in acute coronary syndromes: underused treatments could save thousands of lives. *Cleve Clin J Med* **66**: 615–628.
3. Yusuf S, Sleight P, Pogue J *et al.* (2000) Effects of an angiotensin-converting-enzyme inhibitor, ramipril, on cardiovascular events in high-risk patients. The Heart Outcomes Prevention Evaluation Study Investigators. *N Engl J Med* **342**: 145–153.
4. EUROPA Investigators (2003) *Lancet* **362**: 782–788.

5. Rossouw JE, Lewis B, Rifkin BM (1990) The value of lowering cholesterol after myocardial infarction. *N Engl J Med* **323**: 1112–1119.
6. LRosa JC, Cleeman JI (1992) Cholesterol lowering as a treatment for established coronary heart disease. *Circulation* **85**: 1229–1235.
7. Stenestrand U, Wallentin L (2001) Early statin treatment following acute myocardial infarction and 1-year survival. *JAMA* **285**: 430–435.
8. Schwartz GG, Olsson AG *et al.* (2001) Effects of atorvastatin on early recurrent ischemic events in acute coronary syndromes: the MIRACL study: a randomized controlled trial. *JAMA* **285**: 1711–1718.
9. Eric J Topol (2004) Intensive Statin Therapy — A Sea Change in Cardiovascular Prevention. *N Engl J Med* **350**: 15.
10. Randomized trial of cholesterol lowering in 4444 patients with coronary heart disease: the Scandinavian Simvastatin Survival Study (4S) (1994). *Lancet* **344**: 1383–1389.
11. HOPE Study Investigators (2000) Effects of angiotensin converting enzyme inhibitor, Ramipril, on cardiovascular vents in high risk patients. *N Engl J Med* **342**: 145–153.
12. Vasan RS, Larson MG, Leip EP *et al.* (2001) Assessment of frequency of progression to hypertension in nonhypertensive paticipitants in the Framingham Heart Study: a cohort study. *Lancet* **358**: 1682–1686.
13. The World Health Report (2002) Reducing Risks, Promoting Healthy Life. Geneva, Switzerland: World Health Organization 58.
14. The Seventh Report of the Joint National Committee on Prevention, Detection, Evaluation, and Treatment of High Blood Pressure — the JNC-7 Report (2003). *JAMA* **289**: 2560–2572.
15. The ALLHAT Officers and coordinators for the ALLHAT Collaborative Research Group. Major outcomes in high risk hypertensive patients randomized to angiotensin-converting enzyme inhibitor or calcium channel blocker vs diuretic: the Anti hypertensive and lipid lowering Treatment to Prevent Heart Attack Trial (ALLHAT) (2002). *JAMA* **228**: 2981–2997.
16. Adams HP Jr, Adams RJ, Brott T *et al.* (2003) Guidelines for the early management of patients with ischemic stroke: A Scientific Statement from the Stroke Council of the American Stroke Association. *Stroke* **34**(4): 1056–1083.
17. Karen K Fields, Steven C Goldstein *et al.* (1997) Decision Making in Oncology In. Benjamin Djulbegovic & DM Sullivan (eds.) p. 253, Churchill Livingstone, New York.
18. Schapira AHV (2003) Disease-modifying strategies and challenges in PD. *Neurology* **61** (Suppl 3): S56–S63.
19. Olanow CW, Watts RL, Koller WC. An algorithm (decision tree) for the management of Parkinson's disease (2001): treatment guidelines. *Neurology.* 2001: **56**(suppl 5): S1–S88.
20. Tarsy D, Vitek JL, Lozano AM (eds.) (2003) Surgical Treatment of Parkinson's Disease and Other Movement Disorders. Totowa, NJ: Humana Press.
21. Jordan KM, Arden NK, Doherty M *et al.* (2003) EULAR Recommendations 2003: an evidence based approach to the management of knee osteoarthritis: Report of a Task Force of the Standing Committee for International Clinical Studies Including Therapeutic Trials (ESCISIT). *Ann Rheum Dis* **62**: 1145–1155.
22. Manek NJ (2001) Medical management of osteoarthritis. *Mayo Clin Proc* **76**: 534.

23. Povelka K, Gatterova J, Olejarova M, Machacek S, Giacovelli G, Rovati LC (2002) Glucosamine sulfate use and delay of progression of Knee osteoarthritis: a 3 year, randomised, placebo-controlled, double blind study. *Arch Intern Med* **162**: 2113–2123.
24. Reginster J, Deroisy R, Rovati, Lee R, Lejeune E, Bruyere O, *et al.* (2001) Long term effects of glucosamine sulfate on osteoarthritis progression. *Lancet* **357**: 252–256.
25. Trish R Freeman (2003) Teriparatide: A Novel Agent That Builds New Bone. *Am Pharm Assoc* **43**(4): 535–537.
26. Cranney A, Guyatt G *et al.* (2002) Summary of meta-analyses of therapies for post-menopausal osteoporosis. *Endocrine Reviews* **23**(4): 570–578 571.
27. Snellingen T, Evans JR, Ravilla T, Foster A (2004) Surgical interventions for age-related cataract (Cochrane Review). In: The Cochrane Library, Issue 2, Chichester, UK: John Wiley & Sons, Ltd.
28. Godall KL, Ghosh YK (2004) Total Opacification of Intraocular Lens Implant After Uncomplicated Cataract Surgery: A Case Series. *Arch Ophthalmol* **122**: 782–784.
29. The Age-Related Eye Disease Study Research Group: A randomized, placebo-controlled, clinical trial of high-dose supplementation with vitamins C and E, beta carotene, and zinc for age-related macular degeneration and vision loss. *Arch Ophthalmol* 2001. **119**: 1417–1436.
30. Friberg TR (2002) *Modulating the disease? Laser treatment to Drusen. Retina Subspecialty Day.* Program and abstracts of the American Academy of Ophthalmology 2002 Annual Meeting; (2002) October 18–19, Orlando, Florida.
31. Kuppermann BD (2002) *The Role of Steroid Implants and Devices. Retina Subspecialty Day.* Program and abstracts of the American Academy of Ophthalmology 2002 Annual Meeting; October 18–19, Orlando, Florida.

14

Indian Ayurvedic Medicine in Aging Prevention and Treatment

Bhupinder P. S. Vohra* and Sanjeev K. Gupta†

*Department of Pediatrics, Molecular Biology & Pharmacology
Washington University School of Medicine, Box 8208
Saint Louis, MO 63110, USA; and †Department of Zoology,
Kurukshetra University, Kurukshetra -136119, India.

Ayurveda, the ancient Indian system of healthcare is considered to be the oldest system of medicine in the world.[1] The word Ayurveda is a Sanskrit word, derived from two roots: *ayur*, which means life, and Veda, knowledge. The basic concepts of Ayurveda have continuously being refined since its evolution in Vedic periods and *Charka Samhita* (~900 b.c.),[2] is the first recorded treatise that is fully devoted to the concepts and practice of Ayurveda. The other important treatise of Ayurveda is Sushruta *Samhita* (~600 b.c.),[3] which is devoted to surgery.

Despite the long history of successful Ayurvedic medicinal practice in India, Ayurveda appears unscientific to the scientists and has attracted lot of criticism. The reason for this is that the terms and concepts Ayurveda uses to describe the types and functioning of the human body, to classify levels of health, and to describe the onset of disease, have not yet been related to modern scientific biology and this criticism is also based on the supposed paucity of *in vivo* studies to support the efficacy and safety of individual ayurvedic plant or plant mixtures.

The purpose of this chapter is to outline and correlate the basic ayurvedic concepts to the modern biological science and to access the efficacy of various ayurvedic anti-aging remedies in the light of modern scientific research.

According to ayurvedic concepts the functional aspect of the body is governed by three biological humors (doshas) called *Vata*, *Pitta* and *Kapha* and the root cause of imbalance, or disease, is an aggravation of *dosha*, caused by a wide variety of internal and external factors. Studies have reveled that Vata, Pitta, and Kapha can differentially regulate neuroimmuno-endocrine/cellular integration, which may further regulate the predisposition to various systemic and neuropsychiatric disorders.[4]

Ayurveda adopts a holistic approach to medication. It uses the whole plant or part of plant, which in reality is a mixture of many constituents, and to modern pharmacologists, using a complex mixture is confusing and this has stood in the way of validation of traditional herbal medicines to some extent. The complex mixtures have wide spectrum of activities, some components may be pharmacologically active, and some may be nutrients while some others may act as antidotes to counter potential side effects and these combinations are based on a comprehensive principle for combining different herbs in a formulation (Samyoga). According to Ayurveda, the place where the herb is grown and the time at which the herbs are collected plays an important role in determining the efficacy of the herbs; the same herb grown at different places and collected at different period of year may act differently.

Ayurveda and Aging

Ayurveda includes geriatrics as one of its eight medical divisions, which is termed as Rasayanatantra. Treating aging and age-related disorders is well documented in Ayurveda and it describes the ways to attain longevity, improve memory and freedom from age-related disorders.[5] Usual aging have been distinguished from successful aging, individuals in the usual aging group are non-pathologic but are at a high risk, and the individuals in the successful aging group are at low risk and exhibit higher functional activity.[6] Successful aging has three main components: (i) low probability of disease and disease-related disability, (ii) high capacity for cognitive and physical functioning, and (iii) active engagement with life including interpersonal relations and productive activity.[7] According to Ayurveda, usual aging and the risk associated with it for adverse health events are preventable and can be reversed. Rasayana or science of rejuvenation is a unique concept and Rasayana therapy is the medical management with the advancement of age. The word Rsa is equivalent to the normalcy of health. It is a process by which body tissues are toned for a healthy long life and to delay the aging process. The principal physiological effect of Rasayana is to improve and revitalize the physiological and endocrine functions of the body, to retard the aging process and to make an individual more responsive and resistant to disease, i.e. to improve body function by strengthening the immune system. To gain maximum benefit of this process it must begin in midlife; it will not be effective when started too late because many of the ayurvedic approaches are preventive than curative. Thus, Ayurveda emphasize both on adding life to years as well as adding years to the life.

It is of interest to note that a number of Ayurvedic plants that are grouped under the general class of *Rasayana* (rejuvenators) are used not only as antistress/adaptogenic agents but as a unique class of drugs which, according to Ayurveda, increase the life span and delay aging by countering the degenerative changes normally associated with aging. Among these are the well-known *Amala* (*Emblica officinalis*), *Haritaki* (*Terminalia chebula*), *Guduchi* (*Tinospora cordifolia*), *Pipali* (*Piper longum*), and *Tulsi* (*Ocimum sanctum*).

Age-related cognitive decline is not as severe as dementia, but the worst cases of age-related cognitive decline are at higher risk for dementia.[8,9] The dementias, as with other degenerative diseases, necessitate an integrative-holistic approach which stresses early intervention. Cognitive function is well recognized in Ayurveda and dementia is termed *Cittinasa* (loss of mind). The mind (manus) is thought to be the inner instrument of perception and term *Medhya* refers to all substances/agents that promote or are beneficial to the *Medhya* (i.e. intellect), which essentially encompasses three principles, viz., *Budhi* (intelligence), *Dhritti* (concentration and retention), and *Smriti* (memory). Several Ayurvedic drugs promote memory as well as longevity and general well-being, and these are categorized as *Medhya Rasayana*. The Medhya Rasayans could be somewhat akin to nootropic agents. Plants which posses "antiaging" or "memory-enhancing" effects could also be considered for potential efficacy in disorders now recognized to be associated with cognitive dysfunction, including conditions that feature dementia and Alzheimer's disease.

Table 1. Herbs with Rasayana (Rejuvenators) effect mentioned in Ayurveda.

Longevity-promoting herbs (*Jivniya Rasayana*) *Acorus calamus, Alternanthera sessilis, Argyreia speciosa, Asparagus racemosus, Bacopa monniera, Celastrus paniculatus, Centella asiatica, Clerodendrum infortunatum, Clitoria ternatea, Convolulus pluricaulis, Curculigo orchioides, Dioscorea bulbifera, Enhydra fluctuans, Glycyrrhiza glabara, Jasminum sambac, Leptadenia reticulara Microstylis wallachii, Microstylis mucifera Mucuna pruriens, Nardostachys jatamansi, piper longum, Polygonatum cirrifolium, Polygonatum cirrifolium Royale, Polygonatum verticillatum, Phaseolus trilobus, Phaseolus radiatus, Roscoea procera, Roscoea procera, Terminalia chebula, Terminalia arjuna, Tinospora cordifolia, Valeriana wallichii, Vitex negundo, Withania somnifera.*
Anti-aging herbs (*Vaya sthapna Rasayana*) *Alternanthera sessilis, Argyreia speciosa, Asparagus racemosus, Boerhaavia diffusa Clitoria ternatea, Desmodium gangeticum, Dioscorea bulbifera, Emblica Officinalis, Enhydra fluctuans, Glycyrrhiza glabara, Hydrocotyle asiatica, Tinospora cordifolia, Terminalia chebula, Terminalia arjuna, Pluchea lanceolata, Leptadenia reticulara, Vitex negundo, Withania somnifera.*
Memory-enhancing Herbs (*Medhya Rasayana*) *Abies webbiana, Acorus calamus, Artemisia roxburghiana, Bacopa monniera, Calotropis procera, Callicarpa macrophylla, Celastrus paniculatus, Centella asiatica, Cedrus decodara, Citrullus colocynthis, Clitoria ternatea,Convolvulus pluricaulis, Evolvulus alsinoides, Glycyrrhiza glabra, Litsea glutinosa, Nardostachys jatamansi, Ocimum sanctum, Phytolacca acinosa, Pluchea lanceolata, Pterospermum acerifolium, Rhododendron anthopogan, Picrorhiza kurroa, Salix tetrasperma, Sidha cordifolia Solanum (selinum) tennifolium, Stereospermum suaveolens, Taxus baccata, Terminalia bellerica, Terminalia chebula, Tinospora cordifolia, Valeriana wallichii, Withania somnifera.*

Although all the herbs listed in Table 1 are highly reputed in Ayurvedic texts, and some of these herbs have shown various type of protective activities in different animal models and human trials, many of them lack sound scientific backings to establish their claims. Here we will review the herbs which comes under the category of Medhya Rasayana and whose claims have been verified by the modern scientific research in animals and human trials as well.

Bacopa monniera (Brahmi)

Bacopa is described in Ayurveda for treating age-related mental decline, as well as for improving cognitive processes. The plant has been extensively investigated for its neuro-pharmacological effects. Sedative effects of glycosides of Bacopa were reported as early as in 1959.[10] Alcoholic extract of the plant improved the performance of rats in motor learning[11] and glycoside hersaponin (derived from *B. monniera*) was found better than pentobarbitone in facilitating acquisition and retention of brightness discrimination reaction.[12] The memory enhancing effect of the ethanolic extract was attributed to bacosides A and B[13] and depending upon the different time periods period of collection of *B. monniera* plant, bacoside A predominates in March and April whereas both bacosides A and B are available in May.[14] This fact argues in the favors of the timeliness of collection as described in ayurvedic meteria medica and this also explains the ambiguity in the results of different experiments on the same plant species. Both the extract as well as bacosides were evaluated for their nootropic activity in adult male rats and significant improvement was observed in acquisition, consolidation and retention.[15]

Bacosides attenuated retrograde amnesia produced by electroconvulsive shock, immobilization or scopolamine.[16] *B. monniera* extract was more potent than extract of Ginkgo biloba in counteracting the atropine induced transient amnesia on transfer latency in mice.[10] In a rat model of Alzheimer's disease, oral administration of the *B. monniera* extract markedly reduced the magnitude of memory deficits;[17] the treatment also augmented the acetylcholine concentration, choline acetylase activity and muscarinic receptor binding in hippocampus and frontal cortex. *B. monniera* extract exhibited anxiolytic and the adaptogenic property against acute and chronic stress models in rats. The bacosides showed a mild antidepressant activity as evident by reversal of reserpine syndrome and effect on immobility time in the swimming test in mice.[18,19]

Bacosides exhibited significant anti-stress activity and improved duration and performance in the swimming endurance test, increased survival under hypoxic conditions and protected animals against morphine induced toxicity.[20] Pretreatment with *B. monniera* significantly reduced the acute stress induced increase in the ulcer index, adrenal gland weight, plasma glucose, aspartate aminotransferase, and creatine kinase.[21] Oral administration of the ethanolic extract of the plant in rat increased superoxide dismutase, catalase and glutathione peroxidase activity in the frontal cortex, striatum and hippocampus in a dose dependent manner,[22] Bacosides also effectively protected against stress-induced increase in the level of lipid peroxides in several areas including the hypothalamus and cerebral cortex. There is growing evidence that high concentrations of nitric oxide (NO), generated by activated astrocytes, might be involved in a variety of neurodegenerative diseases, such as Alzheimer's disease, ischemia and epilepsy. Glial cells may produce NO under superoxide radical stimulation by enzyme-independent mechanism. The extract of *B. monniera* inhibited the formation of reactive species and DNA damage in a dose dependent manner in culture of purified rat astrocytes in which toxicity was induced by the nitric oxide donor, S-nitroso-N-acetyl-penicillamine.[23] The bacosides increase protein and serotonin level and decrease in norepinephrine concentration in hippocampus, hypothalamus and cerebral cortex.[16] Immobilization leads to an increased expression of heat shock proteins (HSP) in cerebellum, cerebral cortex and hippocampus and prior administration of the extract, however, markedly reduced the stress-induced expression of HSP, especially in the cerebral cortex and hippocampus.[24] Methanol extract of *B. monniera* showed the dose-dependent free radical scavenging capacity and a protective effect on DNA cleavage.[25]

In a double-blind, randomized trial nineteen attention deficit hyperactive disorder (ADHD) children, aged 8–10 years old, were given 50 mg of *B. monniera* twice daily and after 12 weeks of treatment, a significant improvement in the areas of sentence repetition, logical memory, and pair-associative learning was observed in all the children who took *B. monniera*, since evaluation occurred after four weeks after stopping *B. monniera* usage, indicating that it had a lasting effect.[26] In healthy adults, 12 weeks of *B. monniera* treatment showed improvement in learning and memory, which was accompanied by the significant reduction in anxiety. Chronic effects of *B. monniera* were studied on adults aged between 40 and 65 years in a double-blind randomized, placebo control study and significant effect of the *B. monniera* was observed on a test for the retention of new information and *B. monniera* decreased the rate of forgetting of newly acquired information.[27]

In a contrasting randomized, double-blind, placebo-controlled study, combined *Ginkgo biloba* and *B. monniera* extract did not demonstrate any significant effects on tests investigating a range of cognitive processes including attention, short-term and working memory, verbal learning, memory consolidation, executive processes, planning and problem solving, information processing speed, motor responsiveness and decision making.[28] These findings suggest that treatment of an extract containing *G. biloba* and *B. monniera* for 2 weeks and 4 weeks had no cognitive enhancing effects in healthy subjects. This study can be considered as a classical example of not following the ayurvedic direction and thus the outcome was a negative result. The combination of *G. biloba* and *B. monniera* are not prescribed anywhere in any of the ayurvedic texts and the results in the above study may surprise many but not to an ayurvedic practitioners, and in order to get the expected results one must follow the ayurvedic guideline.

The antioxidant capacity of *B. monniera* explain, the reported anti-stress, immunomodulatory, cognition-facilitating, anti-inflammatory and antiaging effects produced by it in experimental animals and in clinical situations and may justify further investigation of its other beneficial properties. Moreover, these experimental evidences suggests that because of its antioxidant activity, this Ayurvedic drug may also be useful in the treatment of human pathologies in which free radical production plays a key role.

Withania somnifera Dunal (Ashwagandha)

Widely used in Ayurvedic medicine, this plant is an ingredient in many formulations prescribed for a variety of neurological, musculoskeletal conditions and as a general tonic to increase energy, improve overall health and longevity. The various antioxidant activities of this plant have been studied in various experimental paradigms.[29]

There have been numerous studies regarding the cognitive enhancing activities of *W. somnifera*. Sitoindosides IX and X, the steroidal derivatives that have been isolated from *W. somnifera* root augmented learning acquisition and memory in both young and old rats[30] and this effect was due to modulation of cholinergic neurotransmission. The extract containing the sitoindosides VII-X and withaferin A also reversed the ibotenic acid-induced cognitive deficit and reversed the reduction in cholinergic markers (e.g. ACh, ChAT) in rats.[31] The cognition enhancing effects of *W. somnifera* root involve modulations in cholinergic neurotransmission in the cortical and basal forebrain, brain areas involved in cognitive function. Sitoindosides VII-X and withaferin isolated from aqueous methanol extract of roots enhanced acetyl cholinesterase (AChE) activity in the lateral septum and globus pallidus, and decreased AChE activity in the vertical diagonal band.[32] These changes were accompanied by enhanced M1-muscarinic-cholinergic receptor-binding in lateral and medial septum as well as in frontal cortices, whereas the M2-muscarinic receptor-binding sites were increased in a number of cortical regions including cingulate, frontal, piriform, parietal, and retrospinal cortex. Withaferin preferentially affect events in the cortical and basal forebrain cholinergic-signal transduction cascade. *W. somnifera* exhibited a nootropic-like effect in naive and amnesic mice. Thus sitoindosides and withoferins could have potential in Alzheimer's disease therapy. *W. somnifera* improved retention of a passive avoidance task in a step-down paradigm in mice.[33] It also reversed the scopolamine-induced disruption of acquisition and retention and attenuated the amnesia produced by acute treatment with electroconvulsive shock (ECS). *W. somnifera* root and some constituents are also reported to have antioxidant and anti-inflammatory activities, which may also be relevant in Alzheimer's disease therapy. The crude extract is reported to inhibit lipid peroxidation *in vitro* and *in vivo*.[34] The compounds responsible for antioxidant activity include the withanolides,[35–37] but other compounds in the root may also be antioxidants. The glycowithanolides decreased lipid peroxidation in various tissues including the brain in rodents, and both the glycowithanolides and the sitoindosides (VII-X) enhanced catalase and glutathione peroxidase activities in rat frontal cortex and striatum.[38,39] *W. somnifera* has also successfully induced both axonal as well dendritic extensionsin different cell lines.[40,41] If this effect occurred in the CNS, treatment of Alzheimer's disease patients with the root extract may promote synaptic formation, which involves neurite outgrowth; thus cholinergic function may be enhanced. Recently Gupta *et al.*,[42] has shown that *W. somnifera* effectively inhibited both copper induced and iron-induced lipid peroxidation and also attenuated the decreased activities of antioxidant enzymes in the spinal cords of aged rats.

W. somnifera significantly attenuated the decline in learning-memory and motor functions in Streptozoticin-induced diabetic mice and this activity was attributed to its antioxidant properties. The combination of extracts of *W. somnifera* and *Aloe vera* was more effective in reducing oxidative damage in brain regions than the supplementation of single plant extract. The combination also lowered the blood glucose level in comparison to STZ diabetic mice.[43] Thus, this study provides ample support for ayurvedic therapeutic approach in which various herbs are combined together to get the maximum effect. In another study, *W. somnifera* extract plays a significant protective role in haloperidol-induced orofacial dyskinesia and it could effectively prevent neuroleptic-induced extrapyramidal side-effects.[44] In the same study *W. somnifera* extract significantly reduced the lipid peroxidation and significantly reversed the decrease in forebrain SOD and catalase levels. *W. somnifera* is also known for its anti-stress adaptogenic activity. Various studies on different stress models have prove its efficacy in alleviating the stress induced ill effects.[45,46] Electron microscopic observation revealed[47] that pre-treatment with alcoholic extract of *W. somnifera* in the stressed rats protects neuronal damage in the hippocampal sub layer (CA1-CA4 and DG), since these regions are involved in memory processing, it further strengthen the claim for usefulness of *W. somnifera* in Alzheimer's disease management. In a model of stroke, rats were pretreated with an extract of *W. somnifera* a significant motor impairment, with elevated levels of MDA, was observed in vehicle-treated middle cerebral artery occluded rats. Treatment with *W. somnifera* for 15 days did not improve motor performance or decrease the elevated levels of MDA. However, when the pretreatment time of *W. somnifera* was increased to 30 days, it prevented motor impairment and significantly decreased the raised levels of MDA compared with vehicle-treated rats.[48] The take-home message from this study is that as mentioned earlier that Rasayana treatment is a preventive approach; in this study the rats were effectively protected if they were pretreated with *W. somnifera* for 30 days and even 15 days of pre-treatment was ineffective. An alkaloid extract also had tranquillizing effects *in vivo* and potentiated barbiturate, ethanol and urethane induced hypnosis in mice.[49] The pharmacological basis for these observations is unknown, but could reflect the reported GABA-mimetic effect of *W. somnifera* extract.[50] *W. somnifera* administration to mice treated with a carcinogen reduced IL-1 and TNF-β levels,[51] which may also be relevant in Alzheimer's disease, considering the possible involvement of these inflammatory mediators in senile plaque formation and neurodegeneration. *W. somnifera* leaves are also reported to have anti-inflammatory activity.[52] The numerous pharmacological activities of *W. somnifera* indicate that this herb may have multiple beneficial effects in various neurological disorders.

Curcuma longa

The herb is more commonly known as "turmeric". In Ayurveda, *C. longa* has been used for various purposes and through different routes of administration and a recent study has reveled that at least in a mouse model Curcumin, isolated from *C. longa* can correct cystic fibrosis defects.[53] The antioxidant activity of Curcumin is well-documented.[54,55] Curcumin is several times more potent than vitamin E as a free radical scavenger[56] and has shown its beneficial effects in number of free radical related disorders. Its pharmacological effects and clinical use has been reviewed elsewhere[57] and we will discuss its effects on neural tissue. Curcumin protects the brain from lipid peroxidation,[58] and scavenges NO-based radicals.[59] Curcumin inhibited lipid peroxide levels and enhanced of glutathione in rat brain and demonstrated neuro-protection against ethanol-induced brain injury *in vivo* following oral administration.[60] By virtue of its antioxidant properties, some compounds from *C. longa*, including curcumin, demethoxycurcumin, bisdemethoxycurcumin and calebin-A (and some of its synthetic analogues), protected PC12 cells from b-amyloid insult *in vitro*.[61,62] Aqueous extract of *C. longa* inhibited brain monoamine oxidase (MAO) A and was found to be antidepressant in mice.[63] Curcumin treatment Reduced amyloid Pathology, oxidized proteins and interleukin-1 in Alzheimer Transgenic Mouse. These changes were also accompanied by the reduction in astrocytic marker GFAP, reduced insoluble and soluble amyloid and soluble Amyloid. Plaque burden was also significantly decreased in treated animals.[64]

Curcumin blocks AD pathogenesis at multiple sites. Curcumin can act as a scavenger of ROS, including NO and peroxynitrite generated by reactive glia and hydroxyl radicals generated by neurons as a result of direct amyloid toxicity. Its Ibuprofen (NSAID) like action can inhibit microglial activation. Curcumin can also limit damage by inhibiting iNOS, cyclooxygenase-2, and inflammatory cytokine production by reactive glia. By blocking NFkβ and reducing IL-1, IL-6, and ApoE, curcumin should reduce proamyloidogenic factors. Finally, curcumin can lower plasma and tissue cholesterol, potentially lowering amyloid production. Thus Curcumin has great potential to be a potent neurojuvenator/neuroprotector agent.

Celastrus paniculatus

The seed oil improved learning and memory and decreased the levels of noradrenalin, dopamine and 5-hydroxytryptamine (5-HT) in the brain.[65] Administration of the seed oil to rats also reversed a scopolamine-induced task deficit.[66] The aqueous extract increased learning-memory, significantly decreased the brain levels of malondialdehyde, with simultaneous significant increases in levels of glutathione and catalase, thus the aqueous extract of *C. paniculatus* seed has cognitive-enhancing properties and can also act as an antioxidant.[67] Methanol extract of *C. paniculatus* have shown to be anti-inflammatory,[68] which may also have some relevance in the management of neurodegenerative disorders. In another report extracts from *C. paniculatus* showed a dose-dependent free radical scavenging capacity and protected DNA cleavage. *C. paniculatus* significantly attenuated H_2O_2-induced neuronal death and when the neuronal cells were treated with extract of *C. paniculatus* prior to H_2O_2 exposure,[69] catalase activity was increased and levels of malondialdehyde were decreased significantly. These antioxidant effects of active principle of *C. paniculatus* may explain, at least in part, the reported anti-stress, immunomodulatory, cognition-facilitating, anti-inflammatory and antiaging effects produced in experimental animal studies and in clinical situations may justify the further investigation of their other beneficial biological properties.

Centella asiatica

It is used as a revitalizing herb.[70] In mice, extract of *C. asiatica* leaf acted as sedative, antidepressant and showed cholinomimetic activity, which was blocked by atropine.[71] Thus, *C. asiatica* may be appropriate to treat depression and anxiety in neurological disorders, and since it enhances cholinergic activity and therefore it may improve cognitive functions too and it was proved that *C. asiatica* leaf improve learning and memory processes in rats, and modulated dopamine, 5-hydroxytryptamine (5-HT) and noradrenalin systems in rat brain *in vivo*.[72] The triterpene asiatic acid and its derivatives have been shown to protect cortical neurons from glutamate-induced excitotoxicity *in vitro*.[73] A Cognitive-enhancing effect of an aqueous extract of *C. asiatica* was found to be associated with antioxidant mechanism in the CNS.[74] *C. asiatica* prevented radiation-induced behavioral changes during clinical radiotherapy.[75] In a double-blind, placebo-controlled study[76] in healthy subjects, it proved its anxiolytic activity but it remains to be seen whether this herb has therapeutic efficacy in the treatment of anxiety syndromes. The total triterpenes from *C. asiatica* has also shown antidepressant activity and the administration of *C. asiatica* decreased seizures and showed improvement in the learning deficit in the pentylenetetrazole (PTZ) kindled rats.[77] In a streptozotocin model of Alzheimer's disease in rats, an aqueous extract of *C. asiatica* showed a dose-dependent increase in cognitive behavior, a significant decrease in MDA, and an increase in glutathione and catalase,[78] thus indicating that an aqueous extract of *C. asiatica* is effective in preventing the cognitive deficits, as well as the oxidative stress, caused by i.c.v. STZ in rats. Further research may prove the clinical potential of *C. asiatica* in Alzheimer's disease and age-related neurodegeneration.

Clitoria ternatea

This plant is known to promote intelligence. In a passive avoidance test and spatial learning T-maze the aqueous root extract of *C. ternatea* enhanced memory in rats.[79] Alcoholic extracts of *C. ternatea* in rats attenuated electroshock-induced amnesia.[80] The extract produced significant memory retention, and the root parts were found to be more effective. *C. ternatea* extract increased rat brain acetylcholine content and acetyl cholinesterase activity in a similar fashion to the standard cerebroprotective drug pyritinol. An aqueous extract of the root also increased acetylcholine (Ach) levels in rat hippocampus, and it was hypothesized that this effect may be due to an increase in the activity of choline acetyltransferase, an ACh synthetic enzymes.[79] Jain *et al.*[81] showed in the elevated plus maze experiment that *C. ternata* possess nootropic, anxiolytic, antidepressant, anticonvulsant and antistress activity. Although preliminary studies have demonstrated that *C. ternata* possesses many neuroprotective properties but to establish its potentials as neuro-rejuvinator, more research is required.

Acorus calamus

This plant has been used to treat memory loss. Alcohol extract of *A. calamus* root are reported to exert sedative effects and potentiate hypnosis *in vivo*.[82,83] The root has also shown antioxidant activity *in vitro*.[84] *A. calamus* root extract protected rats against acrylamide-induced neurotoxicity and reduced the incidence of paralysis.[85] In a very interesting study,[86] GABA level was significantly increased and glutamate content was significantly decreased in mouse brain by preinhalation of the essential oil, thus suggesting the anticonvulsive effect of its oil. Fragrance inhalation progressively prolonged the pentobarbital-induced sleeping time as inhalation time was lengthened and it presumably involved both of its GABAergic modulatory and antioxidant activities.

Terminalia chebula

Also called Haritiki, it is considered as one of the most powerful Rasayana. *T. chebula* has been reported to exhibit a variety of biological activity, including anticancer, antidiabetic, antimutagenic, antibacterial, and antiviral activities. The ripe fruit (unripe fruit is reported to produce different effects) is reputed to retard the aging process and to improve cognitive processes,[87] thus suggesting apparent benefits in age associated neurodegeneration. There is a general lack of research on this herb and only a limited number of studies provide some explanation for the reputed effects. A methanol extract is reported to bind to NMDA and GABA.[88] Another study showed an aqueous extract of *T. chebula* to be antioxidant.[89,90] Extracts and pure compounds of *T. chebula* were demonstrated to exhibit antioxidant activity at different magnitudes of potency.[91] These studies only provide limited data and further investigation should be conducted to gain more conclusive information regarding its therapeutical values.

Ayurveda and Neurodegenerative Diseases

In addition to age-associated memory disorders Parkinson's disease is the most common neurodegenerative disorder of population over age sixty five. Parkinson's disease is referred in Ayurveda as "Kampa Vata".[92] Drugs supplying the brain with L-dopa have been the mainstay of Parkinson's treatment and Additional drugs are given to prevent the catabolism (breakdown) of dopamine. L-dopa is often administered in combination with other drugs.[93] Ayurvedic treatment for this condition veers around the herb atmagupta (*Mucuna Pruriens*, also known as Kappikacchu). Clinical studies have shown its effectiveness on patients diagnosed with Parkinson's disease.[92] A clinical study used concoction in cow's milk of powdered *M. pruriens*, *Hyoscyamus reticulates* seeds, *W. somnifera* and *Sida cordifolia* roots in 18 clinically diagnosed (with a mean Hoen and Yahr value of 2.22) Parkinsonian patients. The patients which underwent both cleansing (for 28 days) and palliative therapy (56 days showed significant improvement in activities of daily living (ADL) and on motor examination as per UPDRS rating and they exhibited better response in tremor, bradykinesia, stiffness and cramps.[94] Patients treated with *M. pruriens* have shown mild side-effects which include headache, dystonia (abnormal muscle tone) and fatigue.[95] L-dopa is the precursor of dopamine, the neurotransmitter which is absent or decreased in Parkinson's disease due to loss of Dopaminergic neurons in the substantia nigra. *M. pruriens* contains Levodopamine or L-dopa within its seeds,[96] and oral administration of *M. pruriens* increases the dopamine content in the nigrostriatal tract.[97] *M. pruriens* possesses antioxidant properties,[98] which along with the presence of L-Dopa, might together be responsible for the protection offered by this herb in Parkinson's disease.

Ayurveda and Longevity

One verse from the primary text on Ayurveda, the *Charka Samhita* states "*Ayurvedo Amritanaam*," that the purpose of Ayurveda is to gain immortality — the longest lifespan of all. Many studies in the past have conclusively shown that feeding antioxidant and manipulation of antioxidant enzymes may extend the lifespan in many experimental paradigms, and, many herbal preparations (Rasayanas) have shown antioxidant activities. Although to the best of our knowledge no longevity study is yet been conducted by using Rasayana but keeping in view the studies performed and antioxidant nature of Rasayana, one can speculate that

Rasayanic preparation may be of use in adding years to life. Investigators have known for decades that caloric restriction (CR) extends life and the duration of good health in all species in which it has been studied, as long as the diet includes enough nutrition for routine maintenance of the body[99] and caloric restriction might have similar effects in humans. Given that few people would ever reduce their food intake enough to lengthen their lives, biologists are now trying to discover the mechanism that underlies the benefits of caloric restriction and to find agents that might mimic those helpful effects in people without forcing them to go hungry. The two important ayurvedic plants have been suggested which can be used as CR mimetics, one is *Gymnema Sylvestre,* which is also known as "gur-mar" or destroyer of sugar, have been used to treat diabetes and its Ayurvedic medicine, which is effective against both diabetes and obesity and most relevant to CR mimicry is probably a reduction in blood sugar concentrations, especially following an oral glucose load. Its derivatives have shown the reductions of approximately 20 to 30% in serum glucose concentrations 30 minutes after oral administration of 100 mg per kg of body weight.[100] The other suggested herb with potential CR mimetic is *Garcinia cambogia,* known for its weight control activity by increasing lipolysis, decreasing lipogenesis, and appetite suppression.[101] It is hoped that these CR mimetics may be of help to obtain the health- and longevity-promoting effects without actually decreasing food intake.[102]

Several complex formulations have been prescribed in Ayurveda as rejuvenation therapy and to promote longevity. Theses preparations are essentially the mixture of various plants (plant parts), along with clarified cow butter and honey. The most famous of these preparations is Chyavanprash, which helps maintain youthfulness by renewing tissues and counterattacking degeneration. Although composed of more than forty herbs, Chyavanprash is considered a single entity and is the best known herbal compound in Ayurveda. Its adaptogenic properties make it an excellent anti-aging and anti-stress tonic. It is a complex herbal concentrate that has been the geriatric tonic par excellence for centuries. Chyavanprash contains a large percentage of Amalaki (*Emblica officinalis* or Indian Gooseberry). Fruits of *E. oficinalis* have been used for thousands of years in the traditional Indian medicine for the treatment of several diseases. For many years, the therapeutic potential of the fruits was attributed to their high content of ascorbic acid: about 1 g of vitamin C per 100 mL of fresh juice.[103] All the studies published after that time were based on comparison between ascorbic acid and *E. oficinalis* and Chyavanprash were discovered to be more effective than ascorbic acid whether *in vitro* or *in vivo*.[104] In the subsequent years it was discovered that *E. oficinalis* fruits do not contain ascorbic acid neither in free nor in conjugated form, but it contains two new hydrolysable tannins with low molecular weight called emblicanin A (2,3-di-o-galloyl-4,6-(S)-hexahydroxydiphenoyl-2-keto-glucono-d-lactone) and emblicanin B (2,3,4,6-bis-(S)-hexahydroxydiphenoyl-2-keto-glucono-d-lactone).[105]

Geriforte is another commercially available ayurvedic formulation, which has shown antistress, adaptogenic properties. Geriforte administration in rats increased the level of SOD, catalase, which was accompanied by a significant decreased level of MDA.[106,107] In a human trial on aged persons, this formulations has shown positive effects in the form of improvement in appetite, digestive functions, ability to work, and a feeling of well-being.[108] The composition of this formulation is very similar to that of Chyavanprash.

Another famous commercial and probably the most researched ayurvedic Rasayana is Maharishi Amrit Kalash (MAK), and careful search of ayurvedic texts revealed that its herbal components are similar to a reputed Rasayanic preparation called Brahma Rasayana. The chemical analysis of MAK has shown that it contains a large number of compounds, namely tannic acid, flavinoids, catecholamines, α-tocopherol, polyphenols, ascorbates, riboflavin, β-carotenes, mucillage, octacosanol, saponins, sphaeranthine, asparagine, glycyrrhizin, camphene, limonene, bacopins, pinene, etc.[109] MAK has shown more antioxidant potency in preventing LDL oxidation than ascorbic acid, alpha-tocopherol, or probucol. MAK has H_2O_2 binding activity, which decreases the availability of H_2O_2 and help in reducing oxidative stress.[110] MAK induced the activities of cholinergic enzymes in the aged guinea pigs;[111] it also augmented the decreased antioxidant enzymes activities in the old guinea pigs,[112] decreased neuronal lipofuscin[113] and retarded the formation of dark neurons.[114] Ultrastructural studies revealed that MAK can protect age associated mitochondrial degenerations.[115] Thus MAK can be useful in various age associated pathological conditions.

Ayurveda and Toxicity

The most popular belief about ayurvedic formulations is that, these are herbal and are natural remedies and beyond conventional toxicity studies but ayurvedic drugs can be toxic. Charaka[2] states "A potent poison also becomes the best drug on proper administration and on the contrary the best drug becomes a potent poison if used badly". Herbal drugs, when used in combination with allopathic drugs may alter their pharmacodynamics and possible drug interactions may take place. Ayurvedic texts classify toxic plants into different categories depending on the part of the plant that is toxic and by the proper processing in the lines of ayurvedic instructions the toxic components can be destroyed and converted not only to safe but further therapeutically useful entities. The most striking example in this case is of *Aconitum heterophyllum*. The roots of this plant are considered toxic (they contain an alkaloid aconitine) and following ingestion of roots, the toxicity manifests in the form of tingling numbness of mouth and throat, abdominal pain, loss of muscle power, visual and auditory disturbances and finally clonic convulsions.[116] However, aconite forms an important constituent of ayurvedic formulations. The aconite used in the formulations is not a crude agent but one, which is processed. This processing involves boiling of roots with 2 parts of cow's urine (7 hours per day) for 2 consecutive days. The roots are then thoroughly washed with water and boiled with 2 parts of cow's milk for the same duration. These are washed again with lukewarm water, cut into pieces, dried and ground. It has been shown that aconite becomes safe only after this elaborate process and all the steps are essential for complete detoxification.[117,118]

Besides toxicology, ayurvedic pharmacology describes in some detail the side effects that can occur with different therapeutically useful drugs. Further more, it also describes ways (which also include manufacturing techniques) to minimize these side effects. Ayurveda gives instructions regarding time of drug administration, the relationship with food, type of food which should be avoided/permitted with the drug etc. The do's and don'ts are clearly enunciated. For example, amalki (*Emblica officinalis)* should be avoided at bedtime to prevent harmful effects.[119] Similarly, *Piper longum* used in asthma should be avoided in patients with peptic ulcer disease and should be consumed with milk.[120]

Tribhuvankirti is a combination of several plants which is very commonly used to treat "cold in the head" and fever. There are clear instructions in Ayurveda that because it contains aconite and it should be used cautiously. When used, it should be taken with tulsi (Ocimum sanctum) juice, ginger juice or honey.[119] Ayurveda specifies that guggul (a hypolipidimic agent, used to loose weight) should be used cautiously in patients with peptic ulcer disease. While on guggul therapy, the patient is advised to avoid sour food, alcohol and heavy exercise.[121] The subject of teratogenecity also figures in Ayurveda. Thus, certain plants like *Terminalia chebula* (harda) are to be avoided in pregnancy. It is a powerful purgative and is supposed to stimulate gastrointestinal motility and would therefore be contraindicated in pregnancy.[122] These facts are not sufficiently publicized. Apart from plants, Ayurveda also includes metals in its formulary. These metals have to be diligently processed before they are suitable for human consumption and there is again a long list of do's and don'ts regarding their use. Ayurvedic textbooks recommend a special pharmaceutical process to detoxify the metals, e.g. lead, which is used in many ayurvedic preparations, has to be processed by first heating over a fire till it glows. It should then be cooled by dipping into a mixture of sesame oil, buttermilk, cow's urine and a decoction of three plants, viz. *E. officinalis, T. bellerica* and *T. chebula* . After repeating this procedure thrice, the lead is heated the fourth time, following which it is dipped into a powder made of the rind of tamarind and *Piper longum.* This lead is then mixed with arsenic sulphide and wrapped in a betel leaf and warmed in a crucible to a fixed temperature. This process is repeated thirty times before processed lead is ready for use.[123] Unfortunately, there are no quality control methods to standardize such metal containing drugs and to find out whether processing of metal is done appropriately so as to render it nontoxic. This thus increases the probability of toxic effects. For example, if a "herbo-mineral" preparation "Mahayograj Guggul" is prescribed for rheumatoid arthritis in ayurvedic texts and contains several plants and metals. The continuous use for well over two years will provide relief from the arthritis but also lead to poisoning. Thus ayurvedic preparations can also be very toxic and may be life-threatening if taken without consultation from a qualified practitioner.

In the light of above discussion, we conclude that Ayurvedic remedies are of use in circumventing the age-associated pathological conditions, but to get maximum benefits one must follow the strict rules laid down in ayurvedic literature.

REFERENCES

1. Chandra L (ed.) (1970) *India's Contribution to World Thought and Culture*. Vivekananda Rock Memorial Committee, Madras, India.
2. Ray P, Gupta HN (eds.) (1965) *Charaka Samahita*. National Institute of Sciences of India, New Delhi, India.
3. Krishnamurthy KH (ed.) (1991) *Wealth of Susruta*. International Institute of Ayurveda Coimbatore, India.
4. Kurup RK, Kurup PA (2003) Hypothalamic digoxin, hemispheric, chemical dominance, and the tridosha theory. *Intern J Neuroscience* **113**: 657–681.
5. Dahanukar SA, Thatte UM (eds.) (1989) *Ayurveda Revisited*. Popular Prakashan, Mumbai, India.
6. Rowe J, Kahn R (1987) Human aging: Usual and successful. *Science* **237**: 143–149.
7. Rowe JW, Kahn RL (2000) Successful aging and disease prevention. *Ad Renal Replace Th* 7: 70–71.
8. Crook TH (1998) Treatment of age-related cognitive decline: effects of phosphatidylserine. In: Klatz R, Goldman R (eds.) *Anti-Aging Medical Therapeutics*. Vol II, pp. 20–29, Marina Del Ray, CA.
9. Crook TH, Tinklenberg J, Yesavage J (1991) Effects of phosphatidylserine in age-associated memory impairment. *Neurology* **41**: 644–649.
10. Malhotra GL, Das PK (1959) Pharmacological studies of *Herpestis monniera* (Brahmi). *Ind J Med Res* **47**: 294–305.
11. Prakash JC, Sirsi M (1962) Comparative study of the effects of Brahmi (*Bacopa monniera*) on motor learning in rats. *J Sci Indust Res* **21**: 93–96.
12. Sinha MM (1971) Some empirical behavioral data indicative of concomitant biochemical reactions. *Proceed Ind Sci Congress*. Part II, Calcutta 1–26.
13. Chatterjee N, Rastogi RP, Dhar ML (1963) Chemical examination *of Bacopa monniera* Wettst. Part I: Isolation of chemical constituents. *Ind J Chem* **1**: 212–215.
14. Rastogi S, Mehrotra BN, Kulshreshta DK (1994) In: Jain SK (ed.) *Proceedings of IV International Congress of Ehtnobiology*. Nov. 17–21, Deep Publications, New Delhi.
15. Singh HK, Dhawan BN (1982) Effect of *Bacopa monniera Linn* (Brahmi) extract on avoidance responses in rat. *J Ethnopharmacol* **5**: 205–214.
16. Singh HK, Dhawan BN (1997) Neuropsychopharmacological effects of the Ayurvedic nootropic *Bacopa monniera Linn* (Brahmi). *Ind J Pharmacol* **29S**: 359–365.
17. Bhattacharya SK, Kumar A, Ghoshal S (1999) Effect of *Bacopa monniera* on animal models of Alzheimer's disease and perturbed central markers of cognition in rats. *Res Comm Pharmacol Toxicol* **4**: 1–12.
18. Bhattacharya SK, Ghoshal S (1998) Anxiolytic activity of a standardised extract of *Bacopa monniera*: An experimental study. *Phytomed* **5**: 95–100.

19. Shankar G, Singh HK (2000) Anxiolytic profile of standardised Brahmi extract. *Ind J Pharmacol* **32**: 152–159.
20. Singh HK, Shankar G, Patnaik GK (1996) Neuropharmacological and anti-stress effects of bacosides: A memory enhancer. *Ind J Pharmacol* **28**: 47–53.
21. Rai D, Bhatia G, Palit G, Pal R, Singh S, Singh HK (2003) Adaptogenic effect of *Bacopa monniera* (Brahmi). *Pharmacol Biochem Behav* **75**: 823–830.
22. Bhattacharya SK, Bhattacharya A, Kumar A, Ghoshal S (2000) Antioxidant activity of *Bacopa monniera* in rat frontal cortex, striatum and hippocampus. *Phytoherap Res* **14**: 174–179.
23. Russo A, Borrelli F, Campisi A, Acquaviva R, Raciti G, Vanella A (2003) Nitric oxide-related toxicity in cultured astrocytes: effect of *Bacopa monniera*. *Life Sci* **73**: 1517–1526.
24. Chowdhuri DK, Parmar D, Kakkar P, Shukla R, Seth PK, Srimal RC (2007) Antistress effects of bacosides of *Bacopa monnieri*: modulation of Hsp70 expression, superoxide dismutase and cytochrome P450 activity in rat brain. *Phytother Res* **16**: 639–645.
25. Russo A, Izzo AA, Borrelli F, Renis M, Vanella A (2003) Free radical scavenging capacity and protective effect *of Bacopa monniera* L. on DNA damage. *Phytother Res* **17**: 870–875.
26. Negi KS, Singh YD, Kushwaha KP, Rastogi CK, Rathi AK, Srivastava JS, Asthana OP, Gupta RC, Lucknow G (2000) Clinical evaluation of memory enhancing properties of Memory Plus in children with Attention Deficit Hyperactivity Disorder. *Ind J Psychiatry* **42**: 21–29.
27. Stough C, Lloyd J, Clarke J, Downey LA, Hutchison CW, Rodgers T, Nathan PJ (2001) The chronic effects of an extract of *Bacopa monniera* (Brahmi) on cognitive function in healthy human subjects. *Psychopharmacology* **156**(4): 481–484.
28. Nathan PJ, Tanner S, Lloyd J, Harrison B, Curran L, Oliver C, Stough C (2004) Effects of a combined extract of *Ginkgo biloba* and *Bacopa monniera* on cognitive function in healthy humans. *Hum Psychopharmacol* **19**: 91–96.
29. Mishra LC, Singh BB, Dagenais S (2000) Scientific basis for the therapeutic use of *Withania somnifera* (ashwagandha): a review. *Altern Med Rev* **5**: 334–346.
30. Ghosal S, Lal J, Srivastava R (1989) Immunomodulatory and CNS effects of sitoindosides IX and X, two new glycowithanolides from Withania somnifera. *Phytother Res* **3**: 201–206.
31. Bhattacharya SK, Kumar A (1995) Effect of glycowithanolides from *Withania somnifera* on an animal model of Alzheimer's disease and perturbed central cholinergic markers of cognition in rats. *Phytother Res* **9**: 110–113.
32. Schliebs R, Liebmann A, Bhattacharya SK, Kumar A, Ghosal S, Bigl V (1997) Systemic administration of defined extracts from *Withania somnifera* (Indian Ginseng) and Shilajit differentially affects cholinergic but not glutamatergic and GABAergic markers in rat brain. *Neurochem Int* **30**: 181–190.
33. Dhuley JN (2003) Adaptogenic activity of a novel withanolide-free aqueous fraction from the roots of *Withania somnifera* Dun. *Phytother Res* **17**: 531–536.
34. Panda S, Kar A (1997) Evidence for free radical scavenging activity of Ashwagandha root powder in mice. *Ind J Physiol Pharmacol* **41**: 424–426.
35. Bhattacharya A, Ghosal S, Bhattacharya SK (2001) Anti-oxidant effect of *Withania somnifera* glycowithanolides in chronic footshock stress-induced perturbations of oxidative free radical scavenging enzymes and lipid peroxidation in rat frontal cortex and striatum. *J Ethnopharmacol* **74**: 1–6.

36. Bhattacharya A, Ramanathan M, Ghosal S, Bhattacharya SK (2000) Effect of *Withania somnifera* glycowithanolides on iron-induced hepatotoxicity in rats. *Phytother Res* **14**: 568–570.
37. Scartezzini P, Speroni E (2000) Review on some plants of Indian traditional medicine with antioxidant activity. *J Ethnopharmacol* **71**: 23–43.
38. Bhattacharya SK, Satyan KS, Ghosal S (1997) Antioxidant activity of glycowithanolides from *Withania somnifera*. *Ind J Exp Biol* **35**: 236–239.
39. Chaurasia SS, Panda S, Kar A (2000) *Withania somnifera* root extract in the regulation of lead-induced oxidative damage in male mouse. *Pharmacol Res* **41**: 663–666.
40. Kuboyama T, Tohda C, Zhao J, Nakamura N, Hattori M, Komatsu K (2002) Axon- or dendrite-predominant outgrowth induced by constituents from Ashwagandha. *Neuroreport* **13**: 1715–1720.
41. Tohda C, Kuboyama T, Komatsu K (2000) Dendrite extension by methanol extract of Ashwagandha (roots of *Withania somnifera*) in SK-N-SH cells. *Neuroreport* **11**: 1981–1985.
42. Gupta SK, Dua A, Vohra BP (2003) *Withania somnifera* (Ashwagandha) attenuates antioxidant defense in aged spinal cord and inhibits copper induced lipid peroxidation and protein oxidative modifications. *Drug Metabol Drug Interact* **19**: 211–222.
43. Parihar MS, Chaudhary M, Shetty R, Hemnani T (2004) Susceptibility of hippocampus and cerebral cortex to oxidative damage in streptozotocin treated mice: prevention by extracts of *Withania somnifera* and Aloe vera. *J Clin Neurosci* **11**: 397–402.
44. Naidu PS, Singh A, Kulkarni SK (2003) Effect of *Withania somnifera* root extract on haloperidol-induced orofacial dyskinesia: possible mechanisms of action. *J Med Food* **6**: 107–114.
45. Kaur P, Sharma M, Mathur S, Tiwari M, Divekar HM, Kumar R, Srivastava KK, Chandra R (2003) Effect of 1-oxo-5beta, 6beta-epoxy-witha-2-ene-27-ethoxy-olide isolated from the roots of *Withania somnifera* on stress indices in Wistar rats. *J Altern Complement Med* **9**: 897–907.
46. Bhattacharya SK, Muruganandam AV (2003) Adaptogenic activity of *Withania somnifera*: an experimental study using a rat model of chronic stress. *Pharmacol Biochem Behav* **75**: 547–555.
47. Shukla SD, Jain S, Sharma K, Bhatnagar M (2000) Stress induced neuron degeneration and protective effects of *Semecarpus anacardium* Linn. and *Withania somnifera* Dunn. in hippocampus of albino rats: an ultrastructural study. *Ind J Exp Biol* **38**: 1007–1013.
48. Chaudhary G, Sharma U, Jagannathan NR, Gupta YK (2003) Evaluation of *Withania somnifera* in a middle cerebral artery occlusion model of stroke in rats. *Clin Exp Pharmacol Physiol* **30**: 399–404.
49. Malhotra CL, Mehta VL, Das PK, Dhalla NS (1965) Studies on Withania-ashwagandha, Kaul V. The effect of total alkaloids (ashwagandholine) on the central nervous system. *Ind J Physiol Pharmacol* **9**: 127–136.
50. Mehta AK, Binkley P, Gandhi SS, Ticku MK (1991) Pharmacological effects of Withania somnifera root extract on GABAA receptor complex. *Ind J Med Res* **94**: 312–315.
51. Dhuley JN (1997) Effect of some Indian herbs on macrophage functions in ochratoxin A treated mice. *J Ethnopharmacol* **58**: 15–20.
52. Sudhir S, Budhiraja RD, Miglani GP, Arora B, Gupta LC, Garg KN (1986) Pharmacological studies on leaves of *Withania somnifera*. *Planta Med* **1**: 61–63.

53. Egan ME, Pearson M, Weiner SA, Rajendran V, Rubin D, Glockner–Pagel J, Canny S, Du K, Lukacs GL, Caplan MJ (2004) Curcumin, a major constituent of turmeric, corrects cystic fibrosis defects. *Science* **304**: 600–602.

54. Das KC, Das CK (2002) Curcumin (diferuloylmethane), a singlet oxygen (O-1(2)) quencher. *Biochem Biophys Res Commun* **295**: 62–66.

55 Priyadarsini KI (1997) Free radical reactions of curcumin in membrane models. *Free Radic Biol Med* **23**: 838–843.

56. Zhao BL, Li XJ, He RG, Cheng SJ, Xin WJ (1989) Scavenging effect of extracts of green tea and natural antioxidants on active oxygen radicals. *Cell Biophys* **14**: 175–185.

57. Miquel J, Bernd A, Sempere JM, Diaz–Alperi J, Ramirez A (2002) The curcuma antioxidants: pharmacological effects and prospects for future clinical use. A review. *Arch Gerentol Geriatr* **34**: 37–46.

58. Selvam R, Subramanian L, Gayathri R, Angayarkanni N (1995) The anti-oxidant activity of turmeric (*Curcuma longa*). *J Ethnopharmacol* **47**: 59–67.

59. Sreejayan Rao MN (1997) Nitric oxide scavenging by curcuminoids. *J Pharm Pharmacol* **49**: 105–107.

60. Rajakrishnan V, Viswanathan P, Rajasekharan KN, Menon VP (1999) Neuroprotective role of curcumin from curcuma longa on ethanol-induced brain damage. *Phytother Res* **13**: 571–574.

61. Park SY, Kim DSHL (2002) Discovery of natural products from *Curcuma longa* that protect cells from beta-amyloid insult: a drug discovery effort against Alzheimer's disease. *J Nat Prod* **65**: 1227–1231.

62. Kim DSHL, Kim JY (2001) Total synthesis of calebin-A, preparation of its analogues, and their neuronal cell protectivity against b-amyloid insult. *Bioorg Med Chem Lett* **11**: 2541–2543.

63. Yu ZF, Kong LD, Chen Y (2002) Antidepressant activity of aqueous extracts of *Curcuma longa* in mice. *J Ethnopharmacol* **83**: 161–165.

64. Lim GP, Chu T, Yang F, Beech W, Frautschy SA, Cole GM (2001) The curry spice curcumin reduces oxidative damage and amyloid pathology in an Alzheimer transgenic mouse. *J Neurosci* **21**: 8370–8377.

65. Nalini K, Karanth KS, Rao A, Aroot AR (1995) Effects of *Celastrus paniculatus* on passive avoidance performance and biogenic amine turnover in albino rats. *J Ethnopharmacol* **47**: 101–108.

66. Gattu M, Boss KL, Terry AV, Buccafusco JJ (1997) Reversal of scopolamine-induced deficits in navigational memory performance by the seed oil of *Celastrus paniculatus. Pharmacol Biochem Behav* **57**: 793–799.

67. Kumar MHV, Gupta YK (2002) Antioxidant property of *Celastrus paniculatus Willd.*: a possible mechanism in enhancing cognition. *Phytomedicine* **9**: 302–311.

68. Ahmad F, Khan RA, Rasheed S (1994) Preliminary screening of methanolic extracts of *Celastrus paniculatus* and Tecomella undulata for analgesic and anti-inflammatory activities. *J Ethnopharmacol* **42**: 193–198.

69. Godkar P, Gordon RK, Ravindran A, Doctor BP (2003) *Celastrus paniculatus* seed water soluble extracts protect cultured rat forebrain neuronal cells from hydrogen peroxide-induced oxidative injury. *Fitoterapia* **74**: 658–669.

70. Asakawa Y, Matsuda R, Takemoto T (1982) Monoterpenoids and sesquiterpenoids from *Hydrocotyle* and *Centella* species. *Phytochemistry* **21**: 2590–2592.
71. Sakina MR, Dandiya PC (1990) A psycho-neuropharmacological profile of *Centella asiatica* extract. *Fitoterapia* **61**: 291–296.
72. Nalini K, Aroor AR, Karanth KS, Rao A (1992) Effect of *Centella asiatica* fresh leaf aqueous extract on learning and memory and biogenic amine turnover in albino rats. *Fitoterapia* **63**: 232–237.
73. Lee MK, Kim SR, Sung SH, Lim DY, Kim H, Choi H (2000) Asiatic acid derivatives protect cultured neurons from glutamate-induced excitotoxicity. *Res Commun Mol Pathol Pharmacol* **108**: 75–86.
74. Kumar MHV, Gupta YK (2002) Effect of different extracts of *Centella asiatica* on cognition and markers of oxidative stress in rats. *J Ethnopharmacol* **79**: 253–260.
75. Russo A, Izzo AA, Cardile V, Borrelli F, Vanella A (2001) Indian medicinal plants as antiradicals and DNA cleavage protectors. *Phytomedicine* **8**: 125–132.
76. Bradwejn J, Zhou Y, Koszycki D, Shlik J (2000) A double-blind, placebo-controlled study on the effects of Gotu Kola (*Centella asiatica*) on acoustic startle response in healthy subjects. *J Clin Psychopharmacol* **20**: 680–684.
77. Gupta YK, Veerendra Kumar MH, Srivastava AK (2003) Effect of *Centella asiatica* on pentylenetetrazole-induced kindling, cognition and oxidative stress in rats. *Pharmacol Biochem Behav* **74**: 579–585.
78. Veerendra Kumar MH, Gupta YK (2003) Effect of *Centella asiatica* on cognition and oxidative stress in an intracerebroventricular streptozotocin model of Alzheimer's disease in rats. *Pharmacol Physiol* **30**: 336–342.
79. Rai KS, Murthy KD, Karanth KS, Nalini K, Rao MS, Srinivasan KK (2002) *Clitoria ternatea* root extract enhances acetylcholine content in rat hippocampus. *Fitoterapia* **73**: 685–689.
80. Taranalli AD, Cheeramkuzhy TC (2000) Influence of *Clitoria ternatea* extracts on memory and central cholinergic activity in rats. *Pharm Biol* **38**: 51–56.
81. Jain NN, Ohal CC, Shroff SK, Bhutada RH, Somani RS, Kasture VS, Kasture SB (2003) *Clitoria ternatea* and the CNS. *Pharmacol Biochem Behav* **75**: 529–536.
82. Vohora SB, Shah SA, Dandiya PC (1990) Central nervous system studies on an ethanol extract of *Acorus calamus* rhizomes. *J Ethnopharmacol* **28**: 53–62.
83. Zanoli P, Avallone R, Baraldi M (1998) Sedative and hypothermic effects induced by b-asarone, a main component of *Acorus calamus. Phytother Res* **12**(Suppl 1): S114–116.
84. Acuna UM, Atha DE, Ma J, Nee MH, Kennely EJ (2002) Antioxidant capacities of ten edible North American plants. *Phytother Res* **16**: 63–65.
85. Shukla PK, Khanna VK, Ali MM, Maurya RR, Handa SS, Srimal RC (2002) Protective effect of *Acorus calamus* against acrylamide induced neurotoxicity. *Phytother Res* **16**: 256–260.
86. Koo BS, Park KS, Ha JH, Park JH, Lim JC, Lee DU (2003) Inhibitory effects of the fragrance inhalation of essential oil from *Acorus gramineus* on central nervous system. *Biol Pharm Bull* **26**: 978–982.
87. Manyam BV (1999) Dementia in Ayurveda. *J Altern Complement Med* **5**: 81–88.
88. Dev S (1997) Ethnotherapeutics and modern drug development: the potential of Ayurveda. *Curr Sci* **73**: 909–928.

89. Naik GH, Priyadarsini KI, Naik DB, Satav JG, Mohan H (2002) Antioxidant activity and phytochemical analysis of the aqueous extract of *Terminalia chebula*. *Free Radic Biol Med* **33** (Suppl 1): 547.
90. Naik GH, Priyadarsini KI, Satav JG, Banavalikar MM, Sohoni DP, Biyani MK, Mohan H (2003) Comparative antioxidant activity of individual herbal components used in Ayurvedic medicine. *Phytochemistry* **63**: 97–104.
91. Cheng HY, Lin TC, Yu KH, Yang CM, Lin CC (2003) Antioxidant and free radical scavenging activities of *Terminalia chebula*. *Biol Pharm Bull* **26**: 1331–1335.
92. Manyam BV, Sanchez-Ramos JR (1999) Traditional and complementary therapies in Parkinson's disease. *Adv Neurol* **80**: 565–574.
93. Jog M (2004) Review: catechol O-methyl transferase inhibitors plus L-dopa and some surgical interventions improve Parkinson disease symptoms. *ACP J* **140**: 42–49.
94. Nagashayana N, Sankarankutty P, Nampoothiri MR, Mohan PK, Mohanakumar KP (2000) Association of L-DOPA with recovery following Ayurveda medication in Parkinson's disease. *J Neurol Sci* **176**: 124–127.
95. Manyam BV (1990) Paralysis Agitans and Levodopa in Ayurveda: Ancient Indian Medical Treatise. *Movement disord* **5**: 47–48.
96. Chaudhri RD (1996) *Herbal Drug Industry. A Practical Approach to Industrial Pharmacognosy*. Eastern Publishers, New Delhi, India.
97. Manyam BV, Dhanasekaran M, Hare TA (2004) Effect of antiparkinson drug HP-200 (*Mucuna pruriens*) on the central monoaminergic neurotransmitters. *Phytother Res* **18**: 97–101.
98. Tripathi YB, Upadhyay AK (2002) Effect of the alcohol extract of the seeds of *Mucuna pruriens* on free radicals and oxidative stress in albino rats. *Phytother Res* **16**: 534–538.
99. Barger JL, Walford RL, Weindruch R (2003) The retardation of aging by caloric restriction: its significance in the transgenic era. *Exp Gerontol* **38**: 1343–1351.
100. Yoshikawa M, Murakami T, Kadoya M, Li Y, Murakami N, Yamahara J, Matsuda H (1997) Medicinal foodstuffs. IX. The inhibitors of glucose absorption from the leaves of *Gymnema sylvestre* R. BR. (Asclepiadaceae): structures of gymnemosides a and b. *Chem Pharm Bull (Tokyo)* **45**: 1671–1676.
101. Hamilton JG, Sullivan AC, Kritchevsky D (1997) Hupolipidemic activity of (-)-hydroxycitrate. *Lipids* **12**: 1–9.
102. Lane MA, Mattison J, Ingram DK, Roth GS (2002) Caloric restriction and aging in primates: Relevance to humans and possible CR mimetics. *Microsc Res Tech* **59**: 335–338.
103. Kapoor LD (1990) *Handbook of Ayurvedic Medicinal Plants*. CRC, Press, Boca Raton.
104. Manjunatha S, Jaryal AK, Bijlani RL, Sachdeva U, Gupta SK (2001) Effect of Chyawanprash and vitamin C on glucose tolerance and lipoprotein profile. *Ind J Physiol Pharmacol* **45**: 71–79.
105. Ghosal S, Tripathi VK, Chauhan S (1996) Active constituents of *Emblica officinalis*: part 1-the chemistry and antioxidative effects of two hydrolysable tannins, Emblicanin A and B. *Ind J Chem* **35B**: 941–948.
106. Pathania V, Syal N, Hundal MK, Khanduja KL (1998) Geriforte stimulates antioxidant defense system. *Ind J Exp Biol* **36**: 414–417.

107. Singh B, Sharma SP, Goyal R (1994) Evaluation of Geriforte, an herbal geriatric tonic, on antioxidant defense system in Wistar rats. *Ann N Y Acad Sci* **30717**: 170–173.
108. Dubey GP, Agarwal A, Agarwal S, Srivastava VK, Udupa KN (1984) Antistress properties of an indigenous compound and its significance in the management of psychosomatic disorders. *Asian Med J* **27**: 723–727.
109. Dwivedi C, Sharma HM, Dobrowski S, Engineer FN (1991) Inhibitory effects of Maharishi-4 and Maharishi-5 on microsomal lipid peroxidation. *Pharmacol Biochem Behav* **39**: 649–652.
110. Cullen WJ, Dulchavsky SA, Devasagayam TP, Venkataraman BV, Dutta S (1997) Effect of Maharishi AK-4 on H2O2-induced oxidative stress in isolated rat hearts. *J Ethnopharmacol* **56**: 215–222.
111. Vohra BPS, Sharma SP, Kansal VK (2001) Maharishi Amrit Kalash, an ayurvedic medicinal preparation, enhances cholinergic enzymes in aged guinea pig brain. *Ind J Exp Biol* **39**: 1258–1262.
112. Vohra BPS, Sharma SP, Kansal VK (1999) Maharishi Amrit Kalash rejuvenates ageing central nervous system's antioxidant defence system: an *in vivo* study. *Pharmacol Res* **40**: 497–502.
113. Vohra BPS, Sharma SP, Kansal VK, Gupta SK (2001) Effect of Maharishi Amrit Kalash an ayurvedic herbal mixture on lipid peroxidation and neuronal lipofuscin accumulation in ageing guinea pig brain. *Ind J Exp Biol* **39**: 355–359.
114. Vohra BPS, James TJ, Sharma SP, Kansal VK, Chudhary A, Gupta SK (2002) Dark neurons in the ageing cerebellum: their mode of formation and effect of Maharishi Amrit Kalash. *Biogerontol* **3**: 347–354.
115. Vohra BPS, Sharma SP, Kansal VK (2001) Effect of Maharishi Amrit Kalash on age dependent variations in mitochondrial antioxidant enzymes, lipid peroxidation and mitochondrial population in different regions of the central nervous system of guinea-pigs. *Drug Metabol Drug Interact* **18**: 57–68.
116. Franklin CA (1988) *Modi's Medical Junspiudence and Toxicology*, 21st ed. NM Tripathi Pvt. Ltd, Mumbai, India.
117. Sen SP, Khosla RL (1968) Effect of Sodhana on the toxicity of aconite (vatsnava). *Current Med Pract* **12**: 694.
118. Thorat S, Dahanulkar SA (1991) Can we dispense with ayurvedic Somskaras? *J Postgrad Med* **37**: 157–159.
119. Gogate VM (1962) *Drvyaguna Vigyan, 1st ed.* Continental Prakashan, Pune, India.
120. Swami B (1985) *Rasadarpan*. Swami Publication, Patiyala, India.
121. Bhavaprakash Nighantu Karpooradi vargu (1969) *Chaulkhamba Sanskrit Samsthan*. Varanasi, India.
122. Gune G (1934) *Ayurvediya Aushadhi Gunadharma Shastra, Siddhaushadhi, part IV, 2nd ed.* Mohan Mandir, Ahmadnagar, India.
123. Dahanulkar SA, Kapadia AB, Karandikar SM (1982) Influence of trikatu on rifampicin bioavailability. *Indian Drugs* **82**: 271–273.

15

Alzheimer's Disease: Current and Future Treatments

Umesh Kumar
Neurology and GI Centre of Excellence for Drug Discovery
GlaxoSmithKline, New Frontiers Science Park, Harlow
Essex, CM19 5AW, United Kingdom
Email: Umesh_2_Kumar@gsk.com

ALZHEIMER'S DISEASE

Alzheimer's disease (AD), first described in 1906 by Alois Alzheimer, is a progressive neurodegenerative disease that accounts for most cases of dementia seen in the elderly population.[1] Clinical manifestations of the disease start with minor lapses in episodic memory. As the disease progresses, problems with general cognitive functions such as intellectual abilities, memory, executive functions and speech become more common. The cognitive deficit leads to severe personality changes characterized by agitation, depression and social withdrawal. Over a period of year's the condition worsens, resulting in complete immobility, with patients becoming totally dependent on their caregivers for social care.

In the absence of a proven biological marker, the diagnosis of AD remains based on the clinical judgment that the patient's cognitive function has declined from the past level of ability. An internationally agreed criterion for clinical diagnosis of AD includes a detailed history, functional measures of decline such as instrumental activity of daily living scales, mental status tests, Clinical Dementia Rating (CDR), Disability assessment for Dementia (DAD), neuropsychological evaluation, neurological and psychiatric examinations, blood tests, and brain imaging. The accuracy of diagnosis of probable AD is now more than 90% based on autopsy confirmation.

The risk factors for AD include age, genetic polymorphism, Down's syndrome, abnormal protein processing, neurotransmitter deficit, oxidative stress, head trauma, and environmental toxins, e.g. heavy metals. The interaction over time of these genetic and nongenetic risk factors with the biology of aging brain leads to development of the AD process.

It is estimated that around 15 million people worldwide are suffering from AD. The figure is expected to increase significantly over the next 50 years due to increasing life expectancy. Every year 1% of the people over the age of 65 years and 6–8% over the age of 85 are diagnosed with AD in the developed world. A higher proportion of women are affected with AD compared to men. Increase in prevalence of AD with age suggests that every person is likely to develop AD should they live long enough.[2,3]

Neuropathological Features of AD

Neuropathological features of AD can only be studied from post-mortem specimens. Pathological hallmarks of AD include senile plaques and neurofibrillary tangles, neuronal atrophy and cortical neurodegeneration.[4] The senile plaques are extracellular proteinaceous deposits of amyloid-beta (Abeta) peptides. The senile plaques are considered to evolve over a long period of time and their fibrillar nature is due to aggregated 40–42 amino acid long amyloid peptides. Besides Abeta peptides, plaques contain several other components. Dystrophic neurites, activated microglia and reactive astrocytes are all seen near the plaques. Aggregated amyloid fibrils and inflammatory mediators secreted by microglial and astrocytic cells contribute to neuronal dystrophy. Neurofibrillary tangles consist of paired helical filaments which are composed of hyperphosphorylated microtubule associated protein tau.[5] Presence of both plaques and tangles is used as a definitive criterion for diagnosis of AD.

Neuronal death seen in brain is another pathologic hallmark of AD. Certain populations of neurons tend to be lost selectively, and it has been proposed that the loss of synaptic density is likely to have a more immediate relationship to dementia in AD than Abeta accumulation.

AD Pathogenesis

Several hypotheses have been proposed to describe the pathogenesis of AD. The pathophysiologic abnormalities at the anatomical, cellular, and molecular levels support the view that a variety of mechanisms may contribute to AD. The clinical phenotype of AD could be a cumulative effect of all these events. There is no compelling evidence that these mechanisms are mutually exclusive, however, over last 10 years a dominant mechanism has been proposed by the "amyloid hypothesis".

According to another school of thought, tau-associated pathology is the underlying cause of AD pathogenesis. Both these hypotheses are described below. The role of amyloid or tau as a primary cause of neurodegeneration has been debated by two rival groups (Baptists and tauists) over several years, however, the controversy has now been settled due to some recent observations suggesting a link between the two hypotheses but several questions still remain unanswered.[6]

Amyloid Hypothesis

According to this hypothesis, accumulation of Abeta in the brain is the primary influence driving AD pathogenesis.[7] Abeta peptide is produced by proteolytic cleavage of a membrane bound amyloid precursor protein (APP) by two proteases called β and γ-secretases. Under certain circumstances Abeta production is enhanced by changes in activities of both β and γ secretases which leads to a cascade of events including neurofibrillary tangles and cell death. The strongest evidence in favour of this hypothesis is provided by familial cases of AD. Autosomal dominant mutations in the genes for APP, Presenilin-1 (PS1) and Presenilin-2 (PS2) cause early onset familial AD (FAD) by directly increasing synthesis of the toxic Abeta42 peptide. Transgenic mice over expressing Abeta display pathological features of AD such as age specific deposition of Abeta in brain. The other neuropathological characteristics of AD, such as astrocytosis, neuritic dystrophy, and microgliosis are also seen in these animals, however, no neurofibrillary tangles or neurodegeneration is observed. Further genetic evidence is provided by chromosome-21 (C21) trisomy seen in patients with Down's syndrome. C-21 harbours APP gene and one extra copy of APP on C21 results in over production of Abeta. Patients with Down's syndrome develop all the characteristic signs of AD earlier in their lives suggesting that the Abeta accumulation is sufficient to cause symptoms.[8,9] Although the majority of AD cases appear to be sporadic, the strong association of Abeta42 with FAD makes a compelling argument for involvement of Abeta in the etiology of all forms of AD. However, the poor correlation between the concentrations and distribution of amyloid depositions and other parameters, including degree of dementia, loss of synapses, abnormalities of the cytoskeleton and neurodegeneration cannot be explained by this hypothesis. According to the Amyloid hypothesis, neurofibrillary tangles develop due to an imbalance between Abeta production and Abeta clearance. High levels of Abeta disrupt neuronal metabolic and ionic homeostasis and cause aberrant activation of kinases and/or inhibition of phosphatases. These alterations in kinase and phosphatase activities ultimately lead to hyperphosphorylation of tau and formation of neurofibrillary tangles.[10]

Tau Hypothesis

According to this hypothesis, neurofibrillary tangles formed primarily from abnormal aggregations of a microtubule-associated protein tau, interfere with nerve cell functioning by impairing axonal transport. The distribution of neurofibrillary tangles spreads as the severity of the AD increases. During the early stages of the disease, neurofibrillary tangles occur predominantly in the entorhinal region. Subsequently, neurofibrillary tangles appear in the hippocampus and nearby regions of the cortex and finally throughout the cortex. These regions possess a concentration of neurons that receive cholinergic input, and also show the greatest degree of degeneration.[11,12] Decreased levels of acetylcholine and other markers of cholinergic function are characteristically found and have been associated with the deficits in learning and memory seen in AD. Neurotransmitters, including serotonin, glutamate, norepinephrine, and somatostatin, are also decreased, and these changes may contribute to the behavioral abnormalities seen in AD.

In addition to the mechanisms described above, other mechanisms of AD pathogenesis include inflammation, oxidative stress, cerebrovascular stress, hypercholesterolemia, metabolic stress, active cell death and lack of neurotrophic support. A variety of genetic and environmental abnormalities can also contribute to AD, e.g. association of apolipoprotein E4 (apoE4) in familial and sporadic AD. In conclusion, AD is a heterogeneous disease with number of underlying mechanisms operating simultaneously, contributing to the ultimate phenotype.

Treatment

AD is a debilitating disease with serious socio-economic consequences. With the increasing number of AD patients world-wide and their high dependency, the social and economic impact of this disease is likely to increase exponentially. At present the direct and indirect cost of care of AD patients runs into billions of dollars in the developed world annually. There is no definitive cure for AD at present. Currently, only symptomatic treatment is given for cognitive benefit to patients with mild to moderate disease. The treatment becomes less effective as the disease progresses and withdrawal of treatment results in rapid cognitive decline. The currently available symptomatic treatments and potential disease modifying opportunities are described below.

Symptomatic Treatments

Cholinesterase Inhibitors

The gradual neuronal loss resulting in learning and memory disability is due to a decline in the cholinergic levels. Acetylcholine plays an important part in cognition and therefore maintaining its levels by reducing its degradation provides cognitive benefit. A number of inhibitors have been developed that can selectively inhibit acetylcholinesterase, an enzyme that degrades acetylcholine. Short-term clinical trials (3–6 months) with several different inhibitors in different patient groups have shown cognitive benefit to patients with mild to moderate AD. There is no evidence to suggest that such inhibitors alter the course of the underlying disease process. Four acetylcholinesterase inhibitors (Tacrine, Donepezil, Rivastigmine and Galantamine) have been approved by the FDA.[13,14]

Tacrine

Tacrine is the first acetylcholinesterase inhibitor that was approved for AD. Tacrine, given twice daily dose, was efficacious and well tolerated as a short-term therapy. Patients on long-term therapeutic doses required less nursing care than the patients treated with sub-optimal dose. However, gastrointestinal adverse effects and liver toxicity were reported as side effects. Due to these adverse effects Tacrine is now rarely used for AD.[14]

Donepezil

Donepezil, a noncompetitive reversible inhibitor of acetylcholinesterase, is given at doses of 5 or 10 mg a day. A number of randomized placebo controlled clinical trials have been conducted, in which the patients with mild to moderate AD were treated with donepezil, for a period ranging from 6 months to 3 years. The outcome measures in these trials were based on a range of clinical assessment protocols, e.g. activity of daily living (ADL), Mini-Mental State Examination (MMSE), Clinical Dementia Rating (CDR), Disability assessment for Dementia (DAD) and caregiver's time. Donepazil treatment resulted in reduction in functional decline, improved cognition and daily living, no change in MMSE score from baseline, and less reduction in global rating score compared to placebo. The main adverse events are mild gastrointestinal symptoms, which disappear eventually.[14] A recent publicly funded randomized double blind trial confirmed previous results that donepezil treatment provides modest benefit in secondary endpoints such as MMSE score and activity of daily living score over a period of first two years, however, it contradicted the previous claims that long-term donepezil treatment provides benefit in primary endpoints such as disability progression, institutionalization, formal care cost, unpaid caregiver time and adverse effects.[15]

Rivastigmine

Rivastigmine, a reversible cholinesterase inhibitor, inhibits both acetylcholinesterase and butyrylcholinesterase. In a 6-month open label as well as in a prospective, randomized, multicentre, double blind, placebo controlled international trial, Rivastigmine was found to be safe, well tolerated and effective.[16] Long-term treatment with Rivastigmine results in persistent inhibition of acetylcholinesterase and butrylcholinesterase and improved cognitive function in patients with mild AD compared to untreated AD or mild cognitive impairment (MCI) patients.[17] Rivastigmine has a possible cerebrovascular effect and is particularly effective in patients with AD and hypertension. This may be due to improved cerebral blood flow which correlates with cognitive status of patient.[18]

Galantamine

Galantamine is a reversible, competitive and selective inhibitor of acetylcholinesterase that allosterically modulates nicotinic acetylcholine receptors. This dual mechanism of action provided the rationale for a phase II trial of galantamine in AD. In several multicentre, randomized, parallel, double-blind, placebo-controlled trials galantamine 18, 24 and 36 mg/day administered for 3–6 months in patients with mild-to-moderate probable AD demonstrated significant improvement in cognition and global function. Galantamine is well tolerated at the lower doses of 18–24 mg/day and cognitive and daily functions are maintained for 12–36 months, suggesting that galantamine slows the clinical progression of AD. In some patients it shows mild, transient effects typical of cholinomimetic agents.[19]

In the absence of any comparative study of these inhibitors it is not possible to identify the most effective drug. In the future, more potent cholinesterase inhibitors with better safety profiles may be developed. Phenserine is a long-acting selective inhibitor of acetylcholinesterase with a preferential uptake by the brain. Its long duration of action, coupled with its short pharmacokinetic half-life, is claimed to reduce the dose and the body drug exposure.[20] Phenserine has a second mode of action whereby it binds to APP gene and reduces APP expression. Phenserine is currently in phase III clinical trial stage.[21]

NMDA Receptor Antagonist

Glutamate is the primary excitatory amino acid in human brain. Under physiological conditions glutamate activates number of receptors including N-methyl-D-aspartate (NMDA) receptor. Activation of NMDA receptor is associated with learning and memory formation. In AD and other pathological conditions excessive activation of NMDA receptor by glutamate may result in neurodegeneration. NMDA receptor antagonists have therapeutic potential in several central nervous system disorders, including neuroprotective treatment in chronic neurodegenerative diseases, and symptomatic treatment in other neurologic diseases. There is considerable evidence suggesting an excitotoxic component in AD pathogenesis. Neurochemical studies of AD brain show degeneration of glutamatergic pathways and decreased expression of NMDA receptor in hippocampal and cortical regions.[22] Targeting the glutamatergic system may help in reducing neurodegeneration and improving cognition.

Memantine

It is a low to moderate affinity, uncompetitive NMDA receptor antagonist, which has been shown to be effective in preclinical studies. Under physiological conditions, Memantine allows normal glutamatergic neurotransmission but under pathological conditions it inhibits excitotoxicity.[23] A number of clinical studies indicates Memantine to be safe, well tolerated and effective in moderate to severe AD. In a clinical trial, Memantine showed cognitive benefit and reduction in dependency in patients with AD and vascular dementia, and in another, reduced the cognitive decline in patients with moderate to severe AD.[24,25]

In a double blind randomized control trial, patients with moderate to severe AD receiving donepezil and memantine resulted in significantly better outcomes than placebo on measures of cognition, activities of daily living, global outcome and behavior, and was well tolerated. Memantine represents a new approach for the treatment of patients with moderate to severe AD.[26]

Nicotinic Receptor Agonists

Nicotinic receptors (NRs) belong to the group of polymeric receptors of the cell membrane and are key elements of cholinergic transmission. Numerous subtypes of NRs exist with the alpha4 beta2 and alpha7 types being encountered most frequently. Alpha 7 NRs have been proposed to exert a direct or indirect action on the mechanism of Abeta toxicity.

Nicotine has been reported to protect against Abeta-induced neuronal toxicity and death in rat cortical neurons. This neuroprotection can be blocked by dihydro-beta-erythroidine, an alpha4beta2 nicotinic receptor antagonist. Furthermore, incubation with cytisine, a selective alpha4beta2 nicotinic receptor agonist, can inhibit Abeta cytotoxicity. Deficiencies in NRs seem to play a role in Alzheimer's disease.[27]

Clinical studies suggest that nicotine may provide cognitive benefit,[28] however, its long-term use may induce desensitization of nicotinic receptors.[29] Allosteric modulation of NR can circumvent desensitization. This allosteric interaction amplifies the actions of ACh at post- and presynaptic NR. Allosteric modulation of NR could therefore produce significant therapeutic benefit in AD.[30] Galantamine, one of the most potent alpha4/beta2 NR allosteric modulators, maintains patient's level of cognitive and daily function for at least 1 year, which has not been reported for other AChE inhibitors.[30]

A number of other receptors like serotonin 5-hydroxytryptamine6 (5HT6),[31] 5HT4,[32] histamine-H3[33] and γ-aminobutyric acid (GABA)[34] are known to play a role in learning and memory. It is likely that a number of agonists, antagonists and allosteric modulators will be available and undergo trials in near future.

Disease Modifying Treatment

According to the amyloid hypothesis Abeta is central to the pathophysiology of AD. High levels of amyloid peptides, especially Abeta42, initiate aggregation and plaque formation in the areas of brain associated with learning and memory. Therapeutic strategies that lower Abeta formation, prevent aggregation, dissolve plaques or promote clearance from the brain should prove beneficial.

Inhibition of Amyloid Formation

Abeta is produced by two sequential cleavages of amyloid precursor protein (APP) by two proteases, called β- and γ-secretase. β-secretase first cleaves APP in the extracellular domain to release a large APP fragment called APP-β and generates a membrane bound carboxy terminal fragment. γ-secretase cleaves the membrane bound fragment within the transmembrane domain to release Abeta peptide. Both these proteases are excellent targets for disease modification.

β-secretase inhibitors

Beta secretase, a membrane-anchored aspartyl protease, initiates the cleavage of APP at the beginning of Abeta peptide. β-secretase knock out mice lack Abeta and are phenotypically normal, suggesting that therapeutic inhibition of β-secretase may be free of mechanism-based side effects. β-secretase null mice overexpressing human APP are rescued from Abeta-dependent hippocampal memory deficit which correlates with a reduction of amyloid peptides.[35] Due to the potential for disease modification in AD, a number of groups have been trying to develop β-secretase inhibitors. Other aspartyl proteases such as Renin and HIV-1 have provided a rich background for the rational design of potent and selective inhibitors. The elucidation of the crystal structure of β-secretase complexed with inhibitors has further helped in designing of several inhibitors. The β-secretase active site is more open and less hydrophobic than that of other aspartyl proteases.[36] Several peptide based β-secretase inhibitors have been described to date, however, all are relatively large molecules and are not drug-like.[37]

γ-secretase inhibitors

Gamma-secretase is a membrane protein complex with aspartyl protease activity that cleaves APP in its transmembrane domain to release Abeta and the APP intracellular domain (AICD). The identity of γ-secretase complex has been controversial. Identification of PS1 and PS2 as the possible active component of complex was established by genetic linkage studies in familial AD (FAD). Cleavage of APP by mutant presenilin results in the overproduction of amyloidogenic Abeta42. These mutations account for the majority of the cases of the FAD.[38] A number of co-factors have been identified such as nicastrin (Nct), a single transmembrane protein, presenilin enhancer protein-2 (PEN-2) and anterior pharynx defective protein-1 (APH-1). APH-1 stabilizes the presenilin holoprotein in the complex, whereas PEN-2 is required for endoproteolytic processing of presenilin and conferring γ-secretase activity to the complex. Nct undergoes a major conformational change during the assembly of the γ-secretase complex. The conformational change is directly associated with γ-secretase function.[39] Recently, various components of γ-secretase complex when co-expressed in yeast that lacks endogenous γ-secretase activity resulted in reconstitution of γ-secretase activity. This outstanding work finally confirmed presenilin (PS), nicastrin (Nct), APH-1 and PEN-2 as essential components of γ-secretase complex.[40] Since several paralogs and alternatively spliced variants of Presenilin and Aph-1 have been identified, γ-secretase may cleave several other membrane proteins, for example, Notch and Erb4, a receptor tyrosine kinase that regulates cell cycle.[41]

Gamma-secretase is an interesting but complex drug target that challenges classical thinking about proteolytic processing. The complete inhibition of γ-secretase activity is likely to result in serious side-effects. Elan Pharmaceuticals reported a novel class of compounds that reduce Abeta production by functionally inhibiting γ-secretase. Oral administration of one of these compounds, N-[N-(3,5-difluorophenacetyl)-L-alanyl]-S-phenylglycine t-butyl ester (DAPT), to mice transgenic for human APP(V717F) reduces brain levels of amyloid in a dose-dependent manner within 3 h.[42] Development of such novel functional γ-secretase inhibitors will enable a clinical examination of the Abeta hypothesis.

Recently retrospective epidemiological studies reported that patients on long-term non-steroidal anti-inflammatory drugs have reduced risk of AD.[43] When tested *in vitro* and *in vivo* for their Abeta lowering activity, 8 out of 13 NSAIDS and the enantiomers of flurbiprofen were found to be effective, indicating that these compounds directly target the γ-secretase complex.[44] R-flurbiprofen is currently in phase II clinical studies.

Rho-Rock pathway inhibitors

Recently, Rho-Rock pathway has been shown to regulate amyloid precursor protein processing *in vitro* and a subset of NSAIDs that inhibit Rho activity, reduce Abeta42. A selective Rock inhibitors (Y-27632) has also been shown to lower brain levels in a transgenic mouse model of AD.[45] Rho-Rock pathway is a novel therapeutic target for AD.

Inhibition of Abeta Aggregation

The aggregation of soluble Abeta peptide is viewed as a critical event in the pathophysiology of AD, preventing, altering, or reversing aggregation may be of therapeutic value.

Metal chelators

Binding of redox active transition metal ions like Cu^{2+} and Zn^{2+} to Abeta is thought to mediate its reversible aggregation and resistance to proteases. These metal ions are elevated in neocortex of AD patients especially in plaques. Chelating agents can inhibit the binding of these ions to Abeta, therefore these agents have potential therapeutic value. Clioquinol, a bioavailable Cu^{2+}/Zn^{2+} chelator, has been tested for its anti-aggregation activity both *in vitro* and *in vivo.* It inhibited and reversed Cu^{2+}/Zn^{2+} mediated aggregation of synthetic Abeta *in vitro* and solubilized Abeta in deposits in post-mortem AD brain samples. In a transgenic mouse model (APP2576) of AD, the oral administration of clioquinol for nine weeks was associated with significantly lower levels of aggregated Abeta accompanied by increased levels of soluble Abeta.[46]

In a clinical trial, 10 patients were given Clioquinol at 20 mg/day dose and 10 more given the same drug at 80 mg/day for 21 days each. Cerebrospinal fluid (CSF) investigations revealed a decrease in Tau protein and growth-associated protein (GAP43). These proteins are increased in AD and considered stable markers. The levels of CSF-Tau protein correlated positively and significantly with the serum levels of copper and also with the serum copper/zinc ratio. Clinical assessment showed slight improvement after 3 weeks treatment with clioquinol in this open study.[47] In another randomized phase II trial, treatment with metal protein attenuating compound (MPAC, clioquinol) showed significant cognitive benefit in treated patients compared to placebo controls. Plasma levels of Abeta 42 decreased in the clioquinol group and increased in the placebo group.[48] The results support targeting the interactions of Cu^{2+} and Zn^{2+} with Abeta as a novel therapeutic approach for the prevention and treatment of AD.

β-sheet breaker peptides

Several neurodegenerative diseases and systemic amyloidosis are thought to arise from the misfolding and aggregation of an underlying protein. In

AD, blockade of the early steps involving the pathological conversion of the soluble peptide into the abnormally folded oligomeric intermediate precursor of the amyloid fibrils is an attractive therapeutic strategy. β-sheet breaker peptides are small synthetic peptides that are homologous to regions involved in the aggregation. In a transgenic mouse model of AD (PDAPP) administration, one such peptide (iAb5p) subcutaneously blocked amyloid aggregation.[49] The central hydrophobic domain of amyloid peptide interacts with glycosaminoglycan (GAG) and the interaction is involved in aggregation and deposition. GAG mimetics have been developed and tested *in vivo* for their anti-aggregation activity. Two such mimetics NC-758 (Cerebril) and NC-531 (Alzemed) are in phase II clinical trials. It is yet unclear whether the inhibition of the defective folding of A-beta peptide is beneficial for the treatment of AD.

Improved Clearance of Abeta

Abnormal accumulation of Abeta in brain is the driving factor for AD pathogenesis therefore its clearance is likely to be of primary therapeutic benefit.

Immunization

Active immunization with aggregated Abeta or peripherally administered anti-amyloid antibodies reduce amyloid associated pathology and cognitive decline in a transgenic mouse model of AD. A numbers of mechanisms have been proposed for clearance of amyloid from brain. One of the proposed mechanisms by which antibodies may reduce brain amyloid is that the antibodies cross the blood brain barrier (BBB), bind to amyloid plaques and activate microglial phagocytosis of immune complexes. Another proposed mechanism is that there is a dynamic equilibrium of Abeta between brain and periphery and the antibodies in the periphery can act as sink, capturing Abeta in the blood stream and indirectly reducing the Abeta burden in the brain by driving the clearance of peptide from brain to plasma. Another proposed mechanism is that anti-Abeta antibodies directed against specific epitopes might protect against neurotoxicity by inhibiting aggregation of Abeta and by disaggregating already established aggregates or plaques.[50]

Recently, active vaccination with Abeta 42 (AN-1792) entered clinical trials for possible treatment of AD. In phase I, a single dose was found to be safe and consequently, a phase II trial of AN1792 was initiated. The trial was terminated after a small percentage of patients developed signs of meningoencephalitis.[51] A post-mortem study of one of the patients, who died due to unrelated causes, revealed presence of activated T-lymphocytes suggesting the adverse effects seen in some patients might be due to the cellular immune response rather than antibody response.[52] In a cohort of 30 patients from AN-1792 trial a significant correlation was found between antibody response and cognitive decline. Vaccination using smaller fragments of Abeta conjugated to T helper epitopes and various routes of administration are still being pursued in pre-clinical models to study safety and efficacy of this approach.[53]

Compounds unrelated to antibodies have been used recently to test the peripheral sink mechanism. Peripheral treatment with gelsolin or ganglioside (GM-1) reduced the level of Abeta in the brain of transgenic mouse model of AD suggesting that sequestration of plasma Abeta could reduce or prevent brain amyloidosis. Gelsolin, a secretory protein and GM-1, a ganglioside, are known to bind with Abeta with high affinity.[54,55] Future studies with high affinity Abeta binding small molecules may provide further validation of this approach and amyloid hypothesis.

Neprilysin

The steady-state level of Abeta represents a balance between its biosynthesis from the APP and its catabolism by a variety of proteolytic enzymes like neprilysin (NEP), endothelin-converting enzyme, insulin-degrading enzyme, angiotensin-converting enzyme and plasmin. Neprilysin (NEP) is a major Abeta peptide-degrading enzyme as shown by higher Abeta peptide levels in hippocampus, cortex, thalamus/ striatum, and cerebellum of an NEP knockout mouse and by reduction in amyloid load in APP transgenic mice treated with viral vector expressing NEP.[56] Expression of Neprilysin is down-regulated with age and correlates negatively with amyloid deposition in APP transgenic mice. Therapeutic strategies aimed at promoting Abeta degradation may provide a novel approach to treat AD.[57]

Insulin degrading enzyme (IDE)

Epidemiological evidence indicates that insulin resistance in type II diabetes is associated with an increased relative risk for AD. Both genetic linkage and allelic association in the IDE region of chromosome 10 have been reported in families with late-onset AD. This may link diabetes with AD.[58] Naturally occurring IDE missense mutations that result in partial loss of function have been shown to associate with increased levels of insulin and Abeta in plasma.[59] Insulin resistance promotes amyloidosis in APP transgenic mice that corresponds with increased γ-secretase activities and decreased insulin degrading enzyme (IDE) activities. Apparent inter-relationship of insulin resistance to brain amyloidosis is due to a functional decrease in insulin receptor (IR)-mediated signal transduction in the brain. Decreased signal transduction positively correlate with the generation of brain C-terminal fragment of APP, an index of γ-secretase activity, in the brain of insulin-resistant transgenic mice.[60]

Receptor for advanced glycation end products (RAGE)

Blood brain barrier (BBB) regulates entry of plasma derived Abeta into central nervous system and clearance of brain derived Abeta into periphery through several receptors or carrier proteins such as low density lipoprotein related protein-1 (LRP-1), megalin, cubulin and receptor for advanced glycation end products (RAGE).[61] Alterations in the permeability of the BBB may lead to accumulation of Abeta in brain. RAGE is up-regulated in AD brain vasculature[62] and increases in transgenic mouse model of AD with age.[63] RAGE mediated transport of circulating Abeta across BBB is an important factor in the pathogenesis of cerebrovascular amyloidosis as shown by lack of Abeta deposition in transgenic mice treated with soluble RAGE.[64] RAGE knock-out mice are viable, suggesting that blockade of RAGE with immunotherapeutic or small molecule inhibitor may be an important therapeutic opportunity for developing a treatment for AD.[65]

Tau Phosphorylation Inhibitors

Tau is a microtubule associated peptide that is involved in axonal transport. This transport involves repeated phosphorylation and dephosphorylation of tau. Neurofibrillary tangle formation may be due to an imbalance of this process. Glycogen synthase-3 (GSK-3) and cyclin dependent kinase-5 (Cdk-5) are current targets to reduce tau phosphorylation. Certain mood stabilizers such as lithium and valproate may have complex neuroprotective effects including inhibition of GSK-3. Lithium was recently found to reduce amyloid in a mouse model of AD. Valproate will be studied in a multicenter clinical trial in patients with AD.[66,67]

Other Therapies

Cholesterol-Lowering Therapies

A number of epidemiological studies suggest that high cholesterol levels increase the risk of AD significantly, however, there are others which did not report a link.[68] Numerous laboratory studies implicate cholesterol in the process of Abeta production and accumulation. Changes in APP processing by cholesterol may explain how ApoE4 allele increases risk of developing AD.[69] Cholesterol is present in the dense cores of senile plaques both in humans and transgenic mice suggesting that cholesterol plays an important role in the formation and/or progression of senile plaques.[70] A cholesterol-rich diet increases intracellular Abeta and levels of Abeta strongly correlate with the levels of cholesterol in plasma and CNS.[71] Free cholesterol in neurofibrillary tangle-bearing neurons is higher than those of adjacent tangle-free neurons.[72] Genetic heterogeneity in ApoE allele is associated with higher risk of AD. People expressing ApoE4 have higher circulating levels of cholesterol and are at greater risk than people with ApoE2 or ApoE3. ApoE4 accelerate amyloid deposition and promotes Abeta aggregation in cholesterol rich lipid rafts.[73] It is now believed that cholesterol-lowering therapies will be of value as disease modifying agents. Epidemiological studies have shown that statins are associated with a decreased risk of developing AD.[74] These observations require both preclinical and clinical validation. The former involves testing statins in one or more animal models of AD in order to establish relative efficacy and disease features affected by treatment. The latter requires prospective, randomized, placebo controlled trials to evaluate the effect of statin treatment on cognitive and AD biomarker outcomes. High doses of simvastatin show a strong and reversible reduction of cerebral Abeta42 and Abeta40 levels in the cerebrospinal fluid and brain homogenate of transgenic and guinea pig models.[75] In most of the clinical trials, statins have shown no effect on Abeta levels in plasma or cerebrospinal fluid. In several randomized, placebo-controlled, double-blind clinical trials, statins such as simvastatin or atorvastatin did not alter cerebrospinal fluid levels of Abeta40 and Abeta42.[76] However, in a double-blind, randomized, placebo-controlled study, lovastatin reduced serum Abeta levels compared to the baseline.[77] Future controlled clinical trials may help in explaining the contradiction seen in epidemiological and most of the clinical studies.

Anti-inflammatory Therapies

Inflammation clearly occurs in AD brain with full complexity of local inflammatory responses.[78] Microglia, the predominant immune cells in the brain are consistently associated with senile plaques in AD brain and may play a pivotal role in neuroinflammation. There is a considerable body of evidence indicating that the microglia activated by β-amyloid, neurofibrillary tangles or degenerating neurons are the primary source of pro-inflammatory cytokines IL1, IL6, TNF α; chemokines such as MIP-1 α, MCP-1, IL5, IL8; superoxide free radicals and neurotoxic substances. Activated microglia help clear Abeta deposits and thus prevent their harmful effects. Nevertheless, chronic activation of microglia may contribute to neurodegeneration. Patients who show Abeta deposition and neurofibrillary tangle, but limited inflammation, have no history of dementia. Transgenic mice overexpressing various inflammatory mediators show AD like pathology as well as cognitive deficit. The animals show decreased acetylcholine production, neurodegeneration, learning deficit and memory impairment in dose and age-related manner. Based on observations from neuropathology in AD and animal experimentation, inflammation has been considered a therapeutic target for AD. Inflammation is not a primary event in AD and cannot be considered to have a causal role, however, it may add to the progression of the disease. Epidemiological evidence suggests that anti-inflammatory therapies may reduce the risk of developing AD.[79] However, clinical trial data are discouraging for patients with established AD. In a randomized controlled trial, rofecoxib or naproxen showed no effect on cognitive decline. One early trial with indomethacin saw some benefit; subsequent trials with rofecoxib, celecoxib, diclofenac, hydroxychloroquine, naproxen and prednisolone have not shown significant benefits.[80] Inflammatory pathways contributing to AD pathology never occur in isolation but act concurrently. Chronic activation of inflammatory responses may necessitate therapies that target more than one pathway simultaneously to achieve clinical benefits. This may explain why clinical trials with anti-inflammatory drugs have not shown any beneficial effects.

Antioxidants

Free-radical oxidative stress, particularly of neuronal lipids, proteins and DNA, is extensive in those AD brain areas in which Abeta is abundant. Abeta-induced oxidative stress leads to neurodegeneration in AD brain. Abeta leads to neuronal lipid peroxidation, protein oxidation and DNA oxidation by means that are inhibited by free-radical antioxidants. Catecholamines involved with oxidation (monoamine oxidase) are abundant in AD brain where as antioxidant enzymes like superoxide dismutase, catalase, glutathione peroxidase and glutathione reductase are reduced. Therefore, the risk of Alzheimer disease might be reduced by intake of antioxidants that counteract the detrimental effects of oxidative stress.[81]

Vitamins and Selegiline

In a population-based, prospective cohort study conducted in the Netherlands, high dietary intake of vitamin C and vitamin E was associated with lower risk of Alzheimer disease.[82] In a cross-sectional and prospective study of dementia, use of vitamin E and vitamin C supplements in combination was associated with reduced prevalence and incidence of AD. One study found that vitamin E from food, but not other antioxidants, may be associated with a reduced risk of AD. Unexpectedly, this association was observed only among individuals without the APOE epsilon 4 allele. However, another study found that the intake of carotenes and vitamin C, or vitamin E in supplemental or dietary (nonsupplemental) form or in both forms, was not related to a decreased risk of AD. In a double-blind, placebo-controlled, randomized, multicenter trial in patients with Alzheimer's disease of moderate severity, a selective monoamine oxidase inhibitor selegiline (10 mg a day) or alpha-tocopherol (vitamin E) 2000 IU a day slowed the progression of disease. A meta-analysis of the published trials on treatment with selegiline showed little evidence of improvement with selegiline in the short-term in cognition and activities of daily living, which was clinically insignificant. Flavonoids, powerful antioxidants present in wine, tea, fruits and vegetables show inverse correlation with the risk of dementia.[83]

Ginkgo biloba

Ginkgo biloba (GbE) extracts have played a crucial role in Chinese herbal medicine for many centuries. Previous studies have suggested the clinical efficacy of GbE in patients with dementia, cerebral insufficiency, or related cognitive decline. However, the effectiveness of GbE in AD is controversial.[84] GbE preparations are approved in many European countries for the treatment of dementia syndromes including AD.

Neurotrophic Factors

Nerve growth factor (NGF)

Transgenic mice expressing anti-NGF antibodies show AD like neurodegenerative phenotype which includes plaques, neuronal loss, cholinergic deficit, and tau hyperphosphorylation, associated with neurofibrillary pathology suggesting a direct link between NGF signaling and abnormal processing of amyloid precursor protein.[85] NGF therapy might reduce degeneration of cholinergic neurons. In a short-term clinical trial, intracerebroventricular infusion of NGF in AD patients resulted in slight cognitive benefit. Long-term therapy may provide clear benefit but association of the intraventricular route of administration with negative side effects appear to outweigh the positive effects.[86] Due to lack of brain penetration of NGF, orally bioavailable NGF synthesis stimulators, like idebenone and propentofylline, have been tested in pre-clinical models and found to restore age-associated NGF loss. The results suggest that the use of NGF synthesis stimulators may provide a novel therapeutic approach to cholinergic dysfunction.[87]

Brain-derived neurotrophic factor (BDNF)

BDNF is a prosurvival factor induced by cortical neurons that is necessary for survival of cholinergic neurons of the basal forebrain, hippocampus and cortex.[88] The reduction of BDNF early on in AD could weaken synaptic encoding strength of hippocampus and make it vulnerable to degeneration. A single nucleotide polymorphism, val66met, in the BDNF gene has been associated with poor episodic memory and abnormal hippocampal activation in a cohort of 641 human subjects.[89]

In a clinical trial, neurotrophic compound cerebrolysin was found to be well tolerated and resulted in significant improvements in the global score and activities of daily living in patients with AD.[90] In pre-clinical studies, Neotrofin, a purine derivative, was found to stimulate neuritogenesis, the production of neurotrophic factors and to have memory enhancing properties.[91] In a phase I, randomized, double blind, placebo-controlled clinical trial, neotrofin was found to be safe and well tolerated in healthy elderly volunteers.[92] Other approaches like intraparenchymal administration, tissue transplantation and use of viral vectors to deliver neurotrophic factors are underway.

Estrogen

Estrogen may have cholinergic, neurotrophic and neuroprotective effects and may enhance cognitive function.[93] Observational studies have suggested that postmenopausal hormone treatment may improve cognitive function, but data from randomized clinical trials have been sparse and inconclusive. Recently, in a randomized controlled clinical trial of postmenopausal women, estrogen plus progestin did not improve cognitive function but increased risk of clinically meaningful cognitive decline.[94] Raloxifene, a selective estrogen receptor modulator (SERM) that produces both estrogen-agonistic effects on bone and lipid metabolism and estrogen-antagonistic effects on uterine endometrium and breast tissue, has been tested for its safety and efficacy in AD patients. In a randomized double-blind osteoporosis treatment trial, Raloxifene showed no cognitive benefit after 12 months treatment.[95]

Conclusion

Alzheimer's disease is a progressive neurodegenerative disease that accounts for most cases of dementia seen in the elderly. The socio-economic burden of the disease is likely to increase due to increasing life-expectancy. Early clinical diagnosis and timely treatment of AD patients can maintain patient's quality of life and prevent high costs associated with it. There is no definitive cure for AD and the currently available symptomatic treatments show limited efficacy.

It is clear that there are number of strategies to try and intervene in the process of AD. A number of mechanistic targets for AD have been validated by using *in vitro* and *in vivo* systems and several approaches of disease modification are being pursued in the pharmaceutical industry as well as in academia. Recently, active and passive immunization have been successful in clearing the amyloid peptide from brain of transgenic mouse model of AD. However, in a clinical trial, serious adverse effects of active vaccination resulted in early termination of the trial. Other therapeutic opportunities are provided by physiological responses observed in patients such as oxidative stress and neuroinflammation. Epidemiological and clinical trial studies with antioxidants and anti-inflammatory agents have been contradictory. Further controlled trials are needed to address these issues. In the absence of any disease modifying therapy, symptomatic treatment targeting the cholinergic system is the only current option for treatment of AD. A combination of multiple agents is likely to be the option for treatment of AD in the future.

References

1. Linn RT, Wolf PA, Bachman DL, *et al.* (1995) The 'preclinical phase' of probable Alzheimer's disease. A 13-year prospective study of the Framingham cohort. *Arch Neurol* **52**: 485–490.
2. Kukull WA, Ganguli M (2000) Epidemiology of dementia: concepts and overview. *Neurol Clin* **18**: 923–950.
3. Hebert LE, Scherr PA, Bienias JL, *et al.* (2003) Alzheimer disease in the US population: prevalence estimates using the 2000 census. *Arch Neurol* **60**: 1119–1122.
4. Dickson DW (1997) The pathogenesis of senile plaques. *J Neuropathol Exp Neurology* **56**: 321–339.
5. Grundke-Iqbal I, Iqbal K, Tung YC, *et al.* (1986) Abnormal phosphorylation of the microtubule associated protein t (tau) in Alzheimer cytoskeletoal pathology. *PNAS USA* **83**: 4913–4917.

6. Mudher A, Lovestone S (2002) Alzheimer's disease-do tauists and baptists finally shake hands? *Trends Neurosci* **25**: 22–26.
7. Hardy JA, Higgins GA (1992) Alzheimer's disease: the amyloid cascade hypothesis. *Science* **256**: 184–185.
8. Tanzi RE, Bertram L (2001) New frontiers in Alzheimer's disease genetics. *Neuron* **32**: 181–184.
9. Phinney AL, Horne P, Yang J, *et al.* (2003) Mouse models of Alzheimer's disease: the long and filamentous road. *Neurol Res* **25**: 590–600.
10. Oddo S, Caccamo A, Kitazawa M, *et al.* (2003) Amyloid deposition precedes tangle formation in a triple transgenic model of Alzheimer's disease. *Neurobiol Aging* **24**: 1063–1070.
11. Mandelkow EM, Mandelkow E (1998) Tau in Alzheimer's disease. *Trends Cell Biol* **8**: 425–427.
12. Goedert M (1996) Tau protein and the neurofibrillary pathology of Alzheimer's disease. *Ann NY Acad Sci* 777: 121–131.
13. Schneider LS, Tariot PN (2003) Cognitive enhancers and treatments for Alzheimer's disease. In: Tasman A, Kay J, Lieberman JA (eds.) *Psychiatry*, 2nd edition John Wiley and Sons, London.
14. Ibach B, Haen E (2004) Acetylcholinesterase inhibition in Alzheimer's Disease. *Curr Pharm Des* **10**: 231–251.
15. Courtney C, Farrell D, Gray R, *et al.* (2004) Long-term donepezil treatment in 565 patients with Alzheimer's disease (AD2000): randomised double-blind trail. *Lancet* **363**: 2105–2115.
16. Bilikiewicz A, Opala G, Podemski R, *et al.* (2002) An open-label study to evaluate the safety, tolerability and efficacy of rivastigmine in patients with mild to moderate probable Alzheimer's disease in the community setting. *Med Sci Monit* **8**: 9–15.
17. Almkvist O, Darreh-Shori T, Stefanova E, *et al.* (2004) Preserved cognitive function after 12 months of treatment with rivastigmine in mild Alzheimer's disease in comparison with untreated AD and MCI patients. *Eur J Neurol* **11**: 253–261.
18. Vennerica A, Shanks MF, Staff RT, *et al.* (2002) Cerebral blood flow and cognitive responses to rivastigmine treatment in Alzheimer's disease. *Neuroreport* **13**: 83–87.
19. Raskind MA, Peskind ER, Truyen Lb, *et al.* (2004) The cognitive benefits of galantamine are sustained for at least 36 months: a long-term extension trial. *Arch Neurol* **61**: 252–256.
20. Greig NH, De Micheli E, Holloway HW, *et al.* (2000) The experimental Alzheimer drug phenserine: preclinical pharmacokinetics and pharmacodynamics. *Acta Neurol Scand Suppl* **176**: 74–84.
21. Shaw KT, Utsuki T, Rogers J, *et al.* (2001) Phenserine regulates translation of beta-amyloid precursor protein mRNA by a putative interleukin-1 responsive element, a target for drug development. *PNAS USA* **98**: 7605–7610.
22. Palmer GC (2001) Neuroprotection by NMDA receptor antagonists in a variety of neuropathologies. *Curr Drug Targets* **2**: 241–271.
23. Parsons CG, Danysz W, Quack G (1999) Memantine is a clinically well tolerated N-methyl-D-aspartate (NMDA) receptor antagonist — a review of preclinical data. *Neuropharmacology* **38**: 735–767.
24. Reisberg B, Doody R, Stoffler A, *et al.* (2003) Memantine in moderate-to-severe Alzheimer's disease. *N Engl J Med* **348**: 1333–1341.

25. Winblad B, Poritis N (1999) Memantine in severe dementia: results of the 9M-Best Study (Benefit and efficacy in severely demented patients during treatment with memantine). *Int J Geriatr Psychiatry* **14**: 135–146.
26. Tariot PN, Farlow MR, Grossberg GT, *et al.* (2004) Memantine treatment in patients with moderate to severe Alzheimer disease already receiving donepezil: a randomized controlled trial. *JAMA* **291**: 317–324.
27. Bourin M, Ripoll N, Dailly E (2003) Nicotinic receptors and Alzheimer's disease. *Curr Med Res Opin* **19**: 169–177.
28. van Duijn CM, Hofman A (1991) Relation between nicotine intake and Alzheimer's disease. *BMJ* **302**: 1491–1494.
29. Marks MJ, Stitzel JA, Collins AC (1987) Influence of kinetics of nicotine administration on tolerance development and receptor levels. *Pharmacol Biochem Behav* **27**: 505–512.
30. Maelicke A (2000) Allosteric modulation of nicotinic receptors as a treatment strategy for Alzheimer's disease. *Dement Geriatr Cogn Disord* **11** (Suppl 1): 11–18.
31. Reavill C, Rogers DC (2001) The therapeutic potential of 5-HT6 receptor antagonists. *Curr Opin Investig Drugs* **2**: 104–109.
32. Lezoualc'h F, Robert SJ (2003) The serotonin 5-HT4 receptor and the amyloid precursor protein processing. *Exp Gerontol* **38**: 159–166.
33. Bongers G, Leurs R, Robertson J, Raber J (2004) Role of H3-receptor-mediated signaling in anxiety and cognition in wild-type and Apoe$^{-}/^{-}$ mice. *Neuropsychopharmacology* **29**(3): 441–449.
34. Maubach K (2003) GABA(A) receptor subtype selective cognition enhancers. *Curr Drug Targets CNS Neurol Disord* **2**: 233–239.
35. Ohno M, Sametsky E, Younkin N, *et al.* (2004) BACE1 deficiency rescues memory deficits and cholinergic dysfunction in a mouse model of Alzheimer's disease. *Neuron* **41**: 27–33.
36. Hong L, Turner RT, Koelsch G, *et al.* (2002) Crystal structure of memapsin 2 (b-secretase) in complex with an inhibitor OM00-3. *Biochemistry* **41**: 10963–10967.
37. Hussain I (2004) The potential for BACE1 inhibitors in the treatment of Alzheimer's disease. *IDrugs* 7: 653–658.
38. Hardy J (2003) Alzheimer's disease: genetic evidence point to a single pathogenesis. *Ann Neurol* **54**: 143–144.
39. De Strooper B (2003) Aph-1, Pen-2, and Nicastrin with Presenilin generate an active gamma-Secretase complex. *Neuron* **38**: 9–12.
40. Edbauer D, Winkler E, Regula JT, *et al.* (2003) Reconstitution of gamma-secretase activity. *Nat Cell Biol* **5**: 486–488.
41. Kopan R, Ilagan M (2004) γ-Secretase: proteosome of the membrane? *Nature Reviews Molecular Cell Biology* **5**: 499–504.
42. Dovey HF, John V, Anderson JP, *et al.* (2001) Functional gamma-secretase inhibitors reduce beta-amyloid peptide levels in brain. *J Neurochem* **76**: 173–181.
43. in t' Veld BA, Ruitenberg A, Hofman A, *et al.* (2001) Nonsteroidal antiinflammatory drugs and the risk of Alzheimer's disease. *N Engl J Med* **345**: 1515–1521.
44. Eriksen JL, Sagi SA, Smith TE, *et al.* (2003) NSAIDs and enantiomers of flurbiprofen target gamma-secretase and lower Abeta 42 *in vivo*. *J Clin Invest* **112**: 440–449.

45. Zhou Y, Su Y, Li B, *et al.* (2003) Nonsteroidal anti-inflammatory drugs can lower amyloidogenic Abeta42 by inhibiting *Rho*. *Science* **302**: 1215–1217.
46. Cherny RA, Atwood CS, Xilinas ME, *et al.* (2001) Treatment with a copper-zinc chelator markedly and rapidly inhibits beta-amyloid accumulation in Alzheimer's disease transgenic mice. *Neuron* **30**: 665–676.
47. Regland B, Lehmann W, Abedini I, *et al.* (2001) Treatment of Alzheimer's disease with clioquinol. *Dement Geriatr Cogn Disord* **12**: 408–414.
48. Ritchie CW, Bush AI, Mackinnon A, *et al.* (2003) Metal-protein attenuation with iodochlorhydroxyquin (clioquinol) targeting Abeta amyloid deposition and toxicity in Alzheimer disease: a pilot phase 2 clinical trial. *Arch Neurol* **60**: 1685–1691.
49. Soto C, Saborio GP, Permanne B (2000) Inhibiting the conversion of soluble amyloid-beta peptide into abnormally folded amyloidogenic intermediates: relevance for Alzheimer's disease therapy. *Acta Neurol Scand Suppl* **176**: 90–95.
50. Morgan D, Gitter BD (2004) Evidence supporting a role for anti-Abeta antibodies in the treatment of Alzheimer's disease. *Neurobiol Aging* **25**: 605–608.
51. Orgogozo JM, Gilman S, Dartigues JF, *et al.* (2003) Subacute meningoencephalitis in a subset of patients with AD after Abeta42 immunization. *Neurology* **61**: 46–54.
52. Nicoll JA, Wilkinson D, Holmes C, *et al.* (2003) Neuropathology of human Alzheimer disease after immunization with amyloid-beta peptide: a case report. *Nat Med* **9**: 448–52.
53. Robinson SR, Bishop GM, Lee HG, Munch G (2004) Lessons from the AN 1792 Alzheimer vaccine: lest we forget. *Neurobiol Aging* **25**: 609–615.
54. Chauhan VP, Ray I, Chauhan A, Wisniewski HM (1999) Binding of gelsolin, a secretory protein, to amyloid beta-protein. *Biochem Biophys Res Commun* **258**: 241–246.
55. Choo-Smith LP, Surewicz WK (1997) The interaction between Alzheimer amyloid beta(1-40) peptide and ganglioside GM1-containing membranes. *FEBS Lett* **402**: 95–98.
56. Marr RA, Guan H, Rockenstein E, *et al.* (2004) Neprilysin regulates amyloid Beta peptide levels. *J Mol Neurosci* **22**: 5–11.
57. Carson JA, Turner AJ (2002) Beta-amyloid catabolism: roles for neprilysin (NEP) and other metallopeptidases? *J Neurochem* **81**: 1–8.
58. Ertekin-Taner N, Allen M, Fadale D, *et al.* (2004) Genetic variants in a haplotype block spanning IDE are significantly associated with plasma Abeta42 levels and risk for Alzheimer disease. *Hum Mutat* **23**: 334–342.
59. Farris W, Mansourian S, Leissring MA, *et al.* (2004) Partial loss-of-function mutations in insulin-degrading enzyme that induce diabetes also impair degradation of amyloid beta-protein. *Am J Pathol* **164**: 1425–1434.
60. Ho L, Qin W, Pompl PN, *et al.* (2004) Diet-induced insulin resistance promotes amyloidosis in a transgenic mouse model of Alzheimer's disease. *FASEB J* **18**: 902–904.
61. Zlokovic BV (2004) Clearing amyloid through the blood-brain barrier. *J Neurochem* **89**: 807–811.
62. Yan SD, Chen X, Fu J, *et al.* (1996) RAGE and amyloid-beta peptide neurotoxicity in Alzheimer's disease. *Nature* **382**: 685–691.
63. Kawarabayashi T, Younkin LH, Saido TC, *et al.* (2001) Age-dependent changes in brain, CSF and plasma amyloid (beta) protein in the Tg2576 transgenic mouse model of Alzheimer's disease. *J Neurosci* **21**: 372–381.

64. Deane R, Yan SD, Submamaryan RK, *et al.* (2003) RAGE mediates amyloid-beta peptide transport across the blood-brain barrier and accumulation in brain. *Nat Med* **9**: 907–913.
65. Sakaguchi T, Yan SF, Yan SD, *et al.* (2003) Central role of RAGE-dependent neointimal expansion in arterial restenosis. *J Clin Invest* **111**: 959–972.
66. Phiel CJ, Wilson CA, Lee VM, Klein PS (2003) GSK-3alpha regulates production of Alzheimer's disease amyloid-beta peptides. *Nature* **423**: 435–439.
67. Tariot PN, Loy R, Ryan JM, *et al.* (2002) Mood stabilizers in Alzheimer's disease: symptomatic and neuroprotective rationales. *Adv Drug Deliv Rev* **54**: 1567–1577.
68. Simons M, Keller P, Dichgans J, Schulz JB (2001) Cholesterol and Alzheimer's disease: is there a link? *Neurology* **57**: 1089–1093.
69. Frears ER, Stephens DJ, Walters CE, *et al.* (1999) The role of cholesterol in the biosynthesis of beta-amyloid. *Neuroreport* **10**: 1699–1705.
70. Mori T, Paris D, Town T, *et al.* (2001) Cholesterol accumulates in senile plaques of Alzheimer disease patients and in transgenic APP(SW) mice. *J Neuropathol Exp Neurol* **60**: 778–785.
71. Shie FS, Jin LW, Cook DG, *et al.* (2002) Diet-induced hypercholesterolemia enhances brain A beta accumulation in transgenic mice. *Neuroreport* **13**: 455–459.
72. Distl R, Meske V, Ohm TG (2001) Tangle-bearing neurons contain more free cholesterol than adjacent tangle-free neurons. *Acta Neuropathol (Berl)* **101**: 547–554.
73. Kawarabayashi T, Shoji M, Younkin LH, *et al.* (2004) Dimeric amyloid beta protein rapidly accumulates in lipid rafts followed by apolipoprotein E and phosphorylated tau accumulation in the Tg2576 mouse model of Alzheimer's disease. *J Neurosci* **24**: 3801–3809.
74. Crisby M, Carlson LA, Winblad B (2002) Statins in the prevention and treatment of Alzheimer disease. *Alzheimer Dis Assoc Disord* **16**: 131–136.
75. Fassbender K, Simons M, Bergmann C, *et al.* (2001) Simvastatin strongly reduces levels of Alzheimer's disease beta-amyloid peptides Abeta 42 and Abeta 40 *in vitro* and *in vivo*. *PNAS USA* **98**: 5856–5861.
76. Hoglund K, Wiklund O, Vanderstichele H, *et al.* (2004) Plasma levels of beta-amyloid(1-40), beta-amyloid(1-42), and total beta-amyloid remain unaffected in adult patients with hypercholesterolemia after treatment with statins. *Arch Neurol* **61**: 333–337.
77. Friedhoff LT, Cullen EI, Geoghagen NS, Buxbaum JD (2001) Treatment with controlled-release lovastatin decreases serum concentrations of human beta-amyloid (A beta) peptide. *Int J Neuropsychopharmacol* **4**: 127–130.
78. Akiyama H, Barger S, Barnum S, *et al.* (2000) Inflammation and Alzheimer's disease. *Neurobiol Aging* **21**: 383–421.
79. Moore AH, O'Banion MK (2000) Neuroinflammation and anti-inflammatory therapy for Alzheimer's disease. *Adv Drug Deliv Rev* **54**: 1627–1656.
80. Rogers J, Kirby LC, Hempelman SR, *et al.* (1993) Clinical trial of indomethacin in Alzheimer's disease. *Neurology* **43**: 1609–1611.
81. Butterfield DA, Drake J, Pocernich C, Castegna A (2001) Evidence of oxidative damage in Alzheimer's disease brain: central role for amyloid beta-peptide. *Trends Mol Med* 7: 548–554.
82. Engelhart MJ, Geerlings MI, Ruitenberg A, *et al.* (2002) Dietary intake of antioxidants and risk of Alzheimer disease. *JAMA* **287**: 3223–3229.

83. Wilcock GK, Birks J, Whitehead A, Evans SJ (2002) The effect of selegiline in the treatment of people with Alzheimer's disease: a meta-analysis of published trials. *Int J Geriatr Psychiatry* **17**: 175–183.

84. Solomon PR, Adams F, Silver A, *et al.* (2002) Ginkgo for memory enhancement: a randomized controlled trial. *JAMA* **288**: 835–840.

85. Capsoni S, Giannotta S, Cattaneo A (2002) Beta-amyloid plaques in a model for sporadic Alzheimer's disease based on transgenic anti-nerve growth factor antibodies. *Mol Cell Neurosci* **21**: 15–28.

86. Eriksdotter Jonhagen M, Nordberg A, Amberla K, *et al.* (1998) Intracerebroventricular infusion of nerve growth factor in three patients with Alzheimer's disease. *Dement Geriatr Cogn Disord* **9**: 246–257.

87. Yamada K, Nitta A, Hasegawa T, *et al.* (1997) Orally active NGF synthesis stimulators: potential therapeutic agents in Alzheimer's disease. *Behav Brain Res* **83**: 117–122.

88. Bimonte-Nelson HA, Hunter CL, Nelson ME, Granholm AC (2003) Frontal cortex BDNF levels correlate with working memory in an animal model of Down syndrome. *Behav Brain Res* **139**: 47–57.

89. Egan MF, Kojima M, Callicott JH, *et al.* (2003) The BDNF val66met polymorphism affects activity-dependent secretion of BDNF and human memory and hippocampal function. *Cell* **112**: 257–269.

90. Panisset M, Gauthier S, Moessler H, Windisch M (2002) Cerebrolysin in Alzheimer's disease: a randomized, double-blind, placebo-controlled trial with a neurotrophic agent. *J Neural Transm* **109**: 1089–1104.

91. Holmes M, Maysinger D, Foerster A, *et al.* (2003) Neotrofin, a novel purine that induces NGF-dependent nociceptive nerve sprouting but not hyperalgesia in adult rat skin. *Mol Cell Neurosci* **24**(3): 568–580.

92. Grundman M, Capparelli E, Kim HT (2003) A multicenter, randomized, placebo controlled, multiple-dose, safety and pharmacokinetic study of AIT-082 (Neotrofin) in mild Alzheimer's disease patients. *Life Sci* **73**: 539–553.

93. Fillit H, Weinreb H, Cholst I, *et al.* (1986) Observations in a preliminary open trial of estradiol therapy for senile dementia-Alzheimer's type. *Psycho-neuroendocrinology* **11**: 337–345.

94. Rapp SR, Espeland MA, Shumaker SA, *et al.* (2003) Effect of estrogen plus progestin on global cognitive function in postmenopausal women: the Women's Health Initiative Memory Study: a randomized controlled trial. *JAMA* **289**: 2663–2672.

95. Scott JA, Da Camara CC, Early JE (1999) Raloxifene: a selective estrogen receptor modulator. *Am Fam Physician* **60**: 1131–1139.

16

Stem Cells, Regenerative Medicine and Aging

Moustapha Kassem

Department of Endocrinology and Metabolism

University Hospital of Odense, DK-5000 Odense C, Denmark

INTRODUCTION

The hallmark of aging is the gradual and progressive failing of tissues and organs of human body due to a multitude of causes. During steady-state conditions, damaged tissues are replaced by continuous recruitment and differentiation from stem cells in the body. However, with aging, the body ability for regeneration becomes impaired. Regenerative medicine is an emerging discipline in clinical medicine that aims at restoring tissue and organs failing functions due to aging or degenerative diseases by cell transplantation or transplantation of tissue engineered-organs created by *ex vivo* cultured cells. The aim of this chapter is to provide an overview over the use of stem cells in the context of regenerative medicine treatment visions for age-related degenerative diseases. Also, results from some of the recent clinical trials that employed stem cell transplantation for non-malignant diseases indications will be reviewed.

Correspondence to: Moustapha Kassem, MD, PhD, DSc, Professor, Department of Endocrinology and Metabolism, University Hospital of Odense, DK-5000 Odense C, Denmark. Tel. +45-6541 1606, Fax +45-6591 9653, E-mail: mkassem@dadlnet.dk or Moustapha.kassem@ouh.fyns-amt.dk

What are Stem Cells?

Stem cells are undifferentiated cells that are defined by their ability to both self-renew and to differentiate to produce mature progeny cells, including both non-renewing progenitors and terminally differentiated effector cells.[1,2] Classically, stem cells can be classified according to their differentiation potential as totipotent (able to give rise to all embryonic and extra-embryonic cell types), pluripotent (able to give rise to all cell types of the embryo proper), multipotent (able to give rise to a subset of cell types that are derived from the same embryonic origin, for example, mesoderm, ectoderm or endoderm), oligopotent or unipotent (able to give rise to a limited number or one mature cell type). Two types of human stem cells are currently available and can be possibly employed in therapy. The first is the pluripotent human embryonic stem cells (hESC) derived from inner cell mass of blastocysts of early embryos[3] or primitive germ cells of around 2-month-old fetal tissues.[4] The second stem cells type is human adult stem cells (hASC) (or better terminology: tissue-specific stem cells) derived through specific isolation procedures from nearly every organ in the body (bone marrow, liver, pancreas, adipose tissue, dental pulp etc.) and able to give rise to progeny of cells corresponding to their tissue of origin (multipotent or oligopotent). From a therapeutic point of view, important differences exist between hASC and hESC. The self-renewal and differentiation capacities of hESC is much wider than hASC and thus it is possible to obtain larger number of cells for therapy.[3]

The Role of Adult Stem Cells in Tissue Repair

It is increasingly recognized that all tissues in the body exhibit a degree of self-renewal and turnover under regenerative pressure that may vary from high (for example, blood, skin, gut) to low (for example, heart). Even some of the tissues that were generally perceived as non-renewing and post-mitotic, for example brain and heart, have been shown to exhibit previously unappreciated cell turnover.[1,5] These indirect observations suggest the existence of stem cells and progenitor cells within all tissues. However, the degree to which a particular tissue depends upon replenishment of mature cells from stem cells and progenitor cells is still unclear. In some injury/repair models, the presence of stem cells responsible for cell regeneration have been demonstrated for example, hematopoietic system, dermal epithelium, intestinal epithelium.[1,5] However, this is not the case in other tissues. For example, in one recent study that employed clearly defined criteria for stem cell regeneration using genetic approaches, it has been demonstrated that the cells recruited for regeneration of insulin-producing β-cells in mouse pancrease are derived from un-injured differentiated insulin-producing β-cells cells and not from stem cells.[6]

Stem Cell Transdifferentiation and its Relevance for Tissue Repair

In the middle of the 19th century, Cohnehim suggested that tissue repairs in mammals is accomplished through circulating cells derived from blood (and in light of subsequent observations, from the bone marrow). This hypothesis is still under debate but some recent studies have provided some support. The hallmark of these studies is the demonstration of the ability of certain types of ASC and in particular bone marrow stem cells to contribute to multiple non-hematopoietic tissue turnover following transfer to mice or humans.[1] Different populations of bone marrow stem cells were marked by a marker gene, for example green fluorescene protein (GFP), and subsequently infused in a donor organism. The marked cells were detected in several tissues of the recipient and found to express marker characteristics of these tissues suggesting the presence of ASC plasticity that is determined by cues from different microenvironments. One challenging concept that emerged from these studies is the ability of stem cell to cross their embryonic lineage fate and to develop to other embryonic cell lineages [for example, differentiation of blood cells (mesoderm) to neuronal cells (ectoderm) or to hepatocytes (endoderm)].[1,5,7] Based on these results, it has been suggested that it may be possible to use bone marrow stem cells for treatment of various age-related degenerative diseases. Interestingly, the presence of similar stem cell populations in umbilical cord blood has increased the interest in possible isolation and long-term storage of umbilical cord blood. However, the concept of stem cell plasticity is still controversial and some alternative explanations for the claimed plasticity have been put forward including transplantation of heterogenous cell populations with different differentiation potential or cell fusion between transplanted cells and locally resident cells.[1]

Does Senescence of Stem Cells Contribute to the Aging Phenotype?

Little information is available regarding the molecular and cellular changes that occur in stem cells during aging, and one of the major limitations of the reported results in the literature is the absence of prospective markers exclusively defining adult stem cell populations in a particular organ. This may explain the controversies existent in the literature since several methodologies have been employed by various investigators to isolate and characterize stem cells of a particular tissue. The presence of an age-dependent decrease in the number of hematopoietic stem cells,[8] neuronal stem cells,[9] osteoprogenitor cells[10] and muscle stem cells[11] has been reported. However, the absence of an age-related decline in stem cell numbers in various tissues has also been reported.[12] We have studied extensively the effects of aging on human bone marrow mesenchymal stem cells (hMSC). These are clonogenic cells that are present in bone marrow stroma and have the capacity to differentiate into mesoderm-type cells (osteoblast, adipocytes, chondrocytes) and also possibly into non-mesoderm-type cells, such as neuronal cells.[13] hMSC differentiation potential can be tested *in vitro* by inducing their differentiation, using various culture conditions, into various mesoderm-type lineages,[14] and *in vivo* by ectopic transplantation in the subcutaneous tissues of immune-deficient mice and demonstration of the formation of bone, adipocytes and bone-marrow supporting stroma.[15] Using a clonogenic assay for counting the number of MSC isolated from bone marrow of a group of young and old donors, we found no donor-age difference in the number of hMSC.[16] We also demonstrated that hMSC obtained from young and old donors maintained their cell differentiation capacity into osteoblasts and adipocytes *in vitro*[16,17] and *in vivo*.[15] However, upon long-term culture, cells obtained from old donors exhibited a reduced life-span and accelerated senescence.[18] These data suggest that decreased cell proliferation capacity of stem cells may be the limiting factor for cell regeneration with age. This hypothesis needs, however, further confirmation in other human stem cell compartments. Also, it is necessary to determine the contribution of inability of stem cell to meet the increased cell proliferation demands to defective tissue regeneration in several age-related diseases (for example, heart disease, osteoporosis, osteoarthritis).

Clinical Approaches for Treatment of Age-related Degenerative Diseases Using Stem Cells

Assuming that stem cells' dysfunction contributes to the age-related degenerative changes and age-related diseases, replacing or "rejuvenating" resident stem cell populations by stem cell transplantation is a potentially useful therapeutic methodology. Stem cells can be used for therapy and transplantation as undifferentiated cells and they can differentiate *in vivo* depending upon the signals present in their microenvironment. Alternatively, it is possible stem cells are induced into specific lineage *in vitro* and then used in therapy. Most of the clinical studies reported in the literature have used autologus adult stem cell with minimal *in vitro* culture or differentiation. Some of the therapies based on autologus stem cell transplantation are currently being evaluated in clinical trials. In principle, the same approaches can be applied to allogenic ASC and ESC transplantations but rational ways for avoiding complications associated with allogenic ASC and ESC transplantation need to be resolved (see below).

Systemic Transplantation or Immobilization of Stem Cell

Auologus or allogenic hematopoietic stem cells (HSC) have been used for many years for systemic transplantation therapy of several malignant diseases after abalative chemotherapy or irradiation. However, as mentioned above, some recent studies have suggested that HSC may have the ability to home and improve the functional activity of damaged organs. Mobilisation of HSC by cytokines injections is under trial as a treatment for degenerative heart disease due to atherosclerosis with the aim of allowing HSC to engraft in the damaged myocardium and differentiate into new vasculature.[19] Systemic transplantation of allogenic MSC has also been tried in 3 children with severe osteogensis imperfecta. Homing of MSC in bone as well as the production of normal collagen by the transplanted MSC have been demonstrated.[20] Also, case reports of well-tolerated allogenic MSC transplantation and some clinical improvement have been reported in patients with Hurler's syndrome and severe idiopathic aplastic anemia.[21,22] MSC seems to be a good candidate for transplantation therapy due to the presence of an accumulating evidence of their hypo-immunogenic nature allowing transplantation between HLA-incompatible individuals.[23]

Local Implantation of Stem Cells

Local implantation of stem cells at sites of damage may be an easier approach to treat certain degenerative diseases. For example, tissue ischemia following myocardial infraction due to age-related atherosclerosis has been treated with local implantation of a heterogenous population of bone marrow stem cells with impressive improvement of myocardial function following transplantation.[24] Similarly, local implantation of bone marrow stem cells in lower extremities in patients with severe arterial insufficiency produced very promising results in a recent clinical study.[25] In both cases, improving of vasculature has been demonstrated. Other local applications of stem cells implantation are under investigation in phase 1 or 2 clinical trials for example, implantation of hMSC for non-healed fracture, implantation of chondrocytes for treatment of cartilage damage, or implantation of dopaminergic neuronal cells in patients with Parkinson's disease.

Tissue Engineering

Tissue engineering may provide alternative ways for obtaining tissues and organs needed for transplantation due to lack of sufficient number of organ donors and limitations attributable to immunological rejection and mismatch of physical dimensions. Tissue engineering may also allow obtaining patients own cells, seeding them on biodegradable scaffolds that allow formation of a particular tissue. These tissues can be subsequently used to repair tissue defects due to degenerative diseases. Also, tissue engineering may allow *ex vivo* engineering of organs by the means of 3-dimensional mature cells or stem cells and cultivated in bioreactors that lead to the formation of complex tissues or organs, such as liver, hearts, cartilage or kidneys.[26] Several scaffolds are currently available and may be classified as either biologically-derived polymers isolated from extracellular matrix, plants and seaweed for example, collagen type I or fibronectin, alginate from brown algae, or synthetically prepared, such as hydroxyapatite (HA), tri-calcium phosphate (TCP) ceramics, polylactide and polyglycolide, or a combination of these in the form of poly

DL-lactic-co-glycolic acid (PLGA). There exists several animal experiments showing the success of using this approach for example, for treatment of large bone defects in animal models[27] and it is expected that transplantation of tissues based on these methods to humans will be achieved in the coming years.

Combining Stem Cell Therapy with Gene Therapy

Gentic modification of stem cells is an attractive target for gene therapy because of their higher proliferative capacity and long term survival compared with other somatic cells. Some ASC, for example MSC, have been demonstrated to be able to express exogenous proteins (for example, factor VIII and IL-3) for extended period of time and to maintain this ability after transplantation *in vitro*.[15] Therapy based on these genetic modified cells is thus possible.

Limitation for the Use of Stem Cells for Treatment of Age-related Organ and Tissue Degeneration

The clinical potential for using stem cells for treatment of organ and tissue degeneration and dysfunction is enormous. However, some limitations has to be overcome before stem cell therapy is a clinical reality. While ESC have extensive self-renewal and differentiation capacity, they can form teratomas (usually benign tumors) when transplanted *in vivo*. They are also immunogenic and methods need to be devised to alleviate problems with graft rejection. One of these methods is somatic cell nuclear transfer (SCNT) also known as therapeutic cloning.[29] In this method, enucleated non-fertilized ovum is fused with the patient's own adult cell and the hybrid cell is allowed to grow to the blastocyst stage where ESC are isolated and used to generate immunologically compatible tissues and organs with the patient own tissues (so-called personalized organs!). Recently, it has been demonstrated that it is possible to generate ESC from therapeutically cloned blastocyst.[30] For adult stem cells, limited proliferative capacity and *in vitro* replicative senescence limit the ability of obtaining large number of cells needed for clinical or tissue engineering protocols. Replicative senescence of cultured cells is caused by several factors including progressive telomere shortening during continuous subculture *in vitro* owing to the absence of telomerase activity.[31–33] We have recently demonstrated that it is possible to overcome the senescence-phenotype of cultured hMSC by forced expression of human telomerase reverse transcriptase gene (hTERT) in hMSC and thus restoring telomerase activity.[32] These telomerized cells exhibit extended life-span and thus can provide the large number of cells needed for therapy. In addition to this enhanced cell growth, the cells maintained their differentiation potential as examined by *in vivo* implantation in immune-deficient mice. Thus, telomerase activation is a potential strategy for obtaining large number of biologically competent cells for clinical use. Unexpectedly, the extensive cell proliferation *in vitro* led to genetic instability and resulted in MSC transformation.[34] It seems that conditional expression of the hTERT gene or intermittent chemical stimulation of its expression is a more appropriate approach. In addition to solving these basic biological problems related to stem cells, we need to develop clinical methods for evaluation of efficacy and safety of these therapies before their widespread use in the clinic.

Stem cell therapy has the potential of providing a causal treatment for several age-related degenerative diseases and will definitely if realized, improve the quality of life of a large segment of the aging population. However, it is doubtful that it will change the course of the aging process itself or prolong the life span of the general population since aging is associated with failure of several homeostatic mechanisms.[35] However, it is claimed that "rejuvenation" by systemic stem cell transplantation or stimulation of locally resident stem cells by chemical or physical agents may emerge as a causal treatment for the aging process itself. Testing these ideas and realization of their full therapeutic potential will be an intensive topic of biomedical research in the coming years.

References

1. Wagers AJ, Weissman IL (2004) Plasticity of adult stem cells. *Cell* **116**: 639–648.
2. Watt FM, Hogan BL (2000) Out of Eden: stem cells and their niches. *Science* **287**: 1427–1430.
3. Odorico JS, Kaufman DS, Thomson JA (2001) Multilineage differentiation from human embryonic stem cell lines. *Stem Cells* **19**: 193–204.

4. Shamblott MJ, Axelman J, Wang SP, Bugg EM, Littlefield JW, Donovan PJ, Blumenthal PD, Huggins GR, Gearhart JD (1998) Derivation of pluripotent stem cells horn cultured human primordial germ cells. *Proc Natl Acad Sci USA* **95**: 13726–13731.
5. Verfaillie CM, Pera MF, Lansdorp PM (2002) Stem cells: hype and reality. In *Hematology*, pp. 369–391. *American Society of Hematology Education Program.*
6. Dor Y, Brown J, Martinez OI, Melton DA (2004) Adult pancreatic beta-cells are formed by self-duplication rather than stem-cell differentiation. *Nature* **429**: 41–46.
7. Springer ML, Brazelton TR, Blau HM (2001) Not the usual suspects: the unexpected sources of tissue regeneration. *J Clin Invest* **107**: 1355–1356.
8. Allsopp RC, Weissman IL (2002) Replicative senescence of hematopoietic stem cells during serial transplantation: does telomere shortening play a role? *Oncogene* **21**: 3270–3273.
9. Maslov AY, Barone TA, Plunkett RJ, Pruitt SC (2004) Neural stem cell detection, characterization, and age-related changes in the subventricular zone of mice. *J Neurosci* **24**: 1726–1733.
10. Kahn A, Gibbons R, Perkins S, Gazit D (1995) Age-related bone loss. A hypothesis and initial assessment in mice. *Clinical Orthop* **10**: 69–75.
11. Sajko S, Kubinova L, Cvetko E, Kreft M, Wernig A, Erzen I (2004) Frequency of M-cadherin-stained satellite cells declines in human muscles during aging. *J Histochem Cytochem* **52**: 179–185.
12. Van Zant G, Liang Y (2003) The role of stem cells in aging. *Experimental Hematology* **31**: 659–672.
13. Bianco P, Riminucci M, Gronthos S, Robey PG (2001) Bone marrow stromal stem cells: nature, biology, and potential applications. *Stem Cells* **19**: 180–192.
14. Rickard DJ, Kassem M, Hefferan TE, Sarkar G, Spelsberg TC, Riggs BL (1996) Isolation and characterization of osteoblast precursor cells from human bone marrow. *J Bone Miner Res* **11**: 312–324.
15. Stenderup K, Rosada C, Justesen J, Al Soubky T, Dagnaes-Hansen F, Kassem M (2004) Aged human bone marrow stromal cells maintaining bone forming capacity *in vivo* evaluated using an improved method of visualization. *Biogerontology* **5**: 107–118.
16. Stenderup K, Justesen J, Eriksen EF, Rattan SI, Kassem M (2001) Number and proliferative capacity of osteogenic stem cells are maintained during aging and in patients with osteoporosis. *J Bone Mineral Res* **16**: 1120–1129.
17. Justesen J, Stenderup K, Eriksen EF, Kassem M (2002) Maintenance of osteoblastic and adipocytic differentiation potential with age and osteoporosis in human marrow stromal cell cultures. *Calcified Tissue Int* **71**: 36–44.
18. Stenderup K, Justesen J, Clausen C, Kassem M (2003) Aging is associated with decreased maximal life span and accelerated senescence of bone marrow stromal cells. *Bone* **33**: 919–926.
19. Couzin J, Vogel G (2004) Cell therapy — Renovating the heart. *Science* **304**: 192–194.
20. Horwitz EM, Gordon PL, Koo WKK, Marx JC, Neel MD, Mcnall RY, Muul L, Hofmann T (2002) Isolated allogeneic bone marrow-derived mesenchymal cells engraft and stimulate growth in children with osteogenesis imperfecta: Implications for cell therapy of bone. *Proc Natl Acad Sci USA* **99**: 8932–8937.
21. Koc ON, Day J, Nieder M, Gerson SL, Lazarus HM, Krivit W (2002) Allogeneic mesenchymal stem cell infusion for treatment of metachromatic leukodystrophy (MLD) and Hurler syndrome (MPS-IH). *Bone Marrow Transpl* **30**: 215–222.

22. Fouillard L, Bensidhoum M, Bories D, Bonte H, Lopez M, Moseley AM, Smith A, Lesage S, Beaujean F, Thierry D, Gourmelon P, Najman A, Gorin NC (2003) Engraftment of allogeneic mesenchymal stem cells in the bone marrow of a patient with severe idiopathic aplastic anemia improves stroma. *Leukemia* **17**: 474–476.
23. Le Blanc K, Gotherstrom C, Tammik C, Ringden O (2003) HLA expression and immunologic properties of differentiated and undifferentiated adult and fetal mesenchymal stem cells. *Bone Marrow Transpl* **31**: S244–S245.
24. Assmus B, Schachinger V, Teupe C, Britten M, Lehmann R, Dobert N, Grunwald F, Aicher A, Urbich C, Martin H, Hoelzer D, Dimmeler S, Zeiher AM (2002) Transplantation of progenitor cells and regeneration enhancement in acute myocardial infarction — (TOPCARE-AMI). *Circulation* **106**: 3009–3017.
25. Tateishi-Yuyama E, Matsubara H, Murohara T, Ikeda U, Shintani S, Masaki H, Amano K, Kishimoto Y, Yoshimoto K, Akashi H, Shimada K, Iwasaka T, Imaizumi T (2002) Therapeutic angiogenesis for patients with limb ischaemia by autologous transplantation of bone-marrow cells: A pilot study and a randomised controlled trial. *Lancet* **360**: 427–435.
26. Stock UA, Vacanti JP (2001) Tissue engineering: Current state and prospects. *Ann Rev Med* **52**: 443–451.
27. Kon E, Muraglia A, Corsi A, Bianco P, Marcacci M, Martin I, Boyde A, Ruspantini I, Chistolini P, Rocca M, Giardino R, Cancedda R, Quarto R (2000) Autologous bone marrow stromal cells loaded onto porous hydroxyapatite ceramic accelerate bone repair in critical-size defects of sheep long bones. *J Biomed Mater Res* **49**: 328–337.
28. Allay JA, Dennis JE, Haynesworth SE, Majumdar MK, Clapp DW, Shultz LD, Caplan AI, Gerson SL (1997) LacZ and interleukin-3 expression *in vivo* after retroviral transduction of marrow-derived human osteogenic mesenchymal progenitors. *Hum Gene Ther* **8**: 1417–1427.
29. Wilmut I, Beaujean N, de Sousa PA, Dinnyes A, King TJ, Paterson LA, Wells DN, Young LE (2002) Somatic cell nuclear transfer. *Nature* **419**: 583–586.
30. Hwang WS, Ryu YJ, Park JH, Park ES, Lee Koo JM, Jeon HY, Lee BC, Kang SK, Kim SJ, Ahn C, Hwang JH, Park KY, Cibelli JB, Moon SY (2004) Evidence of a pluripotent human embryonic stem cell line derived from a cloned blastocyst. *Science* **303**: 1669–1674.
31. Rattan SIS (2003) Aging outside the body: usefulness of the Hayflick system. In: Kaul SC, Wadhwa R (eds.) *Aging of Cells In and Outside the Body*, pp. 1–8. Kuwer Academic Publishers London.
32. Simonsen JL, Rosada C, Serakinci N, Justesen J, Stenderup K, Rattan SI, Jensen TG, Kassem M (2002) Telomerase expression extends the proliferative life-span and maintains the osteogenic potential of human bone marrow stromal cells. *Nature Biotechnol* **20**: 592–596.
33. Zimmermann S, Voss M, Kaiser S, Kapp U, Waller CF, Martens UM (2003) Lack of telomerase activity in human mesenchymal stem cells. *Leukemia* **17**: 1146–1149.
34. Serakinci N, Guldberg P, Burns JS, Abdallah B, Schrodder H, Jensen T, Kassem M (2004) Adult human mesenchymal stem cell as a target for neoplastic transformation. *Oncogene* **23**: 5092–5094.
35. Rattan SIS (2004) Aging, anti-aging, and hormesis. *Mech Age Dev* **125**: 285–289.

17

Principles and Practice of Hormesis as an Aging Intervention

Suresh I. S. Rattan
Laboratory of Cellular Ageing, Danish Centre for Molecular Gerontology
Department of Molecular Biology, University of Aarhus
DK-8000 Aarhus-C, Denmark

INTRODUCTION

Aging is characterized by a decrease in the adaptive abilities due to progressive failure of maintenance.[1–4] Therefore, it is opined that if cells and organisms are exposed to brief periods of *mild* stress so that their stress response-induced gene expression is enhanced and the related pathways of maintenance and repair are stimulated, one should observe anti-aging, health-improving and longevity-promoting effects. The phenomenon in which adaptive responses to low doses of otherwise harmful conditions improve the functional ability of cells and organisms is known as *hormesis*.[5–9] The paradigm for the applicability of hormesis in aging intervention is the well documented beneficial effects of exercise, which at a biochemical level results in the production of various harmful substances such as free radicals, acids and aldehydes.[10,11]

A wide variety of physical, chemical and biological agents exhibit hormetic effects, including heavy metals, pesticides, antibiotics, chemotherapeutic agents, ethanol, aldehydes, chloroform, pro-oxidants, hypergravity and ionizing radiation. Several meta-analyses performed on thousands of research papers published in the fields of toxicology, pharmacology and radiation biology have led to the conclusion that the most fundamental shape of the dose response is neither threshold nor linear, but U- or inverted U-shaped, depending on the endpoint being measured.[5–7]

The key conceptual features of hormesis are the disruption of homeostasis, the modest overcompensation, the reestablishment of homeostasis and the adaptive nature of the process.[5] Traditionally, homeostasis is defined as the maintenance of a constant internal state for the efficient functioning and the performance of the organism. However, convincing arguments have been put forward to replace the term homeostasis with *homeodynamics*, taking into account the dynamic nature of living processes in an ever-changing lifeline.[12] A critical component of the homeodynamic property of living systems is their capacity to respond to stress. In this context, the term "stress" is defined as a signal generated by any physical, chemical or biological factor (stressor), which in a living system initiates a series of events in order to counteract, adapt and survive. Thermoregulation, detoxification, cell proliferation/apoptosis, DNA repair, heat shock protein synthesis, protein turnover and antioxidative responses are some of the crucial homeodynamic responses.[2,9]

Since the harmful effects of severe stress have long shadowed the hormetic effects of low level stress, applying hormesis in aging research and therapy is a relatively recent development.[4,9,13–15] What follows is a brief review of the published literature on various hormetic agents that have been shown to slow down aging and/or prolong the lifespan of cells and organisms. This is followed by a discussion of the possible molecular mechanisms involved in hormesis and of the issues remaining to be resolved before hormesis can be applied in human aging intervention, prevention and therapy.

IRRADIATION

Radiation hormesis was one of the first to be studied in relation to aging and longevity.[16] Whereas high doses of irradiation decrease lifespan,[17] low doses of irradiation enhance mean lifespan in *D. melanogaster*[17,18] and *Musca domestica*.[19] It has been argued that irradiation leads to female sterility and that the lifespan increase is thus an outcome of decreased fecundity.[17] It was also shown that mutant females without ovaries did not exhibit increased lifespan after irradiation. The environmental conditions influenced the lifespan of male houseflies differently after an exposure to low dose of irradiation.[19] The increased lifespan was observed only when animals were reared in groups, which promoted a high locomotor activity. However, if individually reared and assumed to have a low locomotor activity, flies did not show longer lifespans. Furthermore, irradiated flies had longer lifespan than controls only when the latter were kept on sub-optimal rearing conditions.[15] An increase in the lifespan of gamma rays-exposed female mice has also been reported.[20] Hormetic effects of low doses irradiation on the proliferative ability, genomic stability and activation of mitogen-activated protein kinase pathways have been reported for human diploid cells.[21–23] In contrast to this, some studies have failed to show an increase of lifespan after a low dose of irradiation. For instance, deuteron-irradiated mice exhibit higher mortality rates and lower lifespan in both sexes than non-irradiated ones.[15] There is also a report of lack of lifespan extension in *C. elegans* after gamma irradiation.[24]

Hypergravity

Life-long exposure to hypergravity has been shown to slightly decrease the lifespan in different species studied, i.e. rodents and fruitflies.[15] In contrast, a 2-week exposure to 3 or 5 g at the beginning of imaginal life resulted in an increase of 15% in the lifespan of male but not of female *D. melanogaster*.[13] The lifespan increase was larger than the one observed after heat shock. Those results have been replicated in a more thorough study conducted in two different laboratories (in France and in Belgium) with slightly different conditions and in two different strains of flies.[25] The French study reported that group- or individually-reared male flies kept for 2 weeks at 3 or 5 g exhibited an increased lifespan. This was also the case for males subjected for 3 weeks to 3 g, but not to 5 g and individually-reared males exposed for 3 weeks to either 3 g or 5 g. No effect of hypergravity was observed in females, when longer or intermittent exposures were used. The Belgian laboratory showed that the beneficial effect of hypergravity was observed at least up to 7.38 g and in a range of exposure between 14 and 24 days.[25]

Temperature

Temperature stress is extensively used in the study of hormesis, not only because it is easy to implement and gives consistent results, but also because heat stress mainly acts through an evolutionarily highly conserved stress response pathway known as the heat shock (HS) response.[26] Wild-type and *age-1* mutant hermaphrodite *C. elegans* exposed for 3 to 24 hrs to 30°C exhibited a significant increase in mean lifespan compared to controls.[27] Similarly, a 6 hrs exposure at 30°C of wild-type worms induced a 12.5% increase in lifespan, but no effect was found after exposures of 2 or 4 hrs.[28] In a series of studies,[29–31] the purpose of which was to model survival under stress, *C. elegans* worms were subjected to 35°C HS of different durations. Those studies showed that HS not longer than 2 hr produced an extension of lifespan of animals. In contrast, longer HS had either no effect (3-hr HS) or deleterious effects (exposures longer than 3 hrs). In a study of multiple stresses in *C. elegans* an extension of lifespan after 1 and 2 hrs HS at 35°C was reported.[24] Longer HS had either no effect (3 hr) or deleterious effects (4 hrs and more). The same effects of different HS were observed on thermotolerance (survival time at 35°C).

Virgin males of inbred lines of *D. melanogaster* exhibited a 2-day increase in mean lifespan and lower mortality rates during several weeks after a heat treatment of 36°C for 70 min.[32] No beneficial effect of HS was reported in females or in mated flies. It has also been shown that wild-type *D. melanogaster* exposed 5 mins a day, 5 days a week for one week to 37°C live on average 2 days longer than the control flies.[33] Longer exposures had either no effect or negative effect on lifespan. In our studies on *D. melanogaster*, exposure of young flies to four rounds of mild HS at 34°C significantly increased the average and maximum lifespan of female flies and increased their resistance to potentially lethal heat stress.[34]

Studies have also been performed on the effect of subjecting transgenic *D. melanogaster* overexpressing the inducible Hsp70, to 20 min at 36°C in an incubator under saturated humidity.[15,35] In the control "parental" line, such an exposure significantly increased the lifespan of both virgin flies kept in groups and of mated flies. The effect was more pronounced in males than in females. In individually kept flies, the same trend was observed but was statistically not significant. No beneficial effect of this HS has been seen in the transgenic lines.[15]

Other examples of the effects of thermal stress on longevity include the following. A 2-hr HS at 37°C applied before the first division and after the fourth division extended the replicative lifespan of *Saccharomyces cerevisiae* by 10%.[36] The same HS had no effect if applied later in life as well as if applied everyday. Irradiated and non-irradiated mice intermittently cold-shocked showed lower rates of mortality in non-irradiated males as well as in both sexes in irradiated mice. Longer lifespans were observed in thermally stressed non-irradiated males and irradiated females. Finally, rats kept in water set at 23°C, 4 hrs a day, 5 days a week had a 5% increase in average lifespan.[15] In addition, this treatment seemed to diminish the occurrence of certain age-related diseases.

During the last few years, research done in our labs has shown hormetic effects of mild HS on human skin fibroblasts. Using a mild stress regime of exposing serially passaged human fibroblasts to 41°C for 1 hr twice a week throughout their replicative lifespan *in vitro*, we have reported several beneficial anti-aging effects. These effects include the maintenance of youthful morphology, reduced accumulation of oxidatively- and glycoxidatively-damaged proteins,[14,37–39] increased levels of various HS proteins, increased antioxidative abilities, increased resistance to ethanol, hydrogen peroxide and UV-A irradiation,[40] and increased activities of the proteasome and its 11S activator.[41] An important aspect of these studies is the observation that anti-aging and beneficial effects of RMHS on human cells were observed without inducing additional cell proliferation. This has implications in separating the phenomenon of aging from longevity. It appears that the progression of cellular aging *in vitro* as the increased molecular disorder and accumulation of damage can be slowed down without escaping the regulatory mechanisms of cell cycle arrest and replicative senescence. Thus, the quality of life of the cell in terms of its structural and functional integrity can be improved without upsetting the mechanisms controlling the replicative lifespan of cells.[4,42]

Other Stresses

A study of the lifespan of X-ray irradiated rats subjected to starvation for 9 days, desiccation for 6 days, or forced swimming showed that all groups of stressed irradiated males lived longer than unstressed irradiated rats.[13,15] In females, two of the three stressed groups exhibited a longer lifespan than unstressed irradiated females, but all irradiated groups had shorter lifespans than non-irradiated females. Ordy *et al.*[43] have not only studied the effects of deuteron-irradiation and cold shock, but also exposed irradiated and non-irradiated mice to electric shock for 1 hr each day, from 90 days of age onwards for the next 1 to 20 months, either alone or combined with cold shock. Electric shocks decreased the mortality rates and increased the lifespan in non-irradiated females. No significant effect was observed in the other groups. The combination of electric and cold shocks decreased the mortality rates in all groups and increased lifespan more or less strongly, except in irradiated males in which lifespan was shortened. Chronic low-frequency (10 Hz) electric stimulation of young and old male Brown Norwegian rats resulted in more than 2-fold increase in the proportion of IIa slow muscle fibers and in the content of satellite cells.[44]

Larval crowding can induce both nutritional limitation and high concentrations of waste products. Larval crowding can thus be considered as a stress for the larvae. Several studies have reported that raising larvae in such conditions increased the lifespan of adult flies. For instance, an increase in lifespan with increased larval density, between 5 and 100 larvae per 5 cm^3 of food has been reported.[15] It has also been reported that though the developmental time, starvation resistance, relative fat content and lifespan increased with larval density, viability was dramatically decreased from 91% at density 50 to 59% at density 350.[15] The increase of the lifespan in those conditions might thus be due to a selection process at the larval stage. However, it has been shown that larval crowding without an effect on viability can increase lifespan in *D. melanogaster*.[45]

The effects of repeated physical injuries on lifespan have been studied for a marine oligochaete *Paranais litoralis*, capable of posterior regeneration, and of asexual reproduction.[46] Worms were bisected on a segment immediately anterior to the fission zone (growth zone generating offspring in asexual reproduction) either once at 10, 30 or 50 days of age, or three times during lifespan at 10, 30 and 50 days of age. The results showed that worms bisected once exhibited only a slight increase in lifespan, but the extension of lifespan was increased with the age at bisection. However, worms bisected three times have a significantly longer lifespan than controls.[46] Similarly, mechanical stimulation by very low magnitude, high frequency vibrations have been shown to increase the density of trabecular bone in adult sheep.[47]

Hormetic effects of moderate exercise are also well documented and their molecular basis in terms of alterations in gene expression and antioxidant levels are being elucidated.[10,11,48–50] Anti-aging and life-prolonging effects of calorie restriction observed in rodents and other species are also considered to be an example of hormesis due to low level chronic stress.[51,52]

The Practice of Hormesis and Unresolved Issues

Since hormetic effects of mild stress are normally observed to be quite moderate in experimental systems, it may be difficult to envisage the biological significance of hormesis in terms of its application in human aging intervention and prevention. However, it should be pointed out that although the hormetic effects may be relatively small when studied at the level of an individual biochemical step, often the final biological outcome, such as overall stress-tolerance, functional improvement and survival, is much more higher and significant than expected.[7,40] This suggests that hormesis is involved in the *biological amplification* of adaptive responses leading to the improvement in overall cellular functions and performance.

Exercise is a good example of the biological amplification of beneficial effects of mild stress where it is not only the specific muscle targets which gain benefit, but improvements in the immune system, cardiovascular system, sex hormones, libido and mood are also well documented.[10,53] At present we have very little knowledge of the interactive biochemical pathways which, through a process of biological amplification, result in the maintenance and/or improvement of the physiological functions. In the case of human beings, the role of the mental state and psychological stress in modulating various physiological functions such as the immune response, stress hormone synthesis, gene expression, cardiac output and muscle strength are only beginning to be addressed.[54,55]

There are, however, several issues that remain to be resolved before hormesis can be widely used for modulating aging and preventing the onset of age-related impairments and pathologies. Some of these issues are as follows:

1) establishing biochemical and molecular criteria for determining the hormetic levels for different stresses (including psychological stress);
2) establishing the optimal hormetic regime in terms of the intensity, frequency and recovery periods;
3) identifying differences and similarities in stress response pathways initiated by different stressors;
4) quantifying the extent of various stress responses;
5) determining the interactive and pleiotropic effects of various stress response pathways;
6) adjusting the levels of mild stress for age-related changes in the sensitivity to stress; and
7) determining the biological and evolutionary costs of repeated exposure to stress.

Resolution of these issues requires much more research on hormesis than that at present. The "proof of the principle" has already been provided by experiments with a wide variety of biological systems and by using a range of physical, chemical and biological stressors. Two of the main lifestyle interventions, exercise and calorie restriction, both of which bring their beneficial and anti-aging effects through hormesis,[10,11,51,52] are being widely recognized and increasingly practiced as effective means of achieving a healthy old age. Within the next ten years, one could also expect the availability of certain neutriceutical and pharmacological hormetic agents to mimic HS response and calorie restriction. For example, bimoclomal, a nontoxic, hydroxylamine derivative with hsp-inducing activity and cytoprotective effects is under Phase II clinical trials.[56–59] Similarly, various chemical mimetics of calorie restriction, such as 2-deoxy-D-glucose and its analogues,[60] and resveratrol,[61] which is a polyphenol found in red wine, are being tested for their use as anti-aging agents in the near future.

Another small molecule, N^6-furfuryladenine or kinetin, which has been shown to have significant anti-aging,[62,63] and anti-thrombotic[64] effects in human cells, is considered to work both as an antioxidant,[65,66] and possibly as a hormetic agent.[63,67] Although the use of kinetin at present has been limited to being a cosmeceutical ingredient in a range of cosmetics products,[a] the use of kinetin as a hormetic neutriceutical agent is under investigation.

In the case of irradiation as a hormetic agent, epidemiologic studies of the public, medical cohorts, and occupational workers confirm that low doses of radiation are associated with reduced mortality from all causes, decreased cancer mortality, and reduced mutation load observed in aging and cancer.[68] Increasing use of low-dose total body irradiation as an immunotherapy for cancer[69] also has its basis in hormesis, which, in not-so-distant future, will be developed into a safe and preventive strategy against a variety of age-related diseases. Hormesis through mental challenge[54] and through mind-concentrating meditational techniques[70,71] may be useful in stimulating inter- and intra-cellular debris-removal processes, and thus preventing the neuronal loss that leads to the onset of age-related neurodegenerative diseases.

Thus, applying hormesis in slowing down aging from within in preventing the onset of age-related diseases and in maintaining the physical and mental abilities in terms of healthy old age is a real possibility, but it requires refinement and optimization for different hormetic agents. The principles of hormesis have been established, but its widespread practice depends on resolving certain issues, especially with respect to the distinction between mild and severe stress.

Acknowledgments

Research grants from the Danish Medical Council (SSVF), Danish Research Council (SNF), and Senetek PLC are acknowledged.

References

1. Holliday R (1995) *Understanding Ageing*. Cambridge University Press, Cambridge.
2. Rattan SIS (1995) Ageing — a biological perspective. *Molec. Aspects Med* **16**: 439–508.
3. Rattan SIS (2003) Biology of aging and possibilities of gerontomodulation. *Proc Indian Nat Sci Acad* **B69**: 157–164.
4. Rattan SIS (2004) Aging, anti-aging, and hormesis. *Mech Age Dev* **125**: 285–289.
5. Calabrese EJ, Baldwin, LA (2001) Hormesis: a generalizable and unifying hypothesis. *Crit Rev Toxicol* **31**: 353–424.

---

[a] For product details, see http://www.senetekplc.com

6. Calabrese EJ, Baldwin LA (2001) U-shaped dose-responses in biology, toxicology, and public health. *Annu Rev Public Health* **22**: 15–33.
7. Calabrese EJ, Baldwin LA (2003) Toxicology rethinks its central belief. *Nature* **421**: 891–892.
8. Rattan SIS (2001) Hormesis in biogerontology. *Crit Rev Toxicol* **31**: 663–664.
9. Rattan SIS (2001) Applying hormesis in aging research and therapy. *Hum Exp Toxicol* **20**: 281–285.
10. Singh AMF (2002) Exercise comes of age: rationale and recommendations for geriatric exercise prescription. *J Gerontol Med* Sci **57A**: M262–M282.
11. McArdle A, Vasilaki A, Jackson M (2002) Exercise and skeletal muscle ageing: cellular and molecular mechanisms. *Ageing Res Rev* **1**: 79–93.
12. Rose S (1997) *Lifelines: Biology, Freedom, Determinism,* pp. 335, The Penguin Press, London.
13. Minois N (2000) Longevity and aging: beneficial effects of exposure to mild stress. *Biogerontology* **1**: 15–29.
14. Rattan SIS (1998) Repeated mild heat shock delays ageing in cultured human skin fibroblasts. *Biochem Mol Biol Int* **45**: 753–759.
15. Minois N, Rattan SIS (2003) Hormesis in aging and longevity. In: Rattan SIS (ed.) *Modulating Aging and Longevity,* pp. 127–137. Kluwer Academic Publishers, Dordrecht.
16. Calabrese EJ, Baldwin LA (2000) The effects of gamma rays on longevity. *Biogerontology* **1**: 309–319.
17. Lamb MJ (1964) The effect of radiation on the longevity of female *Drosophila subobscura. J Insect Physiol* **10**: 487–497.
18. Sacher GA (1963) Effect of X-rays on the survival of *Drosophila* imagoes. *Physiol Zool* **36**: 295–311.
19. Allen RG, Sohal RS (1982) Life-lengthening effects of gamma-radiation on the adult housefly, *Musca domestica. Mech Age Dev* **20**: 369–375.
20. Caratero A, Courtade M, Bonnet L, Planel H, *et al.* (1998) Effect of continuous gamma irradiation at a very low dose on the life span of mice. *Gerontol* **44**: 272–276.
21. Suzuki K, Kodama S, Watanabe M (1998) Suppressive effects of low-dose preirradiation on genetic instability induced by X rays in normal embryonic cells. *Radiat Res* **150**: 656–662.
22. Suzuki K, Kodama S, Watanabe M (2001) Extremely low-dose ionizing radiation causes activation of mitogen-activated protein kinase pathway and enhances proliferation of normal human diploid cells. *Cancer Res* **61**: 5396–5401.
23. Suzuki M, Yang Z, Nakano K, Yatagai F, *et al.* (1998) Extension of *in vitro* life-span of gamma-irradiated human embryo cells accompanied by chromosome instability. *J Radiat Res* **39**: 203–213.
24. Cypser JR, Johnson TE (2002) Multiple stressors in *Caenorhabditis elegans* induce stress hormesis and extended longevity. *J Gerontol Biol Sci* **57A**: B109–B114.
25. Le Bourg E, Minois N, Bullens P, Baret P (2000) A mild stress due to hypergravity exposure at young age increases longevity in *Drosophila melanogaster* males. *Biogerontology* **1**: 145–155.
26. Verbeke P, Fonager J, Clark BFC, Rattan SIS (2001) Heat shock response and ageing: mechanisms and applications. *Cell Biol Int* **25**: 845–857.
27. Lithgow GJ, White TM, Melov S, Johnson TE (1995) Thermotolerance and extended life-span conferred by single-gene mutations and induced by thermal stress. *Proc Natl Acad Sci USA* **92**: 7540–7544.

28. Yokoyama K, Fukumoto K, Murakami T, Harada S, *et al.* (2002) Extended longevity of *Caenorhabditis elegans* by knocking in extra copies of hsp70F, a homolog of mot-2 (mortalin)/mthsp70/Grp75. *FEBS Lett* **516**: 53–57.
29. Butov A, Johnson TE, Cypser J, Sannikov I, *et al.* (2001) Hormesis and debilitation effects in stress experiments using the nematode worm *Caenorhabditis elegans:* the model of balance between cell damage and HSP level. *Exp Gerontol* **37**: 57–66.
30. Michalski AI, Johnson TE, Cypser JR, Yashin AI (2001) Heating stress patterns in *Caenorhabditis elegans* longevity and survivorship. *Biogerontology* **2**: 35–44.
31. Yashin AI, Cypser JR, Johnson TE, Michalski AI, *et al.* (2001) Ageing and survival after different doses of heat shock: the results of analysis of data from stress experiments with the nematode worm *Caenorhabditis elegans. Mech Ageing Dev* **122**: 1477–1495.
32. Khazaeli AA, Tatar M, Pletcher SD, Curtsinger JW (1997) Heat-induced longevity extension in Drosophila. I. Heat treatment, mortality, and thermotolerance. *J Gerontol Biol Sci* **52A**: B48–B52.
33. Le Bourg E, Valenti P, Lucchetta P, Payre F (2001) Effects of mild heat shocks at young age on aging and longevity in *Drosophila melanogaster. Biogerontology* **2**: 155–164.
34. Hercus MJ, Loeschcke V, Rattan SIS (2003) Lifespan extension of *Drosophila melanogaster* through hormesis by repeated mild heat stress. *Biogerontology* **4**: 149–156.
35. Minois N, Khazaeli AA, Curtsinger JW (2001) Locomotor activity as a function of age and life span in *Drosophila melanogaster* overexpressing hsp70. *Exp Gerontol* **36**: 1137–1153.
36. Shama S, Lai C-Y, Antoniazzi JM, Jiang JC, *et al.* (1998) Heat stress-induced life span extension in yeast. *Exp Cell Res* **245**: 379–388.
37. Verbeke P, Clark BFC, Rattan SIS (2000) Modulating cellular aging *in vitro*: hormetic effects of repeated mild heat stress on protein oxidation and glycation. *Exp Gerontol* **35**: 787–794.
38. Verbeke P, Clark BFC, Rattan SIS (2001) Reduced levels of oxidized and glycoxidized proteins in human fibroblasts exposed to repeated mild heat shock during serial passaging *in vitro. Free Rad Biol Med* **31**: 1593–1602.
39. Verbeke P, Deries M, Clark BFC, Rattan SIS (2002) Hormetic action of mild heat stress decreases the inducibility of protein oxidation and glycoxidation in human fibroblasts. *Biogerontology* **3**: 105–108.
40. Fonager J, Beedholm R, Clark BFC, Rattan SIS (2002) Mild stress-induced stimulation of heat shock protein synthesis and improved functional ability of human fibroblasts undergoing aging *in vitro. Exp Gerontol* **37**: 1223–1238.
41. Beedholm R, Clark BFC, Rattan SIS (2004) Mild heat stress stimulates proteasome and its 11S activator in human fibroblasts undergoing aging *in vitro. Cell Stress Chaperon* **9**: 49–57.
42. Rattan SIS, Gonzales-Dosal R, Nielsen ER, Kraft DC, *et al.* (2004) Slowing down aging from within: mechanistic aspects of anti-aging hormetic effects of mild heat stress on human cells. *Acta Biochimica Polonica* **51**: 481–492.
43. Ordy JM, Samorajski T, Zeman W, Curtis HJ (1967) Interaction effects of environmental stress and deutron irradiation of the brain on mortality and longevity of C57BL/10 mice. *Proc Soc Exp Biol Med* **126**: 184–190.
44. Putman CT, Sultan KR, Wassmer T, Bamford JA, *et al.* (2001) Fiber-type transitions and satellite cell activation in low-frequency-stimulated muscles of young and aging rats. *J Gerontol Biol Sci* **56A**: B510–B519.

45. Sørensen JG, Loeschcke V (2001) Larval crowding in *Drosophila melanogaster* induces Hsp70 expression, and leads to increased adult longevity and adult thermal stress resistance. *J Insect Physiol* **47**: 1301–1307.
46. Martínez DE (1996) Rejuvenation of the disposable soma: repeated injury extends lifespan in an asexual annelid. *Exp Gerontol* **31**: 699–704.
47. Rubin C, Turner AS, Bain S, Mallinckrodt C, *et al.* (2001) Low mechanical signals strengthen long bones. *Nature* **412**: 603–604.
48. Manson JE, Greenland P, LaCroix AZ, Stefanick ML, *et al.* (2002) Walking compared with vigorous exercise for the prevention of cardiovascular events in women. *New Engl J Med* **347**: 716–725.
49. Radák Z, Kaneko T, Tahara S, Nakamoto H, *et al.* (2001) Regular exercise improves cognitive function and decreases oxidative damage in rat brain. *Neurochem Int* **38**: 17–23.
50. Hitomi Y, Kizaki T, Katsumura T, Mizuno M, *et al.* (2003) Effect of moderate acute exercise on expression of mRNA involved in the calcineurin signaling pathway in human skeletal muscle. *IUBMB Life* **55**: 409–413.
51. Masoro EJ (2000) Caloric restriction and aging: an update. *Exp Gerontol* **35**: 299–305.
52. Yu BP, Chung HY (2001) Stress resistance by caloric restriction for longevity. *Ann NY Acad Sci* **928**: 39–47.
53. Venkatraman JT, Fernandes G (1997) Exercise, immunity and aging. *Aging Clin. Exp Res* **9**: 42–56.
54. Bierhaus A, Wolf J, Andrassy M, Rohleder N, *et al.* (2003) A mechanism converting psychosocial stress into mononuclear cell activation. *Proc Natl Acad Sci USA* **100**: 1920–1925.
55. Padgett RW, Glaser R (2003) How stress influences the immune response. *Trends Immunol* **24**: 444–448.
56. Vigh L, Literati PN, Horváth I, Török Z, *et al.* (1997) Bimoclomol: a nontoxic, hydroxylamine derivative with stress protein-inducing activity and cytoprotective effects. *Nature Medicine* **3**: 1150–1154.
57. Vigh L, Maresca B, Harwood JL (1998) Does the membrane's physical state control the expression of heat shock and other genes? *TIBS* **23**: 369–374.
58. Csermely P (2001) Chaperone overload is a possible contributor to "civilization diseases". *Trends Genet* **17**: 701–704.
59. Söti C, Csermely P (2000) Molecular chaperones and the aging process. *Biogerontology* **1**: 225–233.
60. Lane MA, Ingram DK, Roth GS (2002) The serious search for an anti-aging pill. *Sci Amer* **287**: 24–29.
61. Howitz KT, Bitterman KJ, Cohen HY, Lamming DW, *et al.* (2003) Small molecule activators of sirtuins extend *Saccharomyces cerevisiae* lifespan. *Nature* **425**: 191–196.
62. Rattan SIS, Clark BFC (1994) Kinetin delays the onset of ageing characteristics in human fibroblasts. *Biochem Biophys Res Commun* **201**: 665–672.
63. Rattan SIS (2002) N6-furfuryladenine (kinetin) as a potential anti-aging molecule. *J Anti-aging Med* **5**: 113–116.
64. Hsiao G, Shen MY, Lin KH, Chou CY, *et al.* (2003) Inhibitory activity of kinetin on free radical formation of activated platelets *in vitro* and on thrombus formation *in vivo*. *Eur J Pharmacol* **465**: 281–287.

65. Olsen A, Siboska GE, Clark BFC, Rattan SIS (1999) N^6-furfuryladenine, kinetin, protects against Fenton reaction-mediated oxidative damage to DNA. *Biochem Biophys Res Commun* **265**: 499–502.
66. Verbeke P, Siboska GE, Clark BFC, Rattan SIS (2000) Kinetin inhibits protein oxidation and glyoxidation *in vitro*. *Biochem Biophys Res Commun* **276**: 1265–1267.
67. Barciszewski J, Rattan SIS, Siboska G, Clark BFC (1999) Kinetin — 45 years on. *Plant Sci* **148**: 37–45.
68. Pollycove M, Feinendegen LE (2001) Biologic responses to low doses of ionizing radiation: detriment versus hormesis. Part 2. Dose responses of organisms. *J Nucl Med* **42**: 26N–37N.
69. Safwat A (2000) The role of low-dose total body irradiation in treatment of non-Hodgkins lymphoma: a new look at an old method. *Radiother Oncol* **56**: 1–8.
70. Selkoe DJ (1992) Aging brain, aging mind. *Sci Amer* **267**: 135–142.
71. De Nicolas AT (1998) The biocultural paradigm: the neural connection between science and mysticism. *Exp Gerontol* **33**: 169–182.

18

Foreseeable and More Distant Rejuvenation Therapies

Aubrey D.N.J. de Grey

Department of Genetics, University of Cambridge

INTRODUCTION

Period life tables — the set of proportions of people of each age in a given population who die in a given year — have historically been a reasonable way to estimate the expected age at death of an individual of a given current age, because mortality rates at all adult ages have fallen very slowly. In this essay, I explore a scenario that is at present highly controversial, but, I feel, deserving of urgent scrutiny by gerontologists and many others: that the rather near future of humanity's progress in reducing death rates at older ages will recapitulate the decline in infant mortality seen a century ago, thereby making period life tables first invalid and then mathematically undefinable. I suggest that at some point — probably within only a few decades, and with treatments that I have discussed extensively elsewhere and will therefore only summarize here — we will make progress in restoring the health and vigour (and

Address for correspondence: Department of Genetics, University of Cambridge, Downing Street, Cambridge CB2 3EH, UK.
Email: ag24@gen.cam.ac.uk; Tel.: +44 1223 366197; Fax: +44 1223 333992

consequent low risk of death) of the elderly comparable, in terms of healthy years added to their lives, to the (literal) decimation of mortality rates that Pasteur and those who implemented his ideas gave to infants.[1] Thereafter, it seems inevitable that humanity will take the bit between its teeth and continue indefinitely to strive to reduce the incidence of involuntary death (at whatever age) yet further. In the sections that follow, I outline several of the major biomedical and sociopolitical advances that I feel we are likely to make in this endeavor. The later episodes that I describe may at first seem uninterestingly distant, but such nonchalance becomes questionable when it is realized how early in this chain of events we seem likely to reach what I here term "actuarial escape velocity" — a point at which those with access to modern medical care will begin to enjoy a progressively diminishing mortality risk, i.e. an increasing *remaining* life expectancy as time passes. These considerations have previously led me to predict that the average age of death of those born in wealthy nations in AD 2100 will exceed 5000 years; in fact, however, it is far from clear that that will not be true of those born in AD 2000.

Robust Mouse Rejuvenation: Target Date 2012–2015

As described elsewhere,[2–4] I contend that there are only seven broad categories of molecular and cellular difference between older and younger people that we have any reason to believe we need to fix in order to achieve two to three decades of life extension of humans already in middle age. (Hereafter I will refer to these as the "first-generation" rejuvenation therapies.) These are:

- a diminished number of cells in certain tissues;
- an accumulation of unwanted cells of various types;
- an accumulation of nuclear mutations and epimutations;
- an accumulation of mitochondrial mutations;
- an accumulation of random cross-links between long-lived extracellular proteins;
- an accumulation of chemically inert but bulky "junk" in lysosomes; and
- an accumulation of such junk in the extracellular space.

Further, I and leading researchers in the various relevant disciplines have delineated[2,4–7] approaches to substantially repairing (reversing) all these changes. (This reversal would not initially be anywhere near total, a point returned to in detail in a later section.) These approaches have been argued to be already technically feasible, because the underlying precursor technologies which *a priori* constitute the hardest parts of their implementation have already been developed and the remaining work needed to complete them can be described in considerable detail. I have termed[2–4] these projects "Strategies for Engineered Negligible Senescence" (SENS), since their goal is collectively to eliminate from humans the positive correlation between age and risk of death per unit time that is biogerontologists' formal definition of senescence. They are summarized in Table 1.

Inevitably, however, progress in extending human longevity will only occur following very impressive interventions in the laboratory — impressive enough to overcome the hardly intelligible (to gerontologists, anyway) but deeply entrenched apathy on this matter that is presently so ubiquitous.[8] In order to trigger a concerted effort to extend human lifespan substantially, such interventions will need, I predict, to possess three key features:

- they must be in genetically wild-type mammals, of a long-lived strain of their species;

Table 1. Strategies for engineered negligible senescence: the "seven deadly things" that accumulate with age as side-effects of metabolism and promising first-generation therapies to reverse or obviate that accumulation. For details, see refs. 2–7.

Category of lifespan-limiting damage	Feasible strategy to repair or obviate it
Cell death without matching replacement	Stem cell therapy, growth factors, exercise
Unwanted (e.g. visceral fat; senescent) cells	Cell surface marker-targeted cellular toxins
Oncogenic nuclear [epi]mutations	Somatic telomere elongation knockout ("WILT")
Mitochondrial mutations	Allotopic expression of 13 mtDNA-coded proteins
Extracellular protein/protein cross-links	Phenacyldimethylthiazolium chloride (ALT-711)
Extracellular aggregates (e.g. amyloid)	Immune-mediated phagocytosis
Intracellular aggregates (e.g. oxysterols)	Microbe-derived "xenohydrolases"

- they must more than double the remaining healthy lifespan of controls; and
- they must be initiated more than half-way through the animals' control lifespan.

On this basis I suggest a specific target that researchers should consider as their goal in terms of laboratory life extension results, and which I term "robust mouse rejuvenation" (RMR). This is to take a cohort of mice of a strain whose life expectancy is three years (considerably above average for laboratory mouse strains) and increase their average lifespan to five years with an intervention begun at age two years. I claim[9] that, with funding totalling at most $100m/year, this can be achieved within a decade with 90% probability, given what has already been achieved in all relevant areas.

Robust Human Rejuvenation: Target Date 2025–2040

Supposing that RMR is indeed achieved, how long thereafter can we expect it to be before we can do the same in humans (roughly, extend the remaining life expectancy of 55-year-olds from 25 years to 75 years)? I will term that milestone "robust human rejuvenation" (RHR).

It is indisputable that extending human lifespan by a given factor, starting a given proportion through the default life expectancy, will be far harder than the corresponding achievement in mice. However, the list of target components of aging surveyed above (and deemed substantially reversible within a decade) is based on human aging, not mouse aging. The possibility certainly exists that RMR will be achieved a good deal sooner than a decade hence, if it turns out that only one or two of the "seven deadly things" need be robustly reversed and that they can be reversed in mice by much simpler methods than would be needed in humans. One plausible scenario, purely for illustration, is a combination anti-cancer therapy (virally-delivered up-regulation of *p53* plus angiogenesis inhibition, for example) together with bone marrow transplantation to compensate for excessive apoptosis of haematopoietic stem cells. This strategy would only target cancer, but could still be sufficient to reach the RMR milestone. I think it more likely, however, that such "single-target" approaches will not confer more than about one year of extra life, even if the target is cancer, and that at least three or four of the seven targets will need to be thoroughly addressed, which is unlikely in much less than a decade. But either way, since all seven aspects of ***human*** aging listed above seem amenable to development in suitably "humanized" mice (e.g. mice engineered to suffer from atherosclerosis or neurodegeneration — these already exist) within a decade, the relative ease of mouse life extension is less of a biological issue than it might appear.

It is still a very big issue, however, largely because of a highly relevant non-biological difference: safety. It is acceptable to do an experiment on mice that ends up unintentionally killing them prematurely, but not on humans. In predicting, in the title of this section, that the interval between RMR and RHR is likely to be no more than 20 years, I therefore rely not only on the assumption that all the therapies I claim are necessary for RHR will have been developed in mice by the timeframe I put on RMR, but also on the increased effort that society seems certain to make to expedite the development of RHR once its foreseeability is demonstrated in the tangible terms that RMR constitutes. It is inevitable that numerous problems will indeed be encountered at this stage: the differences between mice and humans are legion. This is a case, however, where money can make a big difference: the immense resources that we can expect to be marshalled to this fight — for it will be at that point that we can truly say that the "war on aging" (WOA) has begun — will compensate very substantially for the safety issue. This is mainly because those funds will permit experimentation on non-human primates on a scale vastly exceeding what would be fundable today, which is exactly what will be needed to solve the biological problems not already solved in mice without the large delays inherent in reliance on safe human trials. Moreover, since the therapies being explored will be late-onset ones, positive as well as negative results will be derivable from primate experiments lasting under a decade.

Universal Access: Target Date 2030–2050

I turn now to a sociopolitical aspect of the development and dissemination of rejuvenation therapies. (Some others will be addressed in a later section.) One of the starkest shortcomings of the contemporary sociopolitical landscape, at the national and even more at the international level, is the enormous number of people who lack access to extremely basic preventative medical care that could substantially prolong their lives.[10] This inequality is just as pronounced in the case of more expensive therapies such as AIDS drugs.[11] Rejuvenation therapies, when first developed, are likely to be among the most elaborate medical procedures in existence; the cost of their delivery will be high. This logic has led some to predict that rejuvenation therapies, as and when they materialize, will result in an even more polarized global health predicament than the current situation, with the wealthy enjoying a large and rapidly increasing healthy life expectancy while the poor are denied this opportunity — at least for many decades.[12]

It may not be unrealistically utopian, however, to suggest that contemporary medical care is a poor precedent for rejuvenation therapies. The desire for anything varies with the amount of benefit that would be obtained from it. I readily concede that this is pure speculation, but I feel that the far greater degree of extension of healthy lifespan offered by rejuvenation therapy compared to any contemporary medicine is likely to translate into a stronger popular motivation to agitate to acquire it.

My use of the word "agitate" may, of course, be an example of classic British understatement, and it is this that leads me to predict a very rapid (five to ten years) transition from the availability of true rejuvenation therapies to anyone to their availability to everyone. Policy-makers no longer neglect the possibility that extremists may wreak unprecedented terrorist havoc, and are aware that the risk of such events varies with both the degree of (as the terrorist would call it) provocation and the number of people provoked. The perceived risk of very large-scale violence (by individuals if not by nations) aimed at gaining access to rejuvenation therapies will, I suggest, be high, and reducible only by the diversion of whatever resources it takes into making such therapies available worldwide at an affordable price. The magnitude of these resources is clearly immense, but not so immense as might perhaps be guessed at first: the inherent lack of confidence in such therapies' safety when they are first introduced, coupled with the fact that they are by definition less beneficial to younger people, will limit initial demand to those at greatest near-term risk. Nonetheless, the speed with which these resources can be mobilized will surely depend on how long before they are needed a policy to set them aside is initiated; this is a pressing reason to educate policy-makers now about the likely timescales discussed in this article.

A phenomenon very different from limited availability has been proposed[13] as likely to limit the rate of rise in average age at death of cohorts born in successive years: namely, death from medical causes that we already think of as avoidable. The two major examples are obesity and infectious diseases. These causes of death seem to me, however, to be unlikely to afflict a significant proportion of those in the industrialized world at that time. Obesity is undoubtedly life-shortening at present and its increasingly epidemic status is well known, but excess subcutaneous fat is almost certain to be amenable to a variant of the strategy for visceral fat and other supernumerary cells listed in Table 1 and discussed below. Further, a sharp increase in takeup of the more primitive weight-reduction drugs and lifestyle changes that are already available seems likely to result from the appreciation that so much more life is at stake once aging can be reversed. Infections are likely to be technically more challenging — high transmissibility and long latency can diminish the efficacy of quarantine, while antibiotic resistance remains a grave threat especially in hospitals. However, recent experience, such as with SARS and avian flu, already shows that society worldwide is rapidly increasing its awareness of such issues and is demonstrating the will to act with the alacrity and decisiveness that such emergencies demand. When this concern is progressively amplified as the risk of death from other causes falls and fatal infections threaten to climb sharply as a proportion of all deaths, this welcome trend is sure to accelerate. I suggest that the demographic outlook is therefore unlikely to be severe.

Bootstrapping: More Distant, But Already Foreseeable, Challenges

One sobering point touched on previously can be confidently stated: most of the first-generation rejuvenation therapies will be not only risky and elaborate but also partial. This suffices in the short term (so long as all tissues are partially rejuvenated), but eventually — and iteratively thereafter — such therapies will have to be improved not only in safety and cost but in comprehensiveness.

Some will be partial only in the sense of their delivery to cells and tissues. The clearest example of this is the obviation of mitochondrial mutations by the insertion of suitably modified versions of the 13 protein-coding genes of our mitochondrial DNA into our nuclear genome, which would express these proteins in a manner targeted for import into mitochondria and thereby complement any mitochondrial mutations that may occur.[5] A cell possessing these transgenes is permanently and completely impervious to the accumulation of mitochondrial mutations; hence, the only problem is to populate tissues with such cells. Unfortunately, the easiest cell types to populate by cell therapy — namely, those that are mitotically active — do not appreciably accumulate mitochondrial mutations in the first place, probably because cells that suffer a clonal expansion of such mutations simply die and are replaced. It is in postmitotic cells, which are apparently tuned to withstand loss of aerobic respiratory function, that these mutations mainly accumulate. Somatic gene therapy is the most obvious route to delivery of the necessary transgenes, but it is not necessarily a solution if (as at present and perhaps for a long time to come) there is a high incidence of potentially mutagenic random insertion, nor if there are toxic side-effects of introducing many copies of such genes into a single cell. It is thus more likely, in my view, that first-generation therapies will be cell-based, combining a stimulation of apoptosis in mononucleated cells that are affected (such as neurons, especially in the substantia nigra[14]) with a replacement of lost cells by stem cell therapy and/or growth factor-mediated cell division. The ostensibly missing link in this strategy is delivery to skeletal muscle, in which only short segments of fibres are affected and thus in which stimulating apoptosis may do much more harm than good.[15] Here, however, there is the option to stimulate circulating hematopoietic stem cells to fuse with muscle fibres, a phenomenon currently under detailed study.[16] Thus, though the proportion of cells possessing these nuclear transgenes could only asymptotically approach unity, it would do so merely by repeating the unimproved first-generation therapy again and again.

Cell replacement has much of the same character. We are still at an early stage in learning how to control differentiation of stem and precursor cells to replenish lost cells in poorly autoregenerative tissues, and not much further along in manipulating growth factors for that purpose. But even at this point, the results of much such work are exceptionally encouraging,[17,18] bolstering my prediction that we will have quite versatile cell replacement technologies in mice within a decade. The longer-term prospects are also bright, simply because we have only a few hundred different cell types deriving from perhaps only one or two dozen precursor cell types: this is a clearly finite problem.

Four other of the "seven deadly things" face a greater obstacle, however. (I turn to the seventh and last, nuclear mutations, shortly.) The approaches currently envisaged for combating them are only capable of reversing a subset of the respective category of damage, however comprehensively they are delivered to tissues. This is sufficient for first-generation therapies, because the damage in question will be considerably reduced in all tissues and will only climb slowly back to pre-treatment levels, but by the time it has done, those therapies will have shot their bolt.

One of these categories of damage is extracellular protein-protein crosslinks. Many of these links are laid down by the process of glycoxidation, in which proteins react with sugars in the circulation to form adducts that can rearrange and then undergo oxidative reactions forming a covalent linkage to a neighboring protein.[19] Such crosslinks are eventually harmful to the function of long-lived extracellular structures, especially the artery wall, because they make these structures less elastic and thus more prone to mechanical damage. The most promising first-generation therapy to break such cross-links is a small molecule known commercially as ALT-711 and technically as phenacyldimethylthiazolium chloride, which appears to cleave some such crosslinks catalytically;[20] there is still debate over whether it actually does this, but its physiological effects on hypertension and other parameters[21] have not been satisfactorily explained in any other way. However, it certainly breaks only one class of such link, known as diketone or dicarbonyl bonds.[22] Numerous other classes of crosslink — mostly formed by glycoxidation, but some (such as dityrosine) not involving sugars at all — have been identified in these materials.[23] Worse, many such crosslinks are thermodynamically much more stable than dicarbonyl linkages, giving reason to doubt that catalytically active small molecules could ever cleave them. What might be necessary is an enzymatic approach whereby the highly endergonic cleavage of the link could be coupled to the hydrolysis of ATP; however, even this seems doubtful given the paucity of ATP in the extracellular medium (though the option of shuttling such an enzyme across the cell membrane by endo- and exocytosis should not be overlooked). The alternative of a non-catalytic small molecule that simply cleaves a single bond should not be discounted, since the minimal rate at which cleavage must occur is only that at which such links arise, i.e. very slowly; however, any compound that reacts with something stable is likely to react with other bioactive molecules too, causing perhaps unacceptable side-effects. It may thus be necessary to devise a strategy that possesses the selectivity of enzymes without the unreactivity of non-ATP-dependent catalysts. A possibility is suggested by the remarkable protein ATase, which removes alkyl adducts from guanines in DNA but is thereby irreversibly converted to an inactive state and eventually degraded after performing just one such reaction.[24]

Analogous challenges await the degradation of intracellular and extracellular aggregates ("junk") and the elimination of superfluous and potentially toxic cells: in each case the general concept outlined for first-generation therapies is potentially extensible, but each such extension will again bite off only part of the remaining problem, so the therapy must become ever more elaborate and multi-faceted as we reach ever-greater ages. (Just why there should be an ever-increasing variety, rather than only quantity, of superfluous cells will be discussed shortly.) There is no reason to suppose that this will frustrate us — the rate at which such new enhancements are required will necessarily be modest if each one repairs a substantial proportion of the remaining problem — but this reality is nonetheless worth stressing sooner rather than later.

Particular mention must be made of the seventh strand of first-generation aging-reversal technologies mentioned earlier: the combating of accumulating mutations in our chromosomes. This problem differs from the six mentioned so far in that there is not likely to be a simple progression of therapies similar in concept to the one we need most urgently. That first-generation one is a genuine cure for cancer, and the approach that I — jointly with leading researchers in all the relevant fields — have propounded recently[7] is a progressive conversion of the body to an entirely telomerase-knockout state (together with corresponding elimination of the gene(s) for telomerase-independent telomere elongation, which we have yet to identify). The motivation for such a therapy is that the rather drastic side-effects which such a procedure would bring about can, in principle, be entirely alleviated by stem cell therapy for tissues whose function relies on continuous repopulation by the division of stem cells. Such therapy would probably need to be undergone about once a decade; this has occasioned great concern from some commentators, but should not, since it is only a small factor more often than one would need to repeat (in progressively updated form) most of the other SENS therapies for reasons just outlined. The reason why cancer is (in my view, though not in every researcher's[25]) the only nuclear mutation-based problem that we need to solve in order to live a lot longer is why cancer kills us when just one cell becomes sufficiently insistent on dividing indefinitely, whereas mutations that do not promote cell division are only harmful once they affect a substantial proportion of the cells in a given tissue. Thus, the imperative not to die of cancer until one has reproduced has, for all species susceptible to cancer (which includes all vertebrates), driven the evolution of DNA maintenance and repair. Protection of DNA from random change must be good enough that not one cell runs amok until some time after the organism reaches adulthood. Thus, all genes not related to the cell cycle get a "free ride" — they are preserved "unnecessarily" well as a side-effect of our anti-cancer defence. This logic applies both for bona fide mutations (sequence changes) and for "epimutations" (changes to the adducts on DNA, and on the histones that it is wrapped around, which affect its propensity to be transcribed). However, clearly there would come a time when — in tissue after tissue — aspects of metabolism that are essential for the correct functioning of that tissue begin to fail as a result of the accumulation of cells with mutations in one or another gene whose product participates in that process. At first sight, this seems a very daunting challenge.

In fact, however, this problem should never appear, because it will be subsumed within two other of the seven SENS strands. Cells that become unable to perform a function that their environment and/or gene expression pattern requires them to perform have two ultimate choices: one is to die (typically by apoptosis) and the other is to be stubborn, sitting around unwanted, failing in their required function and possibly being actively toxic in one or another way. If the latter, they fall under the same heading as already concerns us in respect of senescent cells and visceral adipocytes: our task is thus merely to hurry them on their way to oblivion by activating the immune system against them, by administering drugs that selectively enter them (based on cell surface markers) and kill any cells they enter, or even by more simplistic drugs.

Such approaches are already being explored for the two types of unwanted cell just mentioned;[26,27] the same technique should work more generally, though as with aggregates and cross-links there will be a progressively increased variety of such cells and a corresponding variety of necessary treatments in progressively older people. This then reduces the problem to the other possible fate, cell death, which is probably not a problem for adipocytes or senescent cells but may be deleterious to other tissues; here, of course, the solution is growth factors and/or stem cell therapy to replenish the tissue.

Bootstrapping Better: Developing Later-Generation Therapies in Time

For the reasons surveyed in the previous section, I consider that the seven targets of first-generation rejuvenation therapies will permanently cease to be significant risks of death (for those with access to the latest medical care) from the time RHR is first achieved. This is because the rate at which second- and subsequent-generation therapies of the sort just discussed will appear seems very likely to exceed the rate at which they will become needed by the beneficiaries of the first-generation therapies. But thus far I have only covered specific, identifiable second-generation problems; it will not have escaped the reader that these are unimaginatively similar to those that I consider necessary targets of the first-generation therapies. What about things we haven't thought of?

This entirely valid point will, I predict, motivate — more or less as soon as first-generation rejuvenation therapies appear — an open-ended research program the like of which the world has never seen: one that will dwarf the Human Genome Project, the Apollo missions and even the ongoing international collaboration to develop controlled nuclear fusion. Our unarguably limited ability to predict what aging will throw at us next can be addressed in only one way: by metaphorically pressing the fast-forward button. We are exceedingly fortunate that such an option is indeed available to us.

Specifically, I confidently predict that humanity will at that time set up, and maintain indefinitely, a very large colony of non-human primates of several different species — probably numbering tens of thousands of animals in all — and use them as the testbed for life extension therapies of the future. (Unprecedentedly large-scale primate experimentation is likely to begin early in the WOA, as discussed previously, but I anticipate a sharp additional scaling-up at this later juncture.) Primates have three characteristics which, jointly, motivate this action:

- they are biologically extremely similar to us;
- they don't talk, so if the biomedical imperative is sufficient we feel entitled (as a society) to do more or less anything to them; and
- they all age at least twice as fast as us.

Because of this combination of features (of which only the first two are key to the development of first-generation therapies), a large colony of primates maintained under conditions as similar as we can manage to those under which we maintain ourselves — the same range of diets, the same lack of exercise, and of course the same medical care, including all life-extension treatments in use at the time — will be as good as certain to exhibit (in some species if not in all) all the health-threatening characteristics of aging that we ourselves exhibit, at an age at most half that at which it appears in us. These primates will be the testbeds for succeeding generations of rejuvenation therapies, some of which will have unforeseen side-effects that will kill some of them — which is why we will need such a large colony in order to be sure to maintain, indefinitely, a sufficient number of animals that have reached an age sufficiently exceeding half the age of any human yet alive to ensure that our primate experiments succeed before their results are needed. But it is readily seen that this becomes progressively easier as time passes: we may only just have 80-year-old primates before we have 160-year-old humans, but we can expect to have 100-year-old primates substantially before we have 200-year-old humans, and the lead-time improves forever thereafter. This strategy will be our most powerful defence against the presently unforeseeable biomedical challenges with which the attainment of progressively more advanced ages will surely confront us.

Risk Aversion: Personal and Sociopolitical Consequences

Even the logic presented so far, however, may severely understate the rate at which cohort life expectancy (that is, average age at death of those born in a given year) will increase in coming decades. So far I have discussed only deaths from medically preventable (at that time) causes; we must also consider the likely consequences for lifestyle and thence for mortality rates from other causes, especially violent ones (including accidents).

The principal basis for my view here is a simple observation of the way that people with both a respectable remaining life expectancy and an appreciation of it actually behave. Those most inclined to engage in life-threatening activities are the young, who have not fully grasped their own mortality, and the underprivileged, whose remaining life expectancy is always modest on account of the lesser availability of medical care (especially preventative care), the higher incidence of violent crime, and so on. The same is, I believe, true at a global scale: perhaps it is sheer luck that we are approaching the 60th anniversary of the last time that any western European nations were at war with each other or experienced civil war, an interval not previously seen since Roman times,[28] but I strongly suspect that this arises from a sea change in the readiness of the electorate to sacrifice large numbers of their own lives in the interests of national pride. The elimination of the death penalty throughout Europe and the increasingly stringent restrictions on firearms ownership seen in some European countries (particularly the UK) are examples of the same phenomenon, I claim, as is the increasing public hostility to the habit of driving under the influence of alcohol. The same process is occurring throughout the industrialized world, albeit lagging somewhat behind Europe in several respects. It is for this sort of reason — simple extrapolation from the past century — that I predict that society will consistently act to ensure that death from violent causes remains much rarer than death from aging (i.e. from causes that a younger individual can more reliably survive). That will, of course, entail (presuming that what I have suggested in earlier sections is anywhere near correct) a considerable acceleration in the rate at which we alter our lifestyles. I predicted in 1999[29] that once we functionally cure aging driving will be outlawed; I still think that is likely, at least unless cars become much more automated and road accidents thereby made very rare even in the context of severe human error. This is the final component of the logic on the basis of which I subsequently predicted[30] that the average age at death of those born in wealthy nations in the year 2100 will exceed 5000 years, which is roughly five times the value resulting[31,32] from an indefinite enjoyment of the mortality rate of young teenagers in such nations today. That prediction,[30] however, was made cautiously, since it was a spur-of-the-moment judgment made without the detailed analysis which I have undertaken in this article; the timescales estimated here plainly imply that lifespans of that order will be the norm much sooner, perhaps even for those born in 2000.

Conclusion: The Impending Demise of the Period Survival Curve

The analysis given in this article is a basis for a somewhat more esoteric prediction, interesting perhaps only to demographers but relevant to anyone young who might presume that advertised life expectancy values say anything reliable about their expected remaining lifespan: that it is already invalid to discuss life expectancy of those alive today on the basis of period life tables, and that within the next few decades it will become mathematically impossible to do so. The invalidity needs no further discussion, given the very large ratio of the mortality rates of retirees and young adults. The impossibility arises from the fact that a probability of death at a given age in a given year can only be computed if someone in the population was that age at the start of that year. The point on a period survival curve for a given age is computed by multiplying the probabilities of death in the year in question at all ages up to that age, so the curve only reaches precisely zero if there is an age at which everyone alive at the beginning of the year died during it. All current period survival curves very nearly reach zero even when this condition is not met, however, because in any industrialized country our mortality rate is of the order of 50% for many years prior to the maximum observed age. But if mortality rates at all (or almost all) hitherto observed ages fall to 5% or lower, and mostly to below 1%, the resulting survival curve will still be in mid-air at the greatest age yet observed: it will predict a large number of survivors to that age but will say nothing about their life expectancy thereafter. The only use of such a curve then will be its rhetorical ability to depict how different an era we have entered.

References

1. Armstrong GL, Conn LA, Pinner RW (1999) Trends in infectious disease mortality in the United States during the 20th century. *J Am Med Assoc* **281**: 61–66.
2. de Grey ADNJ, Ames BN, Andersen JK, Bartke A, Campisi J, Heward CB, McCarter RJM, Stock G (2002) Time to talk SENS: critiquing the immutability of human aging. *Ann NY Acad Sci* **959**: 452–462.
3. de Grey ADNJ, Baynes JW, Berd D, Heward CB, Pawelec G, Stock G (2002) Is human aging still mysterious enough to be left only to scientists? *BioEssays* **24**: 667–676.
4. de Grey ADNJ (2003) An engineer's approach to the development of real anti-aging medicine. *Sci Aging Knowl Environ.* 2003:vp1.
5. de Grey ADNJ (2000) Mitochondrial gene therapy: an arena for the biomedical use of inteins. *Trends Biotechnol* **18**: 394–399.
6. de Grey ADNJ (2002) Bioremediation meets biomedicine: therapeutic translation of microbial catabolism to the lysosome. *Trends Biotechnol* **20**: 452–455.
7. de Grey ADNJ, Campbell FC, Dokal I, Fairbairn LJ, Graham GJ, Jahoda CAB, Porter ACG (2004) Total deletion of *in vivo* telomere elongation capacity: an ambitious but possibly ultimate cure for all age-related human cancers. *Ann NY Acad Sci* **1019**: 147–170.
8. Miller RA (2002) Extending life: scientific prospects and political obstacles. *Milbank Q* **80**: 155–174.
9. de Grey ADNJ (2002) Outline proposal for an Institute of Biomedical Gerontology. http://www.gen.cam.ac.uk/sens/IBGcase.htm
10. Fischer PR, Bialek R (2002) Prevention of malaria in children. *Clin Infect Dis* **34**: 493–498.
11. Sidley P (2003) Free retroviral drugs could save up to 1.7 million South Africans. *BMJ* **327**: 184.
12. Lieven A (2001) The second fall. *Prospect* **70**: 22–26.
13. Olshansky SJ, Ludwig D, Carnes BA, Brody J, Hayflick L, Butler R (2003) Obesity, infectious diseases, and forecasts of human life expectancy; the demographic impact of immortality. *Biogerontol* **4** (Suppl. 1): 74–75.
14. Itoh K, Weis S, Mehraein P, Müller-Höcker J (1996) Cytochrome c oxidase defects of the human substantia nigra in normal aging. *Neurobiol Aging* **17**: 843–848.
15. McKenzie D, Bua E, McKiernan S, Cao Z, Aiken JM, Wanagat J (2002) Mitochondrial DNA deletion mutations: a causal role in sarcopenia. *Eur J Biochem* **269**: 2010–2015.
16. Camargo FD, Green R, Capetenaki Y, Jackson KA, Goodell MA (2003) Single hematopoietic stem cells generate skeletal muscle through myeloid intermediates. *Nat Med* **9**: 1520–1527.
17. Barton-Davis ER, Shoturma DI, Musaro A, Rosenthal N, Sweeney HL (1998) Viral mediated expression of insulin-like growth factor I blocks the aging-related loss of skeletal muscle function. *Proc Natl Acad Sci USA* **95**: 15603–15607.
18. Ogawa Y, Sawamoto K, Miyata T, Miyao S, Watanabe M, Nakamura M, Bregman BS, Koike M, Uchiyama Y, Toyama Y, Okano H (2002) Transplantation of *in vitro*-expanded fetal neural progenitor cells results in neurogenesis and functional recovery after spinal cord contusion injury in adult rats. *J Neurosci Res* **69**: 925–933.
19. Monnier VM, Cerami A (1981) Nonenzymatic browning *in vivo*: possible process for aging of long-lived proteins. *Science* **211**: 491–493.

20. Asif M, Egan J, Vasan S, Jyothirmayi GN, Masurekar MR, Lopez S, Williams C, Torres RL, Wagle D, Ulrich P, Cerami A, Brines M, Regan TJ (2000) An advanced glycation endproduct cross-link breaker can reverse age-related increases in myocardial stiffness. *Proc Natl Acad Sci USA* **97**: 2809–2813.
21. Kass DA, Shapiro EP, Kawaguchi M, Capriotti AR, Scuteri A, deGroof RC, Lakatta EG (2001) Improved arterial compliance by a novel advanced glycation end-product crosslink breaker. *Circulation* **104**: 1464–1470.
22. Vasan S, Zhang X, Zhang X, Kapurniotu A, Bernhagen J, Teichberg S, Basgen J, Wagle D, Shih D, Terlecky I, Bucala R, Cerami A, Egan J, Ulrich P (1996) An agent cleaving glucose-derived protein crosslinks *in vitro* and *in vivo*. *Nature* **382**: 275–278.
23. Baynes JW (2001) The role of AGEs in aging: causation or correlation. *Exp Gerontol* **36**: 1527–1537.
24. Hobin DA, Fairbairn LJ (2002) Genetic chemoprotection with mutant O^6-alkylguanine-DNA-alkyltransferases. *Curr Gene Ther* **2**: 1–8.
25. Vijg J, Dolle ME (2002) Large genome rearrangements as a primary cause of aging. *Mech Ageing Dev* **123**: 907–915.
26. Campisi J (2003) Consequences of cellular senescence and prospects for reversal. *Biogerontol* **4** (Suppl. 1): 13.
27. Smith SR, Zachwieja JJ (1999) Visceral adipose tissue: a critical review of intervention strategies. *Int J Obes Relat Metab Disord* **23**: 329–335.
28. Cook C (ed.) (1992) *Pears Cyclopedia*, 100th ed. Pelham Books, London, UK.
29. de Grey ADNJ (1999) *The Mitochondrial Free Radical Theory of Aging*. Landes Bioscience, Austin, USA.
30. Richel T (2003) Will human life expectancy quadruple in the next hundred years? Sixty gerontologists say public debate on life extension is necessary. *J Anti-Aging Med* **6**: 309–314.
31. Comfort A (1979) *Biology of Senescence*, 3rd ed. Churchill Livingstone, Edinburgh, UK.
32. Finch CE (1990) *Longevity, Senescence, and the Genome*. University of Chicago Press, Chicago, USA.

19

Anti-aging Medicine and the Quest for Immortality

S. Jay Olshansky and Bruce A. Carnes
University of Illinois at Chicago, and University of Oklahoma, USA.
Emails: sjayo@uic.edu; bruce-carnes@ouhsc.edu

Introduction

In recent years, there has been an abundance of scientific articles, letters, and editorials published in prominent journals devoted to issues associated with the modern rise of an old concept known as anti-aging medicine.[1–7] The history of efforts to intervene in the aging process dates back thousands of years — an account that has been well chronicled in the scientific literature.[8,9] In fact, it appears that a belief and interest in anti-aging interventions coincides with advances in the medical sciences that embolden those who believe that such interventions are possible.[8] This was the case in the early 20th century as progress was made against infectious diseases, and it is therefore not surprising that entrepreneurs selling anti-aging products have surfaced once again given the combination of recent advances in the biomedical sciences and rapid population aging.

The notion of anti-aging medicine as it is currently promoted and sold at what have come to be known as anti-aging or longevity clinics include a combination of traditional preventive medicine, a battery of tests intended to measure biological age, dietary modification, exercise, and the introduction of hormones and nutritional supplements. Added to this combination of sound advice and substances thought to have anti-aging properties is a heavy dose of exaggerated claims based on the idea that aging is little more than a disease that is already amenable to modification. The underlying premise is that if physiological parameters that are believed to measure biological age in individuals can be modified through interventions so that they resemble levels present at younger ages, then it is suggested that aging has been reversed and length and quality of life extended.

In spite of numerous claims to the contrary, there is no empirical evidence to support the claim that aging in humans has ever been modified by any means,[6,10] nor is there evidence that it is possible to measure biological age[11] or that so-called anti-aging products extend the duration of life. The General Accounting Office (GAO) concluded that the modern anti-aging industry is not only making false claims to the public about the products they sell and the promises they make, but there are also serious risks of physical and financial harm associated with this industry.[10]

The irony is that in recent years, researchers from a broad range of scientific disciplines have begun to piece together important elements of the puzzle of aging, leading many to argue that it is only a matter of time before interventions are developed that modulate the rate of aging in humans.[12–17] Some scientists contend that the inevitable demographics of a rapidly aging population, combined with an increased life expectancy, warrants a significant increase of financial resources and acceleration of scientific efforts to develop aging interventions.[13,14,16,18] Others suggest that if successful, interventions that modify the biological rate of aging in humans would change the fabric of human society — leading to questions about whether such interventions should be pursued.[19–21]

What is evident now is that the public is being exposed to two competing messages. From the anti-aging industry, they are being led to believe that the secret to the fountain of youth already exists, and that it is currently available through clinicians that have been trained in anti-aging medicine. From scientists who work in the various fields that inform the study of aging, the public is being told that anti-aging medicines do not currently exist, but that researchers are closing in on an understanding of the biological processes that contribute to aging — perhaps leading in the future to an intervention that may slow down the process.[6]

We contend that one of the best ways for clinicians interested in the health and quality of life of their patients to address these inquiries about anti-aging medicine is to understand the science of aging. In this chapter we provide an overview of the science that underlies current theoretical and empirical developments in the field of aging. Once the science of aging is understood, it will be self-evident why the claims being made by those currently involved in the anti-aging industry are exaggerated or false, and why both clinicians and the general public should be extremely cautious about so-called anti-aging products.

Why do Species Live as Long as They Do?

Why is the average duration of life 1000 days for most strains of mice, 5000 days for beagle dogs, and 29,000 days for humans? Why is the risk of death extraordinarily high at birth for most forms of life, followed by a notable decline in age-specific mortality until puberty, and then an exponential increase from sexual maturity throughout most of the remaining lifespan?

Why are the bodies that carry the immortal genetic instructions — DNA — not themselves immortal? These and other related questions about life, death and the duration of life have not only occupied the minds of the greatest thinkers of every era, but they have led to countless failed efforts to combat aging and forestall death.[8]

One of the more interesting developments in the study of aging and mortality that formed the core of a heated debate among numerous scientists for nearly two centuries was a discovery in 1825 by the British actuary Benjamin Gompertz. He noted that human deaths tend to occur in a predictable age pattern[22] — a seemingly innocuous and now obvious finding that has done nothing less than shape the mathematics of death ever since. Gompertz believed he had discovered a law about the timing of death that was akin to Newton's law of gravity. Gompertz called his equation the "law of mortality". So much attention was paid to Gompertz's law for more than a century that many scientists from a wide range of disciplines devoted their entire research careers attempting to understand why common age patterns of death should exist. Indeed, scientists were so convinced by the biological arguments for a law of mortality that they extended its applicability to all living things — suggesting that just as Newton's law was universal, so too was the law of mortality. Thus began an intensive era of biological and demographic research to understand and quantify the temporal kinetics of senescence.

Ironically, antecedents to a biological understanding of the timing of death and Gompertz's law of mortality existed well before Gompertz was born. Although there is certainly a literature on aging and death that extends as far back as Egyptian times,[8] one of the earliest related ideas was the belief based on the Old Testament that the lifespans of humans and other species are fixed by a supernatural power or by biological laws that apply to all living things. "My spirit will not contend [remain in] man forever, for he is mortal; his days will be a hundred and twenty years" (Genesis 6:3). The famous Italian noble and health "expert" of the 17th century, Luigi Cornaro, in a now classic statement on this topic, suggested that "even the weakest people had enough 'vital principle' to live for 100 years, and those endowed with a stronger constitution could live to the biblical maximum of 120 years".[23]

One of the more interesting theories about the timing of death was devised by the influential 18th century zoologist, Georges Buffon.[24] Buffon suggested that every person has the same allotment of time from birth to death, and that the duration of life depends not on our habits, customs, or quality of food, but rather on **physical laws** [our emphasis] that regulate the number of our years. This belief was based on his observation that species possessed a suite of fixed biological attributes (e.g. gestation period, age patterns of growth, constant physical form). If all biological phenomenon conform to fixed laws like those governing the timing of gestation and sexual maturity, Buffon reasoned, then duration of life must also be fixed. Buffon's interest in lifespan was based on an extensive database of life history characteristics that he collected for a variety of species (e.g. dogs, cats, rabbits, humans, etc.). Based on these data, Buffon reasoned that a species' lifespan was a product of interconnected chains of functional relationships between biological attributes. He envisioned a fixed duration of gestation giving rise to a fixed duration of growth, which in turn, leads to a fixed duration of life. These data supported his hypothesis that the average lifespan of individuals within a population (i.e. life expectancy) should be proportional to the amount of time that is allocated to growth and development. Specifically, Buffon discovered that life expectancy was consistently six to seven times greater than the time required to reach puberty. As you will see from the evolutionary theory of senescence discussed below, Buffon's observation was prophetic, although he made the common mistake of his time in assuming, much like Luigi Cornaro, that everyone had the potential to live the same length of time. In other words, the early thinkers in this area predicted that a fixed biologically-based limit to life existed and applied equally to everyone, while Gompertz discovered, to the contrary, that there is an age pattern of death suggesting that not everyone shares the same chance of living to older ages.

The Evolutionary Model of Senescence

In order to understand why species live as long as they do, it is important to recognize and appreciate the evolutionary theory of senescence. At the heart of this theory is a fundamental biological link between the timing of reproduction and death. According to evolution theory, the force of natural selection begins to decline rapidly once reproduction commences at puberty, approaching negligible levels at the end of the reproductive window (at menopause).[25] What is meant by "the force of natural selection" is the ability of selection to influence the distribution and frequency of alleles in subsequent generations.

According to Medawar,[26] natural selection operates not just on individuals, but also on all of the genes in our bodies. As the force of selection wanes following the onset of puberty, it then becomes possible for harmful alleles to accumulate in the gene pool because under normal living conditions, there is no penalty for detriments to health occurring in older regions of the lifespan where only a few members of the species normally survive.[26–29] Thus, genes that prove to be harmful and which may be associated with aging are figuratively "pushed" by natural selection to later and later ages, where they have less of an effect on reproduction. Over many generations, harmful alleles will tend to have their age of expression accumulate at or near the end of the reproductive window and beyond. Williams[29] later extended this theory by suggesting that natural selection would favor the accumulation of genes that do beneficial things early in life, even if they are known to be harmful later in life. This concept, known as antagonistic pleiotropy, operates under the same premise that there would normally be no penalty for favoring genes with harmful effects expressed later in life, as long as they enhance reproductive fitness.

Why does aging or senescence occur then? Using the poetic words of Medawar, aging is revealed "only by the most unnatural experiment of prolonging an animal's life by sheltering it from the hazards of its ordinary existence". In other words, aging becomes evident only when survival is extended into the post-reproductive region of the lifespan — as is now the case for most people living in low mortality populations.

Empirical tests of these hypotheses have shown that the age trajectory of death is, in fact, a species-specific phenomenon that, as predicted from evolution theory, is calibrated to the onset and length of a species' reproductive window.[30,31] This means that sea turtles, whales, elephants and humans — species that experience an extended period of growth and development, a significant delay in puberty, and a longer time window within which reproduction can occur — live longer than species that reproduce early and for shorter time periods.

It is important to remember that the reproductive window is a genetically determined and fixed attribute that is established as part of every species' life history strategy that was molded by the environment within which species evolved. Thus, all modern evolutionary theories of senescence rely on the premise that selection is blind to the consequences of gene expression in the post-reproductive period of the lifespan. This premise has numerous implications for aging, health, and longevity. First among these is that aging and death genes or programs cannot arise from the direct action of natural selection. Instead, senescent-related diseases and disorders arise from the unintended degradation of processes that are essential to achieving reproductive success (e.g. growth, development, maturation, maintenance, and repair), but whose continued fidelity of function could not be maintained by an evolutionary process indifferent to post-reproductive survival. The good news in this message is that in the absence of a genetic program for aging, its manifestations (e.g. many of the physiological parameters that change with the passage of time and which cause both frailty and death) are inherently modifiable. This is why exercise, diet and some pharmaceuticals have been used successfully to modulate both physical and physiological attributes of our bodies.

The evolutionary view of senescence also implies that aging is not an unnatural disease. This is in direct contrast to the proponents of anti-aging medicine who claim that aging is itself a disease, and that returning physiological parameters of the body back to levels present at younger ages implies that aging has been reversed and that people can actually grow younger.[32] The biological reality is that aging is a natural and inevitable byproduct of survival extended into the post-reproductive period of the lifespan. This also implies that bodies are not designed for indefinite survival.[33] The biological warranty period implied by the certitude of death has important implications for limits on the lifespan of individuals and the life expectancy of populations.[34]

The reality of what is being demonstrated at anti-aging clinics is a fact about human physiology that has been known for thousands of years. It has always been possible to modify and improve human physiology and physical well-being at any age through diet, exercise, and more recently by pharmaceuticals. The proponents of anti-aging medicine confuse the plasticity of physiology with the manifestations of aging. Further, the tenets of evolutionary biology suggest that it is not possible to influence or measure an aging process that does not exist. Thus, while the mainstays of the anti-aging industry (exercise, diet, hormones and antioxidants) cannot make anyone grow younger, there is abundant scientific evidence showing that physical fitness can be improved at any age with dramatic improvements in quality of life as a beneficial byproduct.

Projecting Life Expectancy Using Mathematical Models

Although demographers study populations, the results of their work are often used to make inferences about limits on the lifespan of individuals. The basic logic is that if there is a limit to the life expectancy of a population, then limits must also exist for the lifespans of the individuals who make up that population. Some researchers have argued that if low mortality populations are approaching a limit to life expectancy as claimed by some biodemographers,[30,35,36] then the approach to these limits should be reflected in the behavior of vital statistics. For example, populations approaching a limit should be characterized by a stagnation in the age trend of the oldest prevalent individual.[37] Critics of limit hypotheses also argue that limits imply that there must be an age beyond which there can be no survivors. Documented violations of both of these conditions have led some demographers to conclude that limits on human life expectancy either do not exist or are not yet in sight.[37,38]

Critics of a purely mathematical approach to the study of human lifespan suggest that the validity of limit hypotheses cannot be ascertained without considering biological evidence on senescence.[34] Duration of life is an outcome variable for scientists in the demographic/actuarial sciences. However, from a biological perspective, duration of life is the product of a multidimensional process involving biological, behavioral, environmental, and random forces.[39,40] As such, there is no defensible basis for the claim, based on purely mathematical models of mortality, that there are no biological or demographic reasons why death rates cannot decline to zero[41] and that life expectancy at birth can rise to 100 years or more.[38]

The regularity of the rise in human life expectancy during the 20th century perceived by the advocates of mathematical extrapolation has led them to predict dramatic increases in life expectancy in this century.[38] However, the unprecedented life expectancy gains achieved during the 20th century occurred primarily because of dramatic reductions in death rates among the young. Duplicating these gains in low mortality populations by the same means is no longer possible because the lives of children can only be saved once. The advocates of mathematical extrapolation also ignore the biomedical significance of the profound shift that has occurred in underlying causes of death.[56] Historically, infectious and parasitic diseases (extrinsic mortality) caused the vast majority of deaths in humans. Heart disease, cancer, stroke and diabetes (intrinsic mortality) dominate the mortality schedule today. Biologically, there is no reason to expect that these two fundamentally different categories of death should or would adhere to the same mortality trend.

Contrary to the perception of those advocating extrapolation methods for projecting life expectancy, the time-frame from the past that has been used to predict the future is anything but representative of the historical mortality experience of humans. The quantum leap in life expectancy achieved over the last 100 years is an unprecedented anomaly in a human history better characterized by fluctuating,[42] stagnating, or slowly rising life expectancy.[43] Because the future course of mortality cannot possibly mimic such an episodic anomaly (characterized by declining early age mortality), this unusual time frame should not be used as the basis for predicting the future course of human life expectancy. The public policy implications and practical importance of this recommendation are evident: 1) it is essential that government agencies responsible for assessing the future solvency of their age-based programs incorporate biological reasoning into their long-term forecasts; and 2) claims made by advocates of anti-aging medicine that the duration of human life can now be extended dramatically based on existing technology, are false.

Death is a biological phenomenon of individuals, not a mathematical property of populations, and the biological evidence is undeniable. The pathology burden within individuals clearly exhibits an age dependence.[34] Cancer and cardiovascular disease are symptoms of a complex underlying age-related pathogenesis that causes cells to lose functionality; a functionality that is necessary for the health and well-being of the individual. The molecular repair processes that maintain the functional integrity of cells also degrade over time. Managing the symptoms of age-related disease (geriatric medicine) is not the same as intervening in the underlying processes (biogerontology) that give rise to these manifestations.[20] Although evolution does not and cannot produce genetic programs for aging or death, forces of deterioration that exist at virtually all levels of biological organization (e.g. molecules, cells, tissues, organs) lead to the undeniable conclusion that there is a limit (expiration date) to how long (warranty period) an individual can live. Since every member of a population is operating under their own unique warranty period, then it is equally impossible to deny that limits also exist for the life expectancy of populations.

Aging and death are predictable byproducts of stable reproductive biologies that evolved under environments far less conducive to survival than those experienced today. Although it is likely that anticipated advances in biomedical technology and lifestyle modification will permit life expectancy to continue its slow rise over the short-term, a repetition of the large and rapid gains in life expectancy observed during the 20th century is extremely unlikely. Such gains would require an ability to slow the rate of aging[3,16] — a technological capability that does not exist today, and even if it did, would require implementation on a broad scale in order to have a measurable impact on the vital statistics of a population.[6]

As such, mathematical models that assume the future course of life expectancy over the long-term will continue the trend observed during the 20th century are likely to fail because they ignore the underlying biology that influences duration of life. Further, the predictions of extreme longevity (life expectancy of 100 years or more) produced by these models are not supported by the biological evidence.

The image of a biological warranty period for the duration of life has been used here to capture a universal and undeniable biological reality — indefinite survival is not possible, and the duration of life would remain limited by biological constraints even if every cause of premature death could be eliminated.

CONCLUSIONS

At its core, the fundamental biological explanation for why individuals senesce — the basic question asked by Gompertz in 1825 about the age pattern of death he observed for humans — is predicated on the importance of natural selection and its declining effectiveness relative to the timing and distribution of reproduction (including a period of grandparenting) within the lifespan of organisms. Applying these principles of evolutionary biology in order to explain age patterns of death among and between sexually reproducing species has come to be known as the biodemography of aging — a concept with historical roots in the search for a law of mortality,[45–50] and contemporary interdisciplinary research on aging.[30,51–53] Although a vast number of genetically-based biological processes that exist within organisms to sustain life and maintain functional integrity have been identified and characterized, senescence is not among them. A genetic program for aging is not required for animals to age, just as there is no program for aging required for man-made machines to experience degradation with time.[16,20]

Although unknown to Buffon in the 18th century and Gompertz in the 19th century, it is the underlying and invisible action of genes that control and establish the predictability and temporal regularity of growth, development, reproduction, and physical form that led to their speculations on a fixed lifespan. After all, it does appear on the surface that lifespan is genetically programmed because species tend to live for prescribed durations of time. As it turns out, natural selection favored these life history traits and biological clocks as ways to ensure that genes are passed on through time.[55] These biological phenomena were molded by the environments in which they arose. Their specific forms and functions were not actively designed in the same way that an engineer would draw the plans for creating a machine and then constructing it. Instead, the biological attributes of individuals that influence duration of life are the product of a directionless and ongoing competition among preexisting genetic variants (alleles) whose "victors" are determined by their ability to propagate themselves.

Most of the rise in human life expectancy has come from saving children from infectious and parasitic diseases, and by reducing mortality in women during childbirth. These gains in life expectancy cannot be repeated in developed countries today because the reservoir of potential person-years associated with further declines in these causes of death has been nearly exhausted. Future gains in life expectancy will have to come from saving the lives of older people through the development and use of interventions that alter the fundamental processes of aging. Although not impossible, there are no interventions in existence today that have been demonstrated to modulate the rate of aging.[6] As such, if another quantum leap in human life expectancy is going to occur among today's population, future trends in mortality will have to be fundamentally different from those observed in the past.

Although Buffon and Gompertz lacked access to knowledge about evolution, Buffon's intuition that senescence and species-specific duration of life is related to a fixed period of growth and development, and Gompertz's prediction of a biological basis for a law of mortality, can now both be supported by evolutionary theory and biological evidence.[34] This, in turn, establishes effective constraints on how long individuals can live and how high life expectancy and maximum lifespan can practically rise. Today, aging and death are viewed as the inadvertent but inevitable byproduct of the degradation of biological structures and processes that evolved for growth, development and reproduction rather than extended operation. These structural and functional constraints exist at every level of biological organization (cells, tissues, organs and organ systems) within an individual, and it is their existence that imposes practical (i.e. probabilistic) limits on the lifespan of individuals and the life expectancy of populations.

Given what is known about the biology of aging, the evolutionary theory of senescence, consistent age patterns of death observed across species, and biodemographic and biomechanical constraints on the duration of life, what messages can clinicians convey to their patients when they come calling for therapies to forestall aging? The answer is simple. The biological process of aging cannot as yet be measured; there is no scientific evidence to support the claim that aging can be modified by any means; it is not currently possible to grow younger; there is no known intervention that has been demonstrated to extend the duration of life of humans; nutritional supplements (in particular antioxidants) may in some cases reduce the risk of disease, but they have no demonstrated affect on aging; and hormone supplements should not be administered except in unusual cases of demonstrated clinical hormonal deficiency. The best approaches available for dealing with the manifestations of aging include a dietary regimen based on moderation, lower caloric intake to reduce body fat, regular exercise, and the avoidance of behaviors that increase the risk of diseases and disorders — such as smoking, excessive alcohol consumption, excessive exposure to sun, and obesity. Although scientists are optimistic that interventions will someday be developed to forestall the aging process, such interventions do not currently exist.

Acknowledgments

Funding for this work was provided by the National Institute on Aging (Grant Nos. AG-00894-01 and AG13698-01). Elements of this manuscript have been published in Olshansky SJ (2003) From Michelangelo to Darwin: The Evolution of Human Longevity. *Israel Medical Association Journal* **52**: 316–316; and Carnes BA, Olshansky SJ, Grahn D (2003) Biological Evidence for Limits to the Duration of Life. *Biogerontology* **4**: 31–45.

References

1. Butler RN, Fossel M, Harman SM, *et al.* (2002) Is there an antiaging medicine? *J Gerontol A Biol Sci Med Sci* **57**: B333–338.
2. Binstock R (2003) The war on anti-aging medicine. *The Gerontologist* **43**: 4–14.

3. de Grey AD, Gavrilov L, Olshansky SJ, Coles LS, Cutler RG, Fossel M, Harman SM (2002a) Antiaging technology and pseudoscience. *Science* **296**: 656.
4. Fisher A, Morley JE (2002) Editorial: Antiaging medicine: the good, the bad, and the ugly. *J Gerontol Med Sci* **57**: M636–639.
5. International Longevity Center. (2002) *Workshop Report, Is There an Antiaging Medicine?* Canyon Ranch Series; New York.
6. Olshansky SJ, Hayflick L, Carnes BA, *et al.* (2002) Position Statement on Human Aging. Scientific American (June), http://www.sciam.com/explorations/2002/051302aging/html. Also published in *J Gerontol Biol Sci* **57A**: B1–B6.
7. Wick G (2002) 'Anti-aging' medicine: does it exist? A critical discussion of 'anti-aging health products'. *Exp Gerontol* **37**: 1137–1140.
8. Gruman GJ (1966) A history of ideas about the prolongation of life. *Transactions of the American Philosophical Society* **56**: 1–102.
9. Olshansky SJ, Carnes BA (2001) *The Quest for Immortality: Science at the Frontiers of Aging.* Norton Press, New York.
10. General Accounting Office. (2001) *Antiaging Products Pose Potential for Physical and Economic Harm.* Special Committee on Aging, GAO-01-1129.
11. Butler R, Sprott R, Warner H, *et al* (2004) Biomarkers of Aging: From Primitive Organisms to Man. *J Gerontol Biol Sci* (in press).
12. Arking R (2004) Extending Human Longevity: A Biological Probability. In: Post SG, Binstock R (eds.) *The Fountain of Youth*, pp. 177–200. Oxford University Press.
13. de Grey ADNJ, Baynes JW, Berd D, Heward CB, Pawelec G, Stock G (2002b) Is human aging still mysterious enough to be left only to scientists? *Bioessays* **24**: 667–676.
14. de Grey ADNJ, Ames BN, Andersen JK, Bartke A, Campisi J, Heward CB, McCarter RJM, Stock G (2002c) Time to talk SENS: critiquing the immutability of human aging. *Annals NY Acad Sci* **959**: 452–462.
15. Guarente L (2003) *Ageless Quest.* Cold Spring Harbor Laboratory Press.
16. Miller R (2002) Extending Life: Scientific Prospects and Political Obstacles. *Milbank Quart* **80**: 155–174.
17. Rattan S (2004) Aging intervention, prevention and therapy through hormesis. *J Gerontol Biol Sci* (in press).
18. Butler RN, Warner HR, Williams TF, *et al.* (2004) The aging factor in health and disease: the promise of basic research on aging. *Aging Clin Exp Res* **16**: 104–112.
19. Kass L (2001) L'Chaim and its limits: why not immortality? *First Things* **113**: 17–24.
20. Hayflick L (2000). The Future of Ageing. *Nature* **408**: 267–270.
21. Stock G, Callahan D (2004) Would doubling the human lifespan be a net positive or negative for us either as individuals or as a society? *J Gerontol Biol Sci* (in press).
22. Gompertz B (1825) On the Nature of the Function Expressive of the Law of Human Mortality and on a New Mode of Determining Life Contingencies. *Philosophical Transactions of the Royal Society of London* **115**: 513–585.
23. Cornaro L (1634) Discourses on the Temperate Life. In: Butler W (ed.) *The Art of Living Long.*
24. Buffon GLL (1747) *Histoire Naturelle Générale et Particuli* 're.
25. Charlesworth B (1994) *Evolution in Age-Structured Populations*, 2nd Edition. Cambridge University Press, New York.

26. Medawar PB (1952) *An Unsolved Problem of Biology*. Lewis, London.
27. Hamilton WD (1966) The Moulding of Senescence by Natural Selection. *J Theor Biol* **12**: 12–45.
28. Kirkwood TBL (1977) Evolution of Aging. *Nature* **270**: 301–304.
29. Williams GC (1957) Pleiotropy, Natural Selection, and the Evolution of Senescence. *Evolution* **11**: 298–311.
30. Carnes BA, Olshansky SJ, Grahn D (1996) Continuing the Search for a Law of Mortality. *Popul Dev Rev* **22**: 231–264.
31. Rose MR (1984) Laboratory evolution of postponed senescence in *Drosophila melanogaster*. *Evolution* **38**: 1004–1010.
32. Roizen M (1999) *RealAge: Are you as Young as you can be?* Cliff Street Books, New York.
33. Olshansky SJ, Carnes BA, Butler R (2003) If Humans Were Built to Last. *Scientific American* Special Issue on Evolution, May.
34. Carnes BA, Olshansky SJ, Grahn D (2003) Biological evidence for limits to the duration of life. *Biogerontology* **4**: 31–45.
35. Olshansky SJ, Carnes BA, Cassel C (1990) In Search of Methuselah: Estimating the Upper Limits to Human Longevity. *Science* **250**:634–640.
36. Olshansky SJ, Carnes BA, Désesquelles A (2001) Prospects for Human Longevity. *Science* **291**: 1491–1492.
37. Wilmoth JR (1997) In Search of Limits. In: *Between Zeus and the Salmon: The Biodemography of Longevity*, pp. 38–64. National Academy Press, Washington, D.C.
38. Oeppen J, Vaupel J (2002) Broken limits to life expectancy. *Science* **296**: 1029–1031.
39. Carnes BA, Olshansky SJ (2001) Heterogeneity and its Biodemographic Implications for Longevity and Mortality. *Exp Gerontol* **36**: 419–430.
40. Finch CE, Kirkwood TBL (2000) *Chance, Development, and Aging*. Oxford University Press, Oxford.
41. Wilmoth J (2001) How long can we live? Science dEbate 2001; http://www.sciencemag.org/cgi/eletters/291/5508/1491.
42. Olshansky SJ, Carnes BA (1996) Prospects for Extended Survival: A Critical Review of the Biological Evidence. In: Caselli G, Lopez A (eds.) *Health and Mortality Among Elderly Populations*, pp. 39–58. Oxford, Clarendon Press.
43. McNeill WH (1976) *Plagues and Peoples*. Anchor, Garden City.
44. Brody S (1924). The Kinetics of Senescence. *J General Physiol* **6**: 245–257.
45. Brownlee J (1919) Notes on the Biology of a Life-Table. *J Royal St Soc* **82**: 34–77.
46. Deevey ES Jr (1947) Life Tables for Natural Populations of Animals. *Q Rev Biol* **22**: 283–314.
47. Greenwood M (1928) Laws of Mortality from the Biological Point of View. *J Hyg* **28**: 267–294.
48. Loeb J, Northrop JH (1916) Is There a Temperature Coefficient for the Duration of Life? *Proc Natl Acad Sci USA* **2**: 456–457.
49. Pearl R, Miner JR (1935) Experimental studies on the duration of life. XIV. The comparative mortality of certain lower organisms. *Q Rev Biol* **10**: 60–79.
50. Pearl R 1921.

51. Carnes BA, Olshansky SJ (1993) Evolutionary Perspectives on Human Senescence. *Pop Dev Rev* **19**: 793–806.
52. Olshansky SJ, Carnes BA (1994) Demographic Perspectives on Human Senescence. *Pop Dev Rev* **20**: 57–80.
53. Olshansky SJ, Carnes BA, Grahn D (1998) Confronting the boundaries of human longevity. *Am Sci* **86**: 52–61.
54. Wachter KW, Finch CE (eds.) (1997) *Between Zeus and the Salmon: The Biodemography of Longevity.* National Academy Press, Washington, D.C.
55. Dawkins R (1992) *The Blind Watchmaker: Why the Evidence of Evolution Reveals a Universe Without Design*, W.W. Norton, New York.
56. Cames BA, Holden L, Olshansky SJ, Sieqel JS (2004) Intrinsic mortality: a bridge between biology and mortality analysis. *J Gerontol Biol Sci*, revised, resubmitted.

Index